# Chlorinated Insecticides

## Volume II
## Biological and Environmental Aspects

Author:

### G. T. Brooks

The University of Sussex
Brighton, Sussex
England

T0262880

## CRC Press
Taylor & Francis Group
Boca Raton London New York

CRC Press is an imprint of the
Taylor & Francis Group, an **informa** business

First published 1974 by CRC Press
Taylor & Francis Group
6000 Broken Sound Parkway NW, Suite
300 Boca Raton, FL 33487-2742

Reissued 2018 by CRC Press

A Library of Congress record exists under LC control number: 73090535

Publisher's Note
The publisher has gone to great lengths to ensure the quality of this reprint but points out that some imperfections in the original copies may be apparent.

Disclaimer
The publisher has made every effort to trace copyright holders and welcomes correspondence from those they have been unable to contact.

ISBN 13: 978-1-138-50533-9 (hbk)
ISBN 13: 978-1-138-55784-0 (pbk)
ISBN 13: 978-1-315-15040-6 (ebk)

Visit the Taylor & Francis Web site at http://www.taylorandfrancis.com and the CRC Press Web site at http://www.crcpress.com

# PESTICIDE CHEMISTRY SERIES — PREFACE

The literature on pesticides is voluminous, but scattered among dozens of journals and texts written or edited by experts. Until now, with the publication of *Chlorinated Insecticides* by G. T. Brooks, there has been no attempt to produce a single, comprehensive series on the chemistry of pesticides. CRC Press should be commended for having undertaken this Herculean task.

When asked by the publisher to serve as editor of the Pesticide Chemistry Series, I discussed the idea with some of my colleagues at the International Pesticide Congress in Tel Aviv in 1971. At that time, Dr. Brooks enthusiastically agreed to become the author of the treatise on chlorinated insecticides, which became such a comprehensive work that it is being published in two volumes.

As editor of this series, my goal has been choosing experts in their respective fields who would be willing to write single-authored books, thus assuring uniformity of style and thought for the individual text as well as the entire series. I would like to express my deep appreciation to each author for having undertaken the large task of writing in such a comprehensive manner without the aid of contributors or an editorial board.

*Chlorinated Insecticides* is the first contribution to this series; subsequent volumes under preparation will appear in the near future under titles such as Organophosphate Pesticides; Herbicides; and Fungicides. Looking into the future, we plan to include books on the chemistry of juvenile hormones and pheromones. Suggestions on other titles and possible authors are invited from the reader.

Gunter Zweig
Editor
Pesticide Chemistry Series

## THE AUTHOR

Gerald T. Brooks was formerly in the Biochemistry Department of the Agricultural Research Council's Pest Infestation Laboratory (now the Pest Infestation Control Laboratory of the Ministry of Agriculture, Fisheries, and Food) at Slough, England. He is currently Senior Principal Scientific Officer in the Agricultural Research Council Unit of Invertebrate Chemistry and Physiology at the University of Sussex, Brighton, England. He holds B.Sc. and Ph.D. degrees of the University of London, and is a Fellow of The Royal Institute of Chemistry. His field of work is the chemistry, biochemistry, and toxicology of insecticides and hormones.

To Ann, my wife and former colleague, and my many friends in Insecticide Toxicology whose work is referred to in these pages.

# TABLE OF CONTENTS

# Chapter 1

# INTRODUCTION

Volume I of *Chlorinated Insecticides* considered the technology of these compounds in a historical context, tracing their development up to the present time. The accounts of the numerous applications in insect control given there provide a link with Volume II in that they pose many questions about the interactions of these compounds with living organisms. Some of these questions will be examined in the subsequent chapters.

Following the early spectacular successes of these compounds in insect control there was a gradual failure of many applications due to the development of resistance to them in the target insects. The emergence of resistance has been a major setback in the use of chlorinated insecticides and is a remarkable example of the way in which the use of chemicals can change, often irreversibly, the nature of living organisms. As an example of what might be termed "accelerated evolution," the history of the development and spread of resistance is fascinating and instructive. It represents the darker side of what has been a highly successful period of human effort to reduce the incidence of disease and to increase world food supplies. Nevertheless, the phenomenon of resistance discussed in this volume, Chapter 2, gave an early stimulus to the investigations outlined in later chapters and the information gained in this field has proved useful in the wider contexts of the vertebrate toxicology of these compounds and the study of their residues in the environment. Accordingly, an account of resistance, representing the "inaction" of chlorinated insecticides, is a logical prelude to the topics of metabolism, toxicity, and action which follow. In particular, metabolism studies have already clarified the reasons for some types of resistance.

The discovery that a chemical has toxic effects is quickly followed by questions about its mode of action at the molecular level since such information can be used to design other active chemical structures. However, the relationship between quantity of a toxicant applied to an organism and its action at the molecular level is usually obscured by numerous peripheral events which occur to varying extents in the tissues of all living organisms. In other words, a particular mode of action may be common to all organisms, but the variation between species in regard to these other events can result in a wide disparity in their sensitivity to both acute and chronic toxic effects. Therefore, an understanding of these events, which collectively describe the pharmacodynamics of the toxicant and involve its distribution and metabolic transformation within an organism, is essential in the study of toxicity. For this reason the various aspects of pharmacodynamics and some relevant biochemical phenomena are presented as a background to the eventual discussion of mode of action.

Especially when they relate to the possible effects of organochlorine insecticides on wildlife, some of these areas are highly controversial and will remain so for a long time. The interactions between chlorinated insecticides and the abiotic and biotic elements of the environment are evidently complex and simple answers to many of the questions asked by environmentalists are not possible at our present state of knowledge. Human ingenuity will undoubtedly answer many of these questions in time. Meanwhile, the immediate safeguards outlined in the final chapter on insecticide residues are indicative of the present degree of international vigilance in those areas related to human welfare.

The situation with regard to wildlife is generally less satisfactory and it is proper that attention has been drawn to the hazards attending the widespread use of chemicals in the environment. We must be aware of these hazards and use all reasonable means to minimize them. However, the debate on the use of chemical pest control agents must be viewed in the light of man's need for day to day survival. In the advanced countries of the world this debate has hitherto been conducted on full stomachs. The recent upsurge in the price of some basic food commodities is an unpleasant reminder that this situation could change very quickly. In these circumstances, well tried chemicals such as the chlorinated insecticides can only be abandoned if equally economic and efficient methods are available for the control of pests of basic commodities.

# Chapter 2

# INSECT RESISTANCE TO CHLORINATED INSECTICIDES

## A. RESISTANCE AND ITS MEASUREMENT

It is well known that there are often great differences between animal species in the way in which they respond to challenges by biologically active compounds. This is the basis of the eternal problem that faces pharmacologists and toxicologists who attempt to predict man's response to toxicants from the effects observed on experimental mammals. No convenient species has been found which approximates in behavior to humans in more than a limited number of respects; fewest differences might well be found between man and the higher nonhuman primates such as the gorilla, orang-utan, and chimpanzee, but such work has obvious difficulties.

Apart from economic considerations, the Arthropoda (comprising the classes Insecta, Arachnida, Crustacea, and Myriapoda) provide abundant and convenient material for fundamental studies of the way in which species differ in their response to bioactive substances, especially toxicants. Valuable information regarding the basis for the natural tolerance (that is, the natural ability of a species to withstand the challenge of a compound which is toxic to other species) of insects to insecticides may come from studies on different species within the same genus (intrageneric or interspecific differences). Finally, there are the differences between individuals in a species (intraspecific differences), and it is with these that the present account is mainly concerned.

"Natural tolerance" is an ability common to all members of a tolerant population, regardless of any challenge by a toxicant — it has always been there. Wild populations which are normally susceptible to a given toxicant will contain some individuals that are more susceptible and some that are less susceptible to the poison. For such a population, the median lethal dose (LD50) or concentration (LC50), defines the amount of toxicant required to kill 50% of the individuals in a randomly chosen sample. Accordingly, the exposure of this population to doses at or above the LD50 or LC50 will kill the more susceptible and the averagely susceptible individuals, leaving those having some characteristic which enables them to

resist the poison. Since some of these characteristics are of genetic origin, interbreeding of the tolerant survivors produces a resistant population, and continued "selection" of successive generations with the poison serves to purify the strain by eliminating any individuals of intermediate resistance produced in the early stages of the selection process. This is particularly true if the "selecting" dose is increased to, say, LD90 (the dose that would kill 90% of the normal or susceptible individuals), so that the selection pressure becomes intense. Thus, the development of insect resistance to insecticides is due to the presence in normal populations of variants carrying pre-adaptations, factors or genes for resistance and the screening out of these variants by selection with an insecticide is a process of Darwinian selection.

In 1957, The World Health Organization Expert Committee on Insecticides defined the phenomenon of resistance in the following terms:[42]

Resistance to insecticides is the development of an ability in a strain of insects to tolerate doses of toxicants which would prove lethal to the majority of individuals in a normal population of the same species. The term "behavioristic resistance" describes the development of an ability to avoid a dose which would prove lethal.

There is a very large volume of literature on insect resistance to insecticides. An excellent, compact account of the problem has recently been given by Brown,[373] and a number of reviews are available which provide access to the more specialized aspects of the problem.[42-44] Because of its great importance and fundamental nature, investigations on resistance are to be found in literature dealing with a range of topics from field testing of insecticides, through medical entomology, agricultural entomology, biology, toxicology, genetics, physiology, and biochemistry.[374]

Resistance may be classified under two heads, "behavioristic resistance," previously defined and "physiological resistance," which is the ability to survive contact with a normally lethal dose. Behavioristic resistance of mosquitoes to DDT is frequently referred to in the literature; it appears that the insects are so irritated by contact with a surface impregnated with the toxicant that they are stimulated to fly before they can pick up a lethal dose. In many cases it has not been easy to determine whether this response is a natural characteristic of the species that was present in all individuals before they actually encountered DDT (in which case it would be "natural irritability" or "normal behavioristic avoidance" equivalent to the natural tolerance referred to earlier) or whether it resulted from selecting out with DDT those individuals carrying preadaptations for this response. However, the first situation should result in avoidance of the toxicant from the moment of its first use, whereas the second would take a few generations of selection before it became manifest. A number of cases falling into the first category are on record; the situation regarding "behavioristic resistance" is less clear, although there are suspected cases in the Americas. It is generally held that a third possibility, in which the insecticide might produce a mutational change in the population that is then passed on to subsequent generations, does not occur in insects.[43]

The interpretation of these apparent behavioristic changes in the presence of the insecticide is complicated by natural phenomena such as the existence of animal-loving (zoophilic) and man-loving (anthropophilic) forms of the same species in the same area.[50] The residual spraying of house interiors with DDT would eliminate the anthropophilic, interior-loving (endophilic) form, leaving the zoophilic and probably exophilic (outside-loving) form, and, in the absence of information about population numbers, there would be a false indication of a behavioristic change from endophily to exophily due to DDT. Furthermore, selection with DDT is known to produce some housefly strains that are resistant to its knock-down effects (the stage of poisoning where the insect can no longer stand on its legs). If the fly becomes so irritated that it leaves the deposit before this prolonged time to knock-down has elapsed, it will survive and on casual observation the reason may appear to be a behavioristic one. If the apparent tolerance is purely behavioristic, an insect exhibiting it should die when compelled to remain in contact with the insecticidal deposit long enough to acquire a normally lethal dose. This is, in fact, what happens in many of the recorded cases, showing that such insects are normally susceptible to the toxicant.

Insect resistance mainly involves "physiological resistance," and it is with this type that the present account is mainly concerned; in other words, it is concerned with what happens in the insect which can actually withstand contact with lethal amounts of the insecticide. The basis of physiological resistance has been mentioned already, namely, that the toxicant acts as a kind of genetic sieve to separate out for survival those few individuals in a population that are already predisposed toward resistance so that they become the founders of a new population of resistant individuals (susceptible individuals will be referred to as S- and resistant ones as R-). The development of resistance depends on the presence of genes for resistance, and these are not present in all populations, so that it may develop in some geographical areas and not in others.

Resistance to an insecticide is measured by comparing the response of an insect strain under test (frequently in terms of the LD50 or LC50, although this is not always entirely satisfactory, as will be seen) with the response of a normal strain which has not been subjected to pressure with the insecticide. If the amount of toxicant needed for equivalent mortality in the test strain has increased over that required by the normal strain, then the former is regarded as being "resistant." In practice, a tenfold or even smaller increase in resistance of houseflies or mosquitoes to organochlorine compounds can result in unsatisfactory control.[43] A statistically significant increase in resistance of less than fivefold has often been described as "vigor tolerance," but there seems to be no really sound reason to distinguish between resistance levels in this way and it is now evident that certain single genes (for example, the factor *Pen,* resulting in reduced toxicant penetration in houseflies) confer only two- or threefold resistance. These low levels should not be confused with the "natural tolerance" previously discussed, and which Brown has called "refractoriness." For example, boll-weevils and grasshoppers are naturally refractory to DDT.

In order to appreciate the intricate phenomenon of insect resistance, some understanding is needed of the way in which the toxicity of insecticides is measured.[375] Having acquired the terminology of insect toxicology one can then go on to examine the way in which resistance develops. In addition to the dosage parameters LD50, LC50, LD90, etc., which were defined earlier, other terms with equivalent meanings, such as KD50 and ED50, are found in the literature. KD50 is the dose of toxicant required to bring half the group of test individuals to the stage of poisoning when they can no longer stand on their *legs* (knock-down), and is obviously applicable mainly to the mature stages of insects. Since knock-down may occur much sooner than eventual death, or may be completely reversible, it should not be confused with LD50. ED50 (Effective Dose 50) is a general term applicable to the dose required to produce any chosen effect in 50% of the individuals treated. A further term, LT50, is the time required to kill 50% of a batch of individuals exposed to the toxicant in some convenient form such as a deposit on paper or glass, or as a vapor. These dosage parameters may be determined in a variety of ways, the methods chosen usually depending on the facilities and the number of insects available. Thus, large numbers of insects may be divided into batches and each batch sprayed with one of a series of concentrations of the toxicant in a suitable solvent, the dead then being counted when some suitable period (frequently 24, 36, or 48 hr) has elapsed after the treatment. Another convenient method is to dip groups of insects into solutions of the toxicant, and then to observe their behavior as before.

In the laboratory, where experimental conditions can be fairly carefully standardized, the usual procedure is to dose insects cultured under defined conditions of diet, temperature and humidity, and segregated with regard to age and sex before use, so that the biological material is as homogeneous as possible. The insects assigned to a particular group for dosing should be chosen randomly from the culture so that there is no bias in the dosage-mortality relationship through the biased selection of individuals. Given these conditions, groups as small as 15 or 20 individuals, and often even 10 for a preliminary estimation, are used for the dosage-response tests. The groups may then receive their dose of toxicant by exposure to a vapor, to a film of toxicant on glass, or to a paper impregnated with the toxicant for a given time; or they may be treated by direct application (topical application) of small drops of solutions of the toxicant in suitable solvents such as acetone (volatile) or mineral oil (nonvolatile). With exposure to films, the amount of toxicant taken up by the insect is not known unless subsequently determined by chemical analysis performed upon the insect, but topical application is a way of applying known amounts of toxicants and is very frequently used in the laboratory. With care,

insects can be injected with small amounts of organic solvents; oil solutions or emulsions and toxicants are sometimes given in this way.

Once the approximate LD50 has been determined in preliminary experiments, a range of doses is chosen so that they are as evenly spaced as possible over the mortality range around the LD50, and they should be in geometric progression for convenience, since toxic effect is better related to the logarithm of the dose than to the dose itself.[375] After the insecticide is applied, the various groups are held in a suitable container with water and nourishment so that they can be observed for the chosen period. Similar groups that have been exposed to films or treated with solutions containing no toxicant are set up as controls. For houseflies or mosquitoes, the maximum response to a given dose may be obtained within 24 hr, but this is not always the case; a period of 48 hr is safer and much longer periods may be needed for some insects. The number of individuals dead or moribund at each dose level are then simply noted at the end of the chosen period so that a plot of dosage-mortality can be obtained. In investigations of structure-activity relationships and for other fundamental investigations, it is often useful to record the progress of poisoning throughout the period of the experiment, since such observations reveal differences in the speed of action of closely related toxicants and in the signs of poisoning shown by the insects. This practice may also reveal other phenomena such as knock-down followed by recovery, which are not evident from simple observations of the end point of the poisoning process. If mortality occurs in the controls, this may be allowed for using Abbott's formula.[44]

The World Health Organization has sought for many years to standardize the measurement of toxicity as a diagnostic for resistance, and a modification of the Busvine-Nash test is now widely used for such determinations. Although designed more specifically for use with mosquitoes, it involves exposure of the insects to papers impregnated with toxicant in a suitable oil, and these can be used for other purposes. For the determination of susceptibility to DDT, papers impregnated with it at concentrations in mineral oil of 0.25, 0.5, 1.0, 2.0, and 4.0%, respectively are used, and for dieldrin the concentrations are 0.05, 0.1, 0.2, 0.4, 0.8, 1.6, and 4.0%. In the standard test, mosquitoes are transferred to holding containers after exposure to the papers for 1 hr, and mortalities are observed after 24 hr.[44,375]

The natural variability of insect populations is such that the dosage response curve obtained by plotting percentage kill against linear dose has an asymmetrical sigmoid shape; there is at first little increase in toxicity as the dose increases, then a rapid increase with dose, followed by a range in which large increments of dose produce only a small increase in mortality. For this reason, the dosage-mortality data are best transformed into the log-dosage-probit mortality form, in which the logarithm of the dose is plotted against percent mortality converted into units (probits) derived from the normal curve of distribution.[376] On the probit scale, the mean (50% effect) has a convenient value of 5.0, the probit transformation allows for the natural variation among individuals, and the data plotted in this way give a straight line with normal, homogeneous insect populations. This straight line, called the regression line, may be fitted by eye or calculated, and the dose required for any particular effect, for example LD50 or LD90, may then be read off directly. The regression line for a normal strain of insects is called the "baseline" in measurements of resistance, and a good deal of difficulty in interpreting resistance phenomena can obviously be avoided if baseline data are available for populations *before* they become resistant; the changes that occur in the nature of the baseline following exposure of a population to an insecticide are of considerable importance for forecasting future trends in susceptibility and for determining the way that resistance has developed.

Some basic propositions and problems are illustrated by reference to Figure 1, which shows log-dosage-probit mortality lines for four strains of the rust red flour beetle, *Triboleum castaneum*, exposed to papers treated with $\gamma$-HCH, the effect being measured by knock-down.[377]

The first point is that slight shifts in KD50 (LD50, LC50, etc.) may occur that are quite independent of the presence of specific genes for resistance. Thus, exposure of populations to insecticides may result in disappearance of the most susceptible genotypes, so that the LD50 will move in the direction of higher dosage without a corresponding increase in survival at the highest dose levels. This is indicated by the crossing of the regression lines for the normal strain ($S_1$) and the strain $S_2$ in Figure 1. Since true resistance requires that some individuals of an R-strain will survive a dose that kills all the individuals of an S-strain, strain $S_2$, although having a higher LD50 than $S_1$ and therefore apparent resistance at that dosage level, is no more resistant than $S_1$ at the 99% mortality level, which is the important consideration from a control point of view.

The regression line $R_1$ (Figure 1), which re-

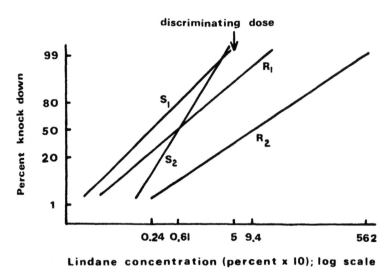

FIGURE 1. Log dosage-percent knockdown lines for two susceptible and two resistant strains of *Tribolium castaneum* exposed to lindane treated papers. In this case, knockdown is used as the criterion for mortality and is plotted as a percentage on the probit scale. (Data adapted from Dyte.[377])

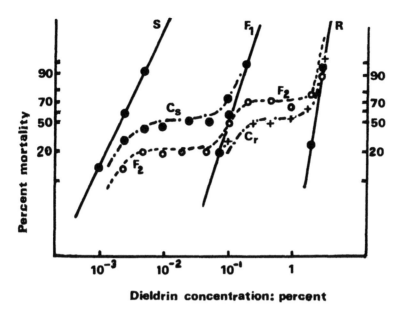

FIGURE 2. Concentration-mortality regression lines for dieldrin treated adult male houseflies (*Musca domestica*). S, normal houseflies; R, dieldrin resistant houseflies; $F_1$ and $F_2$, first and second generation progeny. $C_s$, combined results for various back-crosses between $F_1$ houseflies and S-strain; $C_r$, combined results for various back-crosses between $F_1$ houseflies and R-strain. (Data adapted from Guneidy and Busvine.[381])

presents a hypothetical mixture of $S_1$ and $R_2$, shows an increase in LD50 corresponding to that seen in $S_2$; in this case the regression line does not cross that of $S_1$ and some members of the $R_1$ strain would survive doses sufficient to kill all of the $S_1$ strain. Line $R_1$ has a lower slope than $S_1$ and this tendency may become extreme in heterogeneous populations produced during the selection of resistance. The situation is illustrated for dieldrin resistance in houseflies in Figure 2, which shows that the overall flattening of the line for heterogeneous $F_2$ insects is due to the fact that it is compounded from the individual regression lines of the three genotypes present.

There is a 40-fold difference between $S_1$ and the resistant strain $R_2$ in terms of the LD50s measured from these lines (Figure 1), but because of the flatter line for $R_2$, the LD99 of the $R_2$-strain is more than 100-fold higher than LD99 for the $S_1$-strain, the presence of highly resistant individuals becoming very obvious at this mortality level. The measurement of resistance level at the LD50 therefore underestimates the real problem at the LD99 level, which is the sort of level aimed at in practical control measures

directed at achieving complete kill; it can be seen that LD50 comparisons have little value in measurements of resistance if the slopes of the regression lines being compared are not known.

Small decreases in susceptibility resulting in shifts of the regression lines without changes in their slopes have been attributed to the "vigor tolerance" referred to previously. Thus, strains which have developed a high specific resistance to the toxicant used for selection frequently show small increased tolerances towards other, quite unrelated insecticides that have been attributed to this cause. It has also been invoked to explain why field strains are sometimes more tolerant to insecticides than the same insects reared in the laboratory. The implication of the concept, introduced by Hoskins and Gordon in 1956,[714] is that this "vigor" in the face of challenge from other toxicants is different from specific resistance and is due to the cumulative action of multiple minor genes having slight effects on the physiology which reduce susceptibility. It has been extensively discussed since then, but there is no compelling evidence that the explanation is correct, and opinions about it differ among authorities on resistance.

## B. INHERITANCE AND BIOCHEMICAL GENETICS OF RESISTANCE

### 1. Mode of Inheritance

With the above information available it is possible to consider the genetic origin of organochlorine resistance and its effect on the dosage-mortality regression lines which are normally the only tool available for the examination of resistance. The preadaptations frequently referred to before are genes that are hereditary units carried by the chromosomes contained in the cell nuclei and are passed on from generation to generation in the germ cells. They exist as alleles, such as those producing red or white flowers in peas, and in humans, blue or brown eyes. The DDT-resistance gene is the allele of the normal gene conferring DDT-susceptibility. If the alleles for resistance are not present in a population, resistance cannot develop since insecticides do not cause mutations that might produce them. Resistance can be due to a single gene allele (monofactorial), as is frequently the case when strong resistance to a toxicant has developed, or it can be due to the combined effect of several genes. The two are not always easy to distinguish if pure strains are not available. For example, the development of DDT-resistance is slow because the intermediate stages of resistance (heterozygotes) and the R-homozygotes appear to be less well fitted for survival than the normal susceptible homozygotes, and it seems that supporting or ancillary genes have to be assembled to produce a stable DDT-resistant genotype. For this reason, DDT-resistance in the housefly was once considered to be polyfactorial but is now known to be due to a single main factor frequently associated with one or more others. Examples of monofactorial inheritance of organochlorine resistance in insects are listed in Table 1.[43,373]

In the literature, a genetic character is usually denoted as recessive or dominant if the hybrid is identical with the normal (susceptible) or mutant (resistant) parent, respectively. The immediate offspring (hybrid; $F_1$ generation) of the cross of these two types will show a susceptibility level somewhere between the two extremes, which is often referred to as incompletely dominant resistance, etc., depending on how near the value is to the dominant level. If the level is intermediate between the two (logarithmic average), the resistance is said to be intermediate (also called

TABLE 1

Some Examples of the Monofactorial Inheritance of Resistance to Organochlorine Insecticides

| Dieldrin-resistance[a] | DDT-resistance[a] |
|---|---|
| Musca (i) | Musca: Deh (d) |
| | kdr (r) |
| | md (i) |
| Anopheles (7spp) (i)[b] | Anopheles (5spp) (r) |
| Aedes aegypti (i) | Aedes aegypti (i) |
| Culex fatigans (i) | Culex fatigans (d) |
| Pediculus (i) | Boophilus (r) |
| Cimex (i) | Pediculus (i) |
| Boophilus (d) | Blattella (r) |
| Phaenicia (i) | Euxesta (i) |
| Chrysomyia (i) | Drosophila (d) |
| Blattella (i) | |
| Hylemya (2spp) (i) | |
| Euxesta (i) | |

Data adapted from Brown.[43,373]

[a] d = dominant, i = intermediate, r = recessive.
[b] Dominant in *Anopheles gambiae* from the Ivory Coast.

semidominant). Georghiou has stressed the need for a uniform terminology in this work, and has suggested the terms *completely recessive, incompletely recessive, intermediate, incompletely dominant,* and *completely dominant* for various degrees of resistance that may be observed in the $F_1$ heterozygotes.[378] Table 1 shows that single-gene inheritance has now been established for DDT-resistance in 13 species and for cyclodiene-resistance (dieldrin resistance) in 19 species; the principal gene for DDT-resistance in the housefly (DDT-dehydrochlorinase, DDT-ase; *Deh*) is dominant (incompletely dominant in Georghiou's terminology), but the knock-down resistance gene (*kdr*) is incompletely recessive. In anopheles mosquitoes, except *Anopheles stephensi,* the principal DDT-resistance gene is recessive, but it is intermediate in *Aedes aegypti* and the spotted root maggot (*Euxesta notata*) and dominant in one strain of *Culex fatigans.* Cyclodiene resistance has proved to be intermediate (Table 1), except for its dominance in the cattle tick (*Boophilus microplus*) and in an Ivory Coast strain of *Anopheles gambiae.* The monofactorial inheritance is very sharply revealed in cyclodiene resistance, since the three genotypes are clearly distinct. The RR homozygotes are highly resistant and the Rr heterozygotes are nearly exactly intermediate, so that the three

The Demonstration of Monofactorial Inheritance by Crossing Experiments with Anopheles Mosquitoes

Susceptible killed rr by 0.4%* dieldrin) x Resistant homozygotes RR (survive 4% dieldrin)

Rr($F_1$) Heterozygotes of intermediate resistance survive 0.4% but are killed by 4.0% dieldrin

| rr x Rr | Rr x Rr | Rr x RR |
|---|---|---|
| rr Rr | rr Rr Rr RR ($F_2$) | Rr RR |
| rr killed by 0.4% dieldrin (50%) | rr killed by 0.4% dieldrin (25%)<br>Rr killed by 4.0% dieldrin (50%)<br>RR survive 4.0% dieldrin (25%) | Rr killed by 4.0% dieldrin (50%) |

FIGURE 3. The demonstration of monofactorial inheritance by crossing experiments with anopheline mosquitoes. Asterisk denotes concentration of dieldrin in the oil used for impregnating papers to which insects were exposed. Early experiments were with 0.33% papers; 0.4% later bacame standard. (Data adapted from Davidson.[379,380])

may be distinguished by applying two "discriminating doses" of dieldrin, giving a perfect demonstration of the segregation expected from monofactorial inheritance. In this manner, Davidson[379,380] demonstrated the existence of a single gene allele for dieldrin resistance in a Nigerian strain of *A. gambiae.* The results of crossing experiments between susceptible and homozygous dieldrin resistant adults are shown in Figure 3. Since the doses of dieldrin required to kill all of the susceptible (rr) homozygotes and all of the (Rr) heterozygotes ($F_1$) can be determined, the values obtained can then be used to determine the ratios of the three genotypes present. When back-crossed to each of the parental strains, the $F_1$ heterozygotes gave populations which were shown by use of the respective discriminating dose to contain a 1:1 mixture of heterozygotes and the appropriate parental type, while the response of the progeny of the $F_1$ generation to the discriminating doses showed that they consisted of the three genotypes in the ratio 1:2:1, findings characteristic of monofactorial inheritance.

With a situation as clear-cut as this, the effect of the monofactorial inheritance on the dosage-mortality regression lines can be easily predicted and in the reverse situation, the type of resistance can be deduced from the nature of the regression lines for parents and offspring. As an example, the regression lines of Figure 2 show the monofactorial nature of dieldrin resistance in adult male houseflies, *Musca domestica vicina.*[381] The $F_1$ heterozygotes have intermediate resistance and show a distinct steep regression line midway between the lines for the normal and resistant homozygotes. The $F_2$ generation, obtained (Figure 2) from the heterozygotes, contains susceptibles, resistant heterozygotes and resistant homozygotes in the ratio 1:2:1, so that for $F_2$ the dosage-mortality regression line exhibits plateaus corresponding to complete mortality of the susceptible (rr) at 25% and the heterozygotes (Rr) at 75%. The discriminating dose required to kill each of these types selectively can be read off from the middle of the appropriate plateau. Henceforth, the higher discriminating dose can be used as a check for the development of resistance of that type, since its application to a strain suspected of resistance will kill all individuals except fully resistant homozygotes (RR), whose presence in the population will then be revealed. The $F_2$ regression line represents the response to toxicant of a heterogeneous population, and in less clear-cut circumstances might be taken to be a scatter of points about a single straight line of low slope such as is frequently seen in the development of specific resistance. Thus, the situation is that as selection of a normal population proceeds, the frequency of resistant heterozygotes and homozygotes increases; susceptibles produced by interbreeding of the former are eliminated, and if selection pressure increases, the heterozygotes are finally themselves killed, leaving the fully resistant, homozygous population.

As indicated in Figure 3, the back-cross of heterozygotes (Rr) with the susceptible strain (rr) or with the resistant strain (RR) should give in each case a population consisting of a 1:1 mixture of heterozygotes and the particular parental type of the mating. This is again illustrated for *M. domestica vincina* in Figure 2, in which the plateaus corresponding to breaks in the dose-response curves of the heterogeneous back-cross populations occur at 50% mortality. Precisely similar results were obtained by Georghiou with a different strain of dieldrin-resistant houseflies (Super Pollard),[382] and it seems likely that indications of polyfactorial inheritance obtained by earlier workers were due to the use of mass crosses with genetically impure strains. When the $F_1$ heterozygotes (Rr) are back-crossed to the susceptible parent (rr), the susceptibles in the 1:1 mixture can be killed with a selective dose of insecticide and the back-cross repeated with the surviving heterozygotes. If this back-cross again produces a 1:1 mixture and the process can be repeated several times, good evidence is provided for monofactorial resistance. In the experiments just discussed, the closely similar results obtained with progeny from pairs of reciprocal back-crosses of $F_1$ flies to normal or resistant parents show that the resistance is autosomal, being transmitted to the offspring by either parent (no evidence of sex linkage), and there is no significant extra-chromosomal inheritance (cytoplasmic inheritance). Although the discriminating dose technique is highly useful for detecting such resistance in the field, it is possible that the proportions for apparent monofactorial inheritance could arise by chance, so that a full determination of the regression lines as shown in the figures is desirable for confirmation.

The monofactorial inheritance of intermediate resistance to dieldrin in some of the insects listed in Table 1 represents ideal cases. If the resistance is completely dominant, or nearly so, the resistance of the hybrids (heterozygotes) approaches that of the resistant homozygotes and the regression lines for the two types lie close together. In this case, only the back-cross of heterozygotes with the susceptible parent (since it produces 50% of susceptible rr individuals) will distinctly reveal the 1:1 segregation that indicates monofactorial inheritance; the backcross to the R-parent will give a

1:1 ratio of Rr and RR which may not be readily distinguishable in this case. This is seen in the dosage-mortality relationships established (Figure 4) for a DDT-R strain of *Culex fatigans* from Rangoon, which when crossed with a Rangoon-S strain shows monofactorial inheritance of nearly completely dominant DDT-resistance.[383] The hybrid ($F_1$) regression line lies near to the R-strain regression line; back-crossing of the $F_1$ hybrids to the S-strain gives 50% of hybrids (as shown by the plateau) whose regression line lies in the same area, and 50% of clearly distinguishable susceptibles. Conversely, monofactorial inheritance of completely, or nearly completely, recessive resistance means that the $F_1$ hybrids are as susceptible, or nearly so, as the susceptible homozygotes, so that the regression lines for these types lie close together. In this case only the progeny of the back-cross $F_1$ x RR, consisting of 50% of RR-homozygotes with a distinct regression line displaced to higher doses, will reveal monofactorial inheritance.

The general conclusion from the information available is that the development of strong resistance to a specific type of insecticide usually involves allelism in a single main gene. However, the resistance may derive from several genes in the initial stages of its development, so that it may not at first be apparent that the character is mainly determined by only one of them.[384]

## 2. Biochemical Genetics
### a. Houseflies

In more complicated types of inheritance, inspection of dosage-mortality regression lines may not be very helpful, and other aids are needed. Genetical analysis has been greatly assisted by the location of some of the major genes for resistance on particular chromosomes, aided by crossing experiments with marker strains in which the chromosomes carry mutant genes associated with visible characters. In the early days of investigations on mechanisms of resistance, one problem was to determine whether some particular characteristic of the resistant strain, such as enhanced metabolism of the toxicant, was a cause or a consequence of the resistance, a situation complicated by the fact that the resistant strain used was not always derived from the susceptible strain used for comparison. The situation was sometimes complicated, as in the case of DDT, or HCH, by

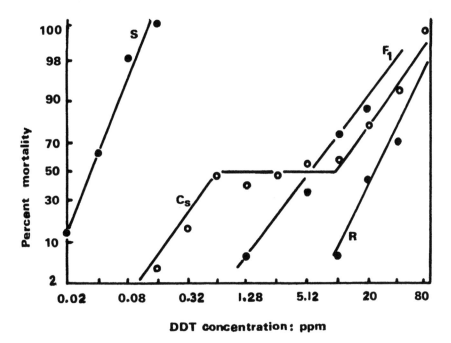

FIGURE 4. Dosage-mortality relationships for normal (S-) and DDT-resistant (R-) strains of mosquitoes (*Culex fatigans*), their first generation progeny (F₁) and the back-cross (C$_S$) between F₁ and S. The resistance is nearly completely dominant and only the susceptibles in the back-cross progeny (C$_S$) are clearly distinct from the resistant homozygotes (R). (Data adapted from Tadano and Brown.[383])

co-existence in a single strain of more than one resistance mechanism for the toxicant.

At this point, it is convenient to introduce some definitions of resistance which will be used later. The term *cross-resistance* has always been used when selection for resistance by one compound produces simultaneous resistance to one or more other compounds to which the insect has never been exposed.[385] From the point of view of genetic origins, it can be subdivided into what might be called "true cross resistance" or "pleiotropic resistance," arising when a single gene controls a mechanism that simultaneously confers resistance to several different insecticides, and *co-existing resistance* (meaning genes co-existing on the same chromosome), which occurs when the genes controlling resistance to different insecticides are close together on the same chromosome so that they are apparently selected together. Pleiotropic resistance is usually found with related structural types, for example, the resistance conferred by the enzyme DDT-dehydrochlorinase to DDT and its analogues that can be dehydrochlorinated, or cyclodiene-resistance (dieldrin-resis-

tance), which confers cross-resistance to all structurally related compounds as well as to HCH and toxaphene. A good example of co-existing resistance is afforded by the proximity of genes *a* (for organophosphate resistance) and *Deh* (for DDT-resistance) on chromosome-2 in the housefly (Figure 5); selection for organophosphate resistance frequently produces DDT-resistance due to *Deh,* although oddly, the reverse situation does not seem to occur.

Double, treble, or *multiple resistance* refers to the simultaneous existence in an insect of several different mechanisms, each protecting against a different type of poison; these mechanisms are usually controlled by genes on different chromosomes (except that co-existing resistance as defined above can be regarded as a special case of multiple resistance) and may arise quite separately, as is seen, for example, when the selection of insects with DDT then dieldrin, produces first DDT-resistance, then dieldrin-resistance. The main gene for dieldrin resistance in houseflies is on chromosome-4, and is quite distinct from the genes for DDT-resistance found on other chromo-

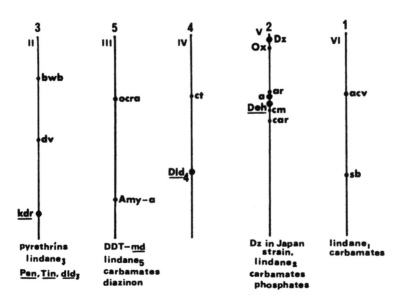

FIGURE 5. Location of genes for visible mutations and for insecticide resistance and metabolism in the five autosomes of the housefly. *kdr*, DDT knockdown resistance; *Deh*, DDT kill resistance (DDT-dehydrochlorinase); Dld₄ (Dieldrin resistance); *a*, organophosphorus insecticide resistance; DDT-*md*, microsomal detoxication; *Ox*, various oxidations.

Mutant marker genes; *bwb*, brown body; *ocra* (ochre eye); *ct*, cut wing tips; *ar*, aristapedia; *acv*, anterior crossveinless; *dv*, divergent; *sb*, subcostal- break; *car*, carnation eye; *cm* carmine eye; *Amy*-a, amylases.

Roman numerals, Hiroyoshi numbering system of linkage groups; Arabic numerals, Wagoner numbering system of chromosomes. (Data adapted from Brown[374] with additional information from Tsukamoto,[386] Sawicki and Farnham,[391] and Plapp.[503])

somes; resistance to DDT does not automatically accompany dieldrin resistance or vice versa (except in one or two cases of coexisting resistance in certain mosquitoes, where the genes for each resistance are distinct but apparently occur close together on the same chromosome). Of course, each of the mechanisms present in multiple resistance may confer resistance to other compounds of that particular type, so that, for example, a strain of houseflies selected for the full armory of genes may display resistance to DDT and its analogues, to dieldrin and its analogues (also HCH and toxaphene), and to a range of organophosphates and carbamates.

Finally, duplicate, or *multiplicate* resistance is said to occur when two or more mechanisms conferring resistance to the *same* type of poison are present together, as can happen with DDT or γ-HCH.

The great value of biochemical genetic studies with visably marked mutants is that they enable these complicated patterns to be unraveled so that individual mechanisms can be tracked down to particular genes. A considerable amount of information regarding the genetic origins of resistance mechanisms in houseflies and mosquitoes has been assembled using this sort of approach to the resistance problem. The housefly has six pairs of chromosomes, of which the five autosomes are shown in Figure 5, the sex chromosome being omitted since it is not normally involved.[384] The locations of some of the mutant genes conferring visible characters are shown, and a full tabulation of these is given in a review by Tsukamoto.[386] Some consideration of the present status of biochemical genetics is now worthwhile, since, as the name implies, it provides a most important link between the biological and the biochemical investigations of resistance. These investigations are most advanced for the housefly and the mosquito; some work has been done on cockroaches also, but the field of agricultural insects is largely unexplored. The early investigations of DDT-resistance in the housefly suggested that its origins differed from strain to strain; some of the work indicated it to be associated with a single, recessive gene, while other findings suggested dominant factors, or resistance of polygenic origin. A major mechanism for DDT-resistance in a number of insects involves its dehydrochlorination to DDE. This conversion is effected by an enzyme called DDT-dehydrochlorinase or DDT-ase, which is controlled in the housefly by a gene of incomplete dominance on chromosome-2 (Figure 5) located between the mutant markers ar and cm.[387] The gene is now usually called Deh (meaning dehydrochlorinase) because in several instances it has been possible to relate it to the enzyme which effects the detoxication.[388] In houseflies, for instance, it has been found in crossing experiments between susceptible and DDT-resistant homozygotes, that the Rr heterozygotes produce about half as much DDE from DDT as the resistant homozygotes do, and this correlation is stable through the usual series of test back-crosses. The housefly enzyme is also able to dehydrochlorinate DDD, so conferring cross-resistance (pleiotropic resistance) to this compound, but there is no cross-resistance to o-chloro-DDT, which is not easily dehydrochlorinated, and only slight resistance to methoxychlor, another molecule that is more refractory towards dehydrochlorination. There is evidence (Vol. II, Chap. 3B.3b) for the existence of variants of DDT-dehydrochlorinase which differ from each other in both activity and substrate specificity, and may be controlled by alleles of Deh. In vitro preparations of DDT-ase from different strains have been found to have varying capacities for detoxication, and Oppenoorth has suggested that the selection of alleles with intermediate levels of activity would present the possibility of secondary mutation to higher levels of activity.[388] One of the classical proofs of DDT-ase involvement in DDT-resistance is the fact that when it is inhibited in vivo, which can be done by mixing DDT with DMC (1,1-bis(p-chlorophenyl)methyl carbinol), some other DDT relatives or WARF-antiresistant, there is pronounced potentiation (synergism) of DDT-toxicity towards the resistant strains.[389] Some of the genes responsible for resistance to organophosphorus and carbamate insecticides are also located on chromosome 2 (Figure 5), and selection for resistance to malathion (an organophosphorus insecticide), Isolan® or carbaryl (carbamate insecticides) produces a tenfold increase in DDT-ase and a considerable increase in resistance to DDT.[390] Using the second chromosomal mutant marker stw (stubby wing), DDT-resistance in the Isolan selected strain was shown to be due to dominant inheritance of Deh. In the "SKA" strain of houseflies investigated at the Rothamsted Experimental Station, moderate levels of resistance to organophosphates contributed by gene a are associated in part of the population with high DDT-resistance due to Deh.[391] Thus it appears that selection with one type of compound has produced resistance to another operating by a different mechanism. This "cross-resistance" to DDT arises simultaneously with resistance to the selecting agent and these cases appear to be examples of co-existing resistance, especially since gene a for organophosphate resistance and Deh are close together. The situation is not quite so clear for the carbamates. In any case, the result is multiple resistance to organophosphates, carbamates, and DDT.

The gene kdr (knock-down resistance) is associated with delayed knock-down of flies exposed to DDT and was first characterized in an Italian strain from Latina Province; it is located on chromosome 3 at crossover ratios of 45.8 from mutant marker dv and 48.6 from bwb.[378] The mechanism of the resistance conferred by kdr is unknown, it does not involve dehydrochlorination (there is no synergistic effect with DDT-ase inhibitors) but it does enhance the expression of Deh, and is itself enhanced by the presence of Deh.[389] The gene is incompletely recessive with regard to delayed knock-down, but intermediate in its effect on "kill resistance." Houseflies have been shown to be resistant to DDT either when kdr is present in the homozygous condition, or when kdr in the heterozygous condition is combined with at least one low activity Deh gene, although the latter alone may confer no appreciable resistance. When acting in conjunction with Deh, it seems likely that the effect of kdr is to increase the toleration of DDT by decreasing the sensitivity of the insect nervous system, so that more time is available for it to be degraded by DDT-ase.[389]

The third chromosome in houseflies appears to be the site of a number of factors that may modify the effects of genes on other autosomes. Thus, another gene for DDT-resistance, called r-DDT, has been assigned to chromosome 3 in a Japanese strain of houseflies called JIR, which is highly resistant to DDT and DDT/DMC combinations. Like *kdr*, this gene is completely recessive; its precise locus is unknown, and it is not clear whether it is allelic to *kdr*. Further examination of the JIR-strain indicated that low nerve sensitivity to DDT is inherited as an incompletely recessive character controlled by a single gene on chromosome 3 which is probably the same as *r-DDT*. If rate of knock-down is taken to be a measure of nerve sensitivity (in the absence of complicating factors such as metabolism), it seems likely that genes *r-DDT* and *kdr* are allelic or identical.[378] A further third chromosomal gene conferring incompletely recessive resistance is *kdr-O*, which is responsible for resistance to DDT knock-down in the American Orlando strain of houseflies. This gene and *kdr* are not allelic, but reinforce one another when present together in the heterozygous condition. *Kdr-O* apparently confers high cross-resistance to *o*-chloro-DDT, which is not enzymically dehydrochlorinated, so that it resembles *kdr* in this respect.[392] Genetic manipulations in which chromosomes 2 and/or 3 of the Orlando strain were substituted by the corresponding marked chromosomes from susceptible marker strains, showed that recessive resistance to methoxychlor, Dilan®, and pyrethrins was also under the control of a gene or genes on chromosome 3. When the *kdr-O* gene was subsequently transferred from the Orlando strain into a DDT-susceptible strain with concurrent selection with DDT or pyrethrins/piperonyl butoxide mixture, resistance to Dilan, *o*-chloro-DDT and methoxychlor was associated with the transferred resistance to DDT and pyrethrins, suggesting that *kdr-O* is responsible for a pleiotropic resistance to several compounds through some mechanism other than their metabolic detoxication.[393]

As indicated previously, the SKA strain of houseflies has gene *a* (chromosome 2) controlled resistance to organophosphates associated with simultaneous DDT (*Deh*) resistance, and it also exhibits delayed penetration of several insecticides through the cuticle. Diazinon, dieldrin, and DDT penetrate at about half the rate found with normal flies, and there is a low resistance (12-fold) to tributyl tin acetate, these phenomena being associated with a single gene on chromosome 3 which has been called the penetration factor *Pen*.[391] This gene appears to be allelic with an incompletely recessive gene having similar properties characterized in a highly parathion

resistant strain of houseflies by Plapp, and called *tin*[394] (Figure 5), since it conferred up to tenfold resistance to tributyl tin chloride. DDT and dieldrin penetrate more slowly into strains containing the *tin* gene[391] and knock-down with these compounds, and also with parathion, is slower than in normal strains which lack the gene. These are significant findings, because although such genes may confer no resistance to kill by themselves, they appear to act as important intensifiers of resistance when they are present with other mechanisms such as metabolic detoxications. For example, a 30-fold resistance to parathion associated with gene *a* combined with a threefold resistance associated with *tin* becomes 100-fold resistance to parathion when the *tin* gene is incorporated into the parathion resistant strain (see Table 3). The gene *Pen* confers slight resistance to diazinon, dieldrin, DDT, and methoxychlor; when it is combined in the homozygous condition with a chromosome 5 factor which confers tenfold resistance to DDT due to a microsomal detoxication mechanism, the flies become immune to DDT.[391]

Also on chromosome 3 is a minor factor for dieldrin resistance, referred to as *dld*$_3$ to distinguish it from the major factor or chromosome 4 (Dld$_4$).[391,395] This factor delays the onset of signs of poisoning by dieldrin because it delays dieldrin penetration when either heterozygous or homozygous; however, it gives no resistance to kill when heterozygous and no more than a twofold protection when homozygous. It appears to be incompletely recessive or intermediate in character, since although the heterozygote is practically indistinguishable from the susceptible parent in several respects, it is nevertheless killed more slowly than the susceptible parent by a number of unrelated insecticides and differs from it in regard to penetration and rate of knock-down. Sawicki and Farnham regard *Pen*, *dld*$_3$, and *tin* as being the same gene, or alleles of it.[391] It is not known whether this gene interacts with the main dieldrin-resistance gene Dld$_4$. Available evidence suggests that the genes *tin* and *kdr-O* are not identical; *kdr-O* delays knock-down by DDT more efficiently than *tin* does, but the latter gene retards DDT penetration more effectively, so that the two actions are apparently different.

Supporting evidence for the role of non-metabolic mechanisms in DDT-resistance comes from work with DDT analogues. Thus, houseflies that were highly resistant to Prolan® (Vol. I; Figure

2, structure 17), which is not subject to the dehydrochlorination mechanism, were also highly resistant to DDT but converted less of it into DDE than did other, less resistant strains.[396] This strain was also resistant to methoxychlor, o-chloromethoxychlor and o-chloro-DDT (derived from methoxychlor and DDT, respectively, by introducing an o-chloro-atom into one of the benzene rings) and deutero-DDT in which the tertiary hydrogen in DDT is replaced by deuterium. It converted less deutero-DDT to DDE than other less resistant strains did, and while the o-chloroderivative was absorbed just as well by this strain as by others, it was hardly metabolized during a 24-hr period. Ortho-chloro-DDT is refractory to both chemical and enzymic dehydrochlorination and is therefore usually toxic to houseflies whose resistance mechanism is DDT-dehydrochlorinase. This resistance to these analogues, especially o-chloro-DDT, shown by the Prolan-R strain is evidently not associated with dehydrochlorination.

In the course of experiments designed to correlate levels of DDT-ase in various housefly strains with resistance to DDT, Oppenoorth compared a strain L from the University of Illinois, highly resistant to DDT and originally used by Lipke and Kearns to isolate DDT-ase, with a strain $F_c$ of Danish origin and resistant to both DDT and diazinon.[397] The diazinon resistance of the original strain F had earlier been shown to be partly associated with gene *a* on chromosome 2 and partly with other, undefined genes. When gene *a* was removed from the original strain F by outbreeding to a susceptible strain, further selection with diazinon then retained the resistance to both diazinon and DDT in the substrain $F_c$. It was soon found that, in contrast to strain L, live $F_c$ flies produced little DDE from DDT, and F-DMC (bis(p-chlorophenyl)trifluoro methylcarbinol), which inhibits DDT-ase and is therefore a good synergist for DDT against strain L, was ineffective as a synergist with $F_c$ flies. On the other hand, the compound sesamex (sesoxane; 2-(3,4-methylenedioxyphenoxy)-3,6,9-trioxaundecane), a well established inhibitor of the oxidative detoxifying enzymes located in the microsomal fraction of various insect tissues, proved to be a powerful synergist for DDT against strain $F_c$, although it was virtually ineffective as a synergist against strain L. Since sesamex also synergized diazinon against the $F_c$ strain, it became apparent that selection with diazinon had

resulted in the development of an oxidative resistance mechanism effective for both compounds, and it was subsequently shown to be associated with a gene on chromosome 5, which was named DDT-*md*. This appears to be an example of pleiotropic resistance, since one gene confers resistance to the two different structural types.

The gene DDT-*md* has so far been definitely found in only two strains of housefly, $F_c$ and the related SKA strain mentioned previously, but other insects can degrade DDT by oxidative processes, as will be seen. The LD50 of DDT for the $F_c$ strain is more than 10 $\mu$g/fly, in contrast to 0.15 $\mu$g for a susceptible strain, giving a resistance factor of more than 67-fold (that is 10/0.15), which is similar to that found with diazinon (2/0.03). For the SKA strain, an LD50 is recorded for DDT of more than 100 $\mu$g/fly (resistance factor >625) and for diazinon of 11 $\mu$g/fly (resistance factor 209).[398] By crossing the SKA strain with a multi-marker susceptible strain *ac; ar; bwb; ocra*-SRS, carrying the markers on chromosomes 1, 2, 3, and 5 respectively, the four triple-marked strains and the quadruple-marked one shown in Table 2, were obtained.[391] Chromosome 4 carries no marker and could be derived from either or both parents in the triple marked strains; in SKA flies it carries the major dieldrin-resistance factor and is derived only from the SKA parent in the quadruple-marked strain, as can be ensured by selection of the flies with dieldrin. In each of these strains, therefore, a different chromosome from SKA has been isolated, and so the separate contribution each chromosome makes to resistance to various compounds can be measured. These contributions are listed under each chromosome in Table 2. DDT-*md* on chromosome 5 is of intermediate dominance, being less protective than *Deh*, which is selected along with gene *a* resistance to organophosphates and confers considerable resistance to DDT. By itself, the DDT-*md* gene isolated from SKA gives slight to moderate resistance to DDT (tenfold) and methoxychlor (20-fold). However, in the intact SKA flies it is reinforced by the penetration factor *Pen* on chromosome 3 and the dehydrochlorination mechanism to *Deh*, which by itself evidently confers strong resistance, is also present in some individuals. In $F_c$, the main resistance factors appear to be *Pen* and DDT-*md*, but here again *Deh* is present and may make some contribution to the

## TABLE 2

Genetic Constitution of Housefly Strains Derived from the Cross SKA x ac; ar; bwb; ocra SRS and the Contribution of Isolated SKA Autosomes to Different Resistance Mechanisms

| Strain | Mutant markers | Autosomes from | | Resistance gene |
|---|---|---|---|---|
| | | Susceptible | SKA | |
| 446.500 | ac; ar; bwb | 1,2,3,(4) | 5(4)[a] | DDT-*md* (Sesamex inhibited) |
| DR | ac; ar; bwb; ocra | 1,2,3,5 | 4 | *Dld*$_4$ (dieldrin resistance) |
| 348 | ac; ar; ocra | 1,2,(4),5 | 3,(4) | *Pen* (delayed penetration) |
| 393 | ac; bwb; ocra | 1,3,(4),5 | 2,(4) | *a* (modified aliesterase) |
| 338 | ar; bwb; ocra | 2,3,(4),5 | 1,(4) | slight chlorthion resistance |

### Resistance factors in SKA (Composite parent-strain)

| Autosome: | 1 | 2 | 3 | 4 | 5 |
|---|---|---|---|---|---|
| Substrain: | 338 | 393 (gene *a*) | 348 (*Pen*) | DR (*Dld*$_4$) | 466.500 (DDT-*md*) |
| | Chlorthion | chlorthion x 14 | chlorthion x 5[b] | Dieldrin (x 700) | diazinon x 9 |
| | Chloroxon | chloroxon x 12 | diazinon x 2[b] | | diazoxon x 4.6 |
| | Malaoxon | diazinon x 13 | dieldrin x 2[b] | | ethyl malaoxon x 3.3 |
| | (x2) | diazoxon x 8 | DDT x 2[b] | | methoxychlor x 20 |
| | | ethyl-chlorthion x 48 | tributyl tin | | DDT x 10 |
| | | ethyl-chloroxon x 14 | acetate (x 12) | | |
| | | ethyl malathion x 4 | | | |
| | | ethyl malaoxon x 6.5 | | | |
| | | malathion x 1.4 | | | |
| | | malaoxon x 6 | | | |
| | | parathion x 15 | | | |
| | | paraoxon x 3.6 | | | |
| | | *Deh*; | | | |
| | | DDT x 1000 | | | |

(Data adapted from Sawicki and Farnham.[91])

a Parentheses indicate that autosome 4 could come from either or both parents. In this book Arabic rather than Roman numerals are applied to the chromosomes (see also legend to Figure 5).

b Factor is a measure of delayed penetration, not resistance to kill, although dieldrin-kill resistance is (X2) when *Pen* is homozygous.

total resistance. However, according to Sawicki,[391] the tenfold resistance to DDT conferred by DDT-*md* in the SKA flies, combined with the factor *Pen* conferring a twofold reduction in penetration for DDT, can confer immunity to DDT when present in homozygous condition. Therefore, a contribution to resistance by *Deh* may not be necessary in either the SKA or the $F_c$ strain; in fact, DDT-ase occurs in less than 20% of SKA flies. The resistance to diazinon conferred by DDT-*md* is of intermediate dominance in strain SKA and it appears to give slightly less protection against this compound than gene *a* does. Gene *a* is reinforced when *Pen* is also present and DDT-*md* can interact to give high resistance to compounds normally unaffected by this mechanism if gene *a* is present simultaneously. The locus of DDT-*md* has been determined to be approximately 30 crossover units from *ocra*.

The SKA strain provides a good illustration of the various types of resistance that may be present simultaneously in one strain. The gene *a*, *Deh* relationship has already been cited as a special example of double resistance due to two genes co-existing on the same chromosome. DDT-*md* exhibits pleiotropic resistance towards DDT, methoxychlor, diazinon, and diazoxon; gene *a* toward a series of organophosphates, and *Pen* toward diverse structural types. The presence of the factors on chromosomes 2, 3, 4, and 5 together in the same strain and embracing different structural types is quadruple (multiple) resistance, and there are three distinct resistance mechanisms for DDT in the same insect (*Deh*, DDT-*md* and *Pen*) comprising triplicate (multiplicate resistance). It is clear from the work of Plapp and of the Rothamsted group that the combined effect of resistance genes is frequently much more than additive. Some of the resistance levels obtained when different factors are present together in houseflies have been listed by Plapp, and are shown in Table 3.

As with other organochlorine compounds, investigations of the causes of resistance to the cyclodiene compounds and $\gamma$-HCH are complicated by lack of knowledge of their precise mode of action. This makes the simplicity of the genetic background of dieldrin resistance as found in many insect species (Table 1) all the more intriguing. The investigations of Busvine and of Georghiou that demonstrated the monofactorial inheritance of dieldrin-resistance in *Musca domestica* have already been referred to and a similar conclusion was reached by Milani at about the same time. While investigating DDT-resistance involving the gene *kdr*, Milani concluded that a moderate dieldrin resistance in this strain was not associated with chromosome 3, nor was it located on chromosome 2 of a strain in which dieldrin resistance had appeared spontaneously, without selection. Crossing experiments between the two strains further confirmed that the factors for dieldrin resistance in these strains were different.[399]

More recent investigations have shown that a major gene ($Dld_4$) conferring dieldrin resistance of intermediate dominance is located on chromosome 4 (Figure 5) and that in some strains minor effects are produced by another gene on chromosome 3 ($dld_3$). In 1966, Oppenoorth conducted crossing

TABLE 3

The Effect of Combinations of Resistance Factors in Houseflies

Gene (individual resistance level conferred)

| Toxicant | | | Combined resistance level |
|---|---|---|---|
| DDT | *Kdr*-O (200) | *Deh* (100) | 2500 |
| DDT | *tin* (2) | DDT-*md* (50) | 900 |
| DDT | *ox* (2)[a] | *Kdr*-O (200) | 450 |
| Dieldrin | *tin* (1·3) | $Dld_4$ (1000) | 2000 |
| Parathion | *tin* (3) | gene[a] (30) | 100 |

(Data adapted from Plapp.[503])

[a] A second chromosomal gene responsible for the oxidation of carbamates and phosphates.

experiments with a Uruguayan strain of houseflies (called strain $U_2$) which had been kept under selection pressure with $\gamma$-HCH but had never been exposed to dieldrin. Using a strain carrying the gene on chromosome 4 for the visible character *ct*, it was shown that the cross-resistance to dieldrin was associated with chromosome 4 in strain $U_2$ but no clear evidence was found for the involvement of secondary factors from other chromosomes in this dieldrin-resistance.[400] Later, Sawicki and Farnham examined the genetic origins of dieldrin-resistance in the SKA strain using the strain with mutant markers for genes 1, 2, 3, and 5 as shown in Table 2 and showed that the major factor for dieldrin-resistance in this strain is also located on chromosome 4.[395] At the same time, they showed that a factor on chromosome 3 called $R_2$, when present in the homozygous condition, produced a small resistance to it (twofold) at the death end point in toxicity determinations. This factor was later equated with the penetration factor *Pen*, is thought to be the same as *tin*, and is referred to as $dld_3$ in Figure 5. The synthetic strain DR, homozygous for $Dld_4$ from the SKA strain, is immune to high doses of dieldrin (up to 50 $\mu$g per fly) for the first 24 hr following treatment, but the end point of toxicity is not reached until 3 days later, the flies being gradually killed during the second and third day to give a straight line dosage-mortality regression line on the fourth day. The susceptibility to dieldrin is then measurable and the LD50 is about 7.0 $\mu$g/female fly, corresponding to a 700-fold resistance level. The influence of $dld_3$ in the intact SKA strain is not clear, but since it delays the penetration of dieldrin in this strain and can confer a twofold resistance when homozygous, it seems likely that it does make some contribution to the total dieldrin resistance. Sesamex, the well known inhibitor of microsomal oxidations (Vol. II, Chap. 3B.3a), has no effect on the toxicity of dieldrin to strain DR, as was found earlier with the homozygous dieldrin-resistant Slough strain of *Musca domestica vicina*, and the resistance mechanism controlled by this gene is unknown. It is likely that $Dld_4$ is the factor responsible for the extensive cross-resistance spectrum shown to other cyclodienes and $\gamma$-HCH by dieldrin-resistant housefly strains (Vol. II, Chap. 3B.3b).

The cross-resistance to $\gamma$-HCH has been an intriguing problem for many years, and there has long been evidence that this molecule is vulnerable to metabolic detoxication processes significant for resistance, in addition to the cyclodiene-resistance of unknown mechanism which appears not to involve metabolism. Before the application of marked strains in this problem, one or two lines of evidence indicated that several genes might be involved in HCH-resistance in houseflies. In particular, crosses of the HCH-resistant Uruguay strain, later used by Oppenoorth, with a normal colony gave hybrids ($F_1$) having an intermediate level of resistance and an $F_2$ with no segregation and few parental phenotypes, suggesting resistance of polygenic origin. Tsukamoto then used the marker gene technique to analyze the HCH-resistance of a Japanese strain and concluded that an incompletely dominant gene on chromosome 5 was responsible for the resistance, with significant contributions by chromosomes 3, 2, and 1. The heterozygous contributions of the chromosomes were rated as $5 > 2 > 3 > 1$, and the homozygous effects as $3 > 5 \simeq 2$. In another Japanese strain, $\gamma$-HCH resistance was conferred mainly by an incompletely dominant gene on chromosome 3.[386]

In the course of his examination of the Uruguayan strain $U_2$ for the locus of dieldrin resistance, using a strain with the marker *ct*, Oppenoorth included that the dieldrin resistance gene $Dld_4$ was partly responsible for $\gamma$-HCH resistance, with additional contributions from other chromosomes, especially the second. Thus, each of the five chromosomes of the housefly has been implicated in $\gamma$-HCH resistance in one strain or another, and it seems that different factors can operate in different strains. From what is now known, the dieldrin-resistance factor $Dld_4$ is always involved in $\gamma$-HCH resistance; in Tsukamoto's experiments with one of the Japanese strains, chromosome 4 was not examined, and therefore any contribution it may have made is not listed.

*b. Mosquitoes*

In anopheline mosquitoes, two main types of organochlorine resistance arise, dependent on at least two genetic factors that are normally separate. One confers a moderate degree of resistance to DDT and its analogues, while the other usually confers a high degree of resistance to dieldrin and related cyclodiene compounds and a moderate degree of cross-resistance to $\gamma$-HCH. The two types of resistance can occur simultaneously in the same individual and this double resistance has become very common.[48]

Several cases of resistance to DDT in anophelines appear to be due to a single recessive or incompletely recessive gene. This was clearly shown for a Javan strain of *Anopheles sundaicus*, having a tenfold resistance to DDT, by Davidson, who crossed it with a susceptible strain from Kuala Lumpur. Unlike the parent Javan strain, the $F_1$ hybrids were all susceptible to 2.5% DDT by the Busvine-Nash test, and 75% of the $F_2$ generation were also killed by this concentration. Since the $F_1$ back-cross to the susceptible parent gave all susceptible progeny, and that to the resistant parent gave 50% of resistant individuals, all the facts indicate monofactorial resistance due to a recessive gene.[43] The situation is generally similar for *A. albimanus*, *A. pharoensis* (although a strain of the latter from the United Arab Republic shows partial dominance), and *A. quadrimaculatus*. It appears, however, that in anophelines the expression of DDT-resistance can be modified significantly by the genetic background of the strain. In a strain of *A. stephensi* from Madras, its inheritance was influenced by factors linked to the male sex chromosome, so that resistance introduced by the male parent was intermediate in the hybrid, while that introduced by the female was incompletely recessive. DDT-resistance in *A. stephensi* appears to be of intermediate dominance, but in a strain from Iran it was nearly completely dominant in the presence of homozygous dieldrin-resistance but had a lower degree of dominance when the dieldrin-resistance was heterozygous. DDT-resistant *A. sacharovi* and *A. sundaicus* produce much DDE from DDT, but the role of dehydrochlorination or other detoxication reactions is unclear in the other species, since the toxicity of DDT towards them was not improved by inhibitors of DDT-dehydrochlorinase (WARF-anti-resistant) or of microsomal mixed function oxidases (piperonyl butoxide).[378]

The inheritance of dieldrin-resistance has been investigated in nine species of Anopheles; *A. gambiae*, *A. albimanus*, *A. quadrimaculatus*, *A. pharoensis*, *A. stephensi*, *A. sundaicus*, *A. pseudopunctipennis*, *A. funestus*, and *A. sacharovi*, and appears to be monofactorial and incompletely dominant, (intermediate) in all cases except for strains of *A. gambiae* from the Ivory Coast, in which it appears to be dominant. The genotypes in these cases can usually be clearly distinguished by

the use of discriminating doses of insecticide (Figure 3). Some crossing experiments that have been carried out between strains of *A. gambiae* having intermediate or dominant types of dieldrin-resistance suggest that the two types of character may be inherited independently, but verification of this requires the use of morphological marker strains, of which few have been described for anophelines.[378]

Resistance to both DDT and dieldrin occurring in the same species has been examined in *A. albimanus*, *A. quadrimaculatus*, *A. stephensi*, and *A. pharoensis*.[401] When crossed with susceptible strains, doubly resistant *A. albimanus* and *A. quadrimaculatus* show the difference in the two resistances in regard to mechanisms of inheritance; dieldrin resistance shows partial dominance, and the recessive character of DDT-resistance is revealed. Further proof in the case of *A. albimanus* comes from the isolation of strains showing DDT-resistance without dieldrin-resistance and vice versa. In *A. pharoensis*, the genes for DDT and for dieldrin resistance are apparently close together on the same chromosome so that a case of "co-existing" resistance occurs, the two factors being selected together by either toxicant.

These findings have some significance for malaria control. A highly efficient insecticide like dieldrin will rapidly select out a resistant population from a natural one containing quite low original frequencies of resistant individuals, and it appears, in fact, that such individuals are by no means uncommon among some populations that have never been treated with this chemical. A less efficient toxicant like DDT (partly because its irritant nature provokes avoidance), when applied against a population containing some naturally resistant individuals whose resistance even in the homozygous state is only moderate and in the heterozygous state is hardly perceptible (since the resistance is recessive), takes much longer to select out a resistant population. This makes DDT the insecticide of first choice in malaria eradication campaigns, because even if the potential for DDT-resistance is present, selection for resistance may be so slow that eradication of the malaria parasite can be achieved before the numbers of resistant individuals is sufficient to maintain its transmission. Another important consideration is that because of the moderate level of the DDT-resistance, this toxicant will still often kill sufficient numbers of a resistant population of an inefficient vector species (although the frequency of control measures and the amounts applied may have to be increased) to prevent malaria transmission by them. With a highly efficient vector species such as *A. gambiae*, on the other hand, it seems that DDT, even in the absence of resistance, is unable to inflict sufficient mortality on the populations to interrupt malaria transmission. For this species, HCH or dieldrin is a better choice, except for the resistance question, although cyclodiene-resistance is uncommon amongst East African *A. gambiae* despite high selection pressure in some areas. Thus, although cyclodienes might still be used successfully in some areas of Africa, the prospect of double resistance, coupled with the inefficient control by DDT with *A. gambiae* makes the possibility of malaria eradication at some future date most uncertain without additional control measures.[401]

*Culex pipiens fatigans* is normally less susceptible to DDT than other mosquitoes and can therefore become decisively resistant to DDT when selected with it. The resistance to DDT (tenfold) in a Northern Indian strain was monofactorial and recessive, whereas in a Philippine strain with 13-fold resistance and a highly resistant strain from Southern India, it was monofactorial and nearly completely dominant. Monofactorial inheritance with nearly complete dominance was also found in a strain from Rangoon that was selected with DDT for 10 to 15 generations, when the larval LC50 became 25 to 40 ppm (concentration in water) compared with 0.15 ppm for the original strain collected in the field.[383] The same field strain was also selected with dieldrin for 11 generations to give a dieldrin-resistant strain, or successively with DDT and then dieldrin to give a strain resistant to both DDT and dieldrin (double resistance). The dieldrin-resistance was found to be monofactorial and of intermediate dominance and crosses were then made with susceptible strains carrying morphological markers to determine the relationships between the loci of the resistance factors and the markers. The DDT-resistance factor was not linked with gene *w* (white eye) on chromosome 1, but with genes *y* (yellow larva) and *ru* (ruby eye) on chromosome 2, while the dieldrin resistance was independent and linked with the gene *kps* (clubbed palpi) on chromosome 3. The DDT-resistance gene is located 20 crossover units from *y* and 45 crossover units from *ru* on chromosome 2, while that for dieldrin-resistance is

35 to 40 from *kps* (Figure 6). In this species, the correlation between DDT-resistance and dehydrochlorination is not so evident as it is in *Ae. aegypti*. Thus, tolerance toward different DDT-analogues varies from strain to strain;[378] a Savannah strain of *Culex fatigans* was found to have a resistance mechanism for *p,p'*-DDT but not for *o*-chloro-DDT or *o,p'*-DDT, while a strain from Delhi had resistance to *p,p'*-DDT which extended to *o,p'*-DDT, *o*-chloro-DDT, deutero-DDT, and DDD. The ortho-substituted compounds are resistant to enzymic dehydrochlorination and therefore are usually more toxic than DDT to the strains having the DDT-ase resistance mechanism, as found with the Savannah strain. Hence, the resistance to them exhibited by the Delhi strain suggests that a different defense mechanism is involved. In support of this possibility, inhibitors of DDT-dehydrochlorinase, such as DMC or WARF anti-resistant, do not show much synergistic activity with DDT against the Delhi strain.

This variability also extends to dieldrin-resistance, which continues, for example, to be very low in California where selection pressure has been present for some years, whereas resistance is high in Upper Volta, Nigeria. It may be that different genes are involved or that the genetic background of the strains influences the expression of a single common gene for dieldrin-resistance, as in the case of DDT-resistance in *Anopheles stephensi*. The physiological mechanism of dieldrin resistance is unknown, as in the case of houseflies, although *C. fatigans* appears to be able to convert dieldrin into the corresponding *trans*-diol by breaking open the epoxide ring, and this conversion appears to be a detoxication mechanism.[402]

Selection of *C. fatigans* with deutero-DDT induces resistance to this compound in parallel with the cross-resistance to DDT (240-fold to both insecticides in eight generations).[403] Conversely, DDT induces rather less cross-resistance to deutero-DDT in this insect (2.5-fold for a tenfold increase in DDT-resistance). The finding that DDT-resistance and dieldrin resistance are associated with genes on different chromosomes in this species is a significant one, for it means that there is no chance that these two resistances can be selected together, as happens with *A. pharoensis* and *Ae. aegypti*. In an Oak Ridge strain of *Culex tarsalis* DDT-resistance was found to be incompletely recessive, but determination of the number

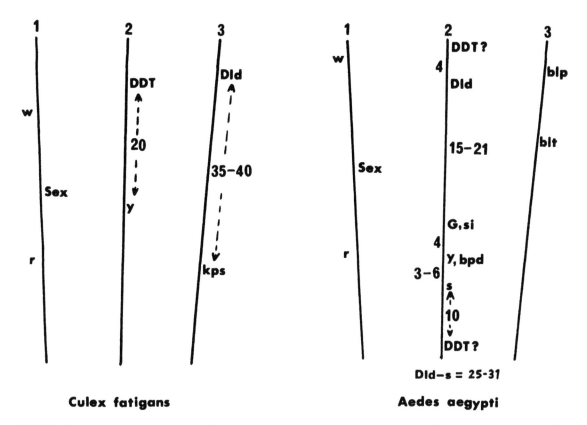

**Culex fatigans**          **Aedes aegypti**

FIGURE 6. Location of genes for dieldrin-resistance and DDT-resistance in the three chromosomes of the mosquitoes *Culex fatigans* and *Aedes aegypti*.

Mutant marker genes; *w*, white eye; *r*, red eye; *y*, yellow larva; *kps*, clubbed palpi; *G*, gold; *si*, silver thorax; *s*, spot abdomen; *bpd*, black pedicel; *blp*, black palp; *blt*, black tarsus. The position of the DDT-resistance gene on chromosome 2 of *Ae. aegypti* is uncertain. (Data adapted from Brown and Pal[43] with data of Lockhart, Klassen, and Brown.[406])

of genes involved was prevented by partial overlap of the susceptibility ranges of the R- and S- strains. *Culex tarsalis* larvae having a 2000-fold resistance to DDT had also appreciable resistance to other compounds which undergo dehydrochlorination and to Prolan, *o*-chloro-DDT and *o*-chloro-DDD in which dehydrochlorination is impossible or difficult.[404] DDT/DMC and DDT/WARF combinations did not overcome the resistance, but compounds such as iso-DDT and DDT hydroperoxide in which oxidation at $C_2$ is prevented or inhibited by substitution largely overcame the resistance. These findings indicate that dehydrochlorination is rather insignificant for the resistance of these larvae. An examination of the fate of DDD, using the compound labeled with carbon-14, showed that it was metabolized by hydroxylation at the tertiary carbon atom, as well as by dehydrochlorination, and polar metabolites could be recovered from both the larvae and the water in which they were kept. It therefore seems likely that oxidative mechanisms are significant for the degradation of DDT and its relatives in this strain.

Most populations of the yellow fever mosquito (*Ae. aegypti*) can develop resistance to DDT on account of a major gene of intermediate dominance located on chromosome 2. The segregation is not as clear cut as in other cases, however, since there is considerable overlapping between the genotypes. In some strains, the DDT-resistance requires the support of certain ancillary genes which are slowly selected during DDT-pressure. The genetic aspects of resistance in *Ae. aegypti* have attracted more attention than is the case with *C. fatigans*, more morphological markers being available for the former. Contrary to normal experience with the two types of resistance, there appears to be a link between DDT-resistance and dieldrin-resistance (which is also monofactorial and of intermediate dominance) in a number of strains of *Ae. aegypti* so far examined. The first DDT-resistant population of this species was found in Trinidad in the middle 1950s, and it proved to be susceptible to dieldrin, but a strain in San Juan, Puerto Rico, developed dieldrin as well as DDT-resistance when it was selected with DDT.[43] When this strain was continually back-crossed to a susceptible strain, either DDT or dieldrin maintained the resistance to both compounds. Since then, many populations of this mosquito in the Caribbean area generally have been found to have developed DDT and dieldrin resistance together. In several cases, interpretation of the situation is complicated by the fact that dieldrin had actually been used, but in Barbados, DDT exposure produced resistance to both compounds although no dieldrin had been applied on the island. Populations of *Ae. aegypti* in Bangkok have also developed dieldrin-resistance along with DDT-resistance, although there has been only sparing use of HCH and dieldrin.[408]

The determination of the location and relationship of the genes for DDT and for dieldrin-resistance in *Ae. aegypti* has proved to be a complex matter. Using susceptible strains with morphological markers, Klassen and Brown showed that neither gene is linked with markers on chromosomes 1 or 3, so that both are evidently on chromosome 2.[405] They could not at first be separated in the Puerto Rico (Isle Verde) strain, either by selection with dieldrin or with DDT coupled with backcrossing to a susceptible strain. Subsequently, the two were separated and Dld-R; DDT-S and Dld-S; DDT-R strains were obtained; crossing experiments between these and between the doubly-resistant and doubly-susceptible strains then indicated that the two genes were only four crossover units apart on chromosome 2 (Figure 6), so that selection for one gene appeared likely to select for the other. The order of genes on chromosome 2 appeared to be DDT-*Dld-G-bpd-y-s,* and the situation appeared to be similar in the Trinidad strain. In contrast, some other work indicated the DDT-gene in the Trinidad strain to be on the opposite side of *s* from *bpd*, and since *bpd* is closely linked with *y*, the order appeared to be *bpd, y-s*-DDT. Recently reported work by Brown and his colleagues involved crossing experiments between five dieldrin R-strains and susceptible marker strains and indicated that the gene order on chromosome 2 is *Dld-si-s* (*si* is silver, very closely linked with *Gold*) and that DDT is on the

opposite side of *s* from *si* or *y* in the Trinidad strain, indicating the orders *si-s*-DDT and *y-s*-DDT, with the total distance *s*-DDT being about ten units. Commenting that no unequivocal proof of the relationship between the resistance genes on chromosome 2 has yet been provided, they conclude that the order in *Ae. aegypti* is probably Dld-*si-bpd, y-s*-DDT.[406] This implies that the total crossover distance between *Dld* and DDT is about ten times greater (Figure 6) than was apparent from the earlier experiments, so that the frequent close association between DDT and dieldrin-resistance in this species is not quite so easily explained on the basis of the most recent findings. Further work with morphological markers should eventually clear up this interesting question. There is evidently also a close relationship between dieldrin and DDT-resistance in *Ae. taeniorhynchus*. In *Ae. aegypti*, the principal mechanism of DDT-resistance is detoxication to DDE, while the mechanism of dieldrin-resistance is unknown. However, there is some indication that *Ae. aegypti*, like *C. fatigans*, can open the epoxide ring of dieldrin to give the *trans*-diol, and adults of both species certainly possess enzymes that are able to open epoxide rings.[235] In the Caribbean strains of *Ae. aegypti*, the DDT-resistance levels appear to be closely correlated with the levels of DDT-dehydrochlorinase,[407] and investigations with the morphological markers suggest that the resistance genes involved in these strains are either allelic or identical, as seems to be the case with the genes responsible for dieldrin-resistance in the various strains of this region. Although *Ae. aegypti* has developed DDT-resistance in the Caribbean region and in Southeast Asia, this has not been the case in West Africa, and the initial response of strains from the last area to DDT-pressure in the laboratory was much slower than with the other strains, although high levels of resistance were eventually attained. The same resistance gene on chromosome 2 appears to be involved in strains from Upper Volta, Bangkok, Trinidad or Penang, but in the Penang strain, DDT-resistance shows more proportionality with the dehydrochlorination of DDD than of DDT, and there is in any case some natural tolerance towards DDT.

The locus of the dieldrin-resistance gene in the Upper Volta strain, which readily developed this resistance, is comparable with that in the Caribbean strains and in a strain from Karachi which also has the second chromosal DDT-resistance.[408]

The DDT-dehydrochlorinase of *Ae. aegypti* larvae can be inhibited by DMC and WARF-antiresistant, and differs from the housefly enzyme in that it acts upon a wider range of substrates; it is more active towards DDD than DDT, and acts also on *o*-chloro-DDT, but not on duetero-DDT. In accordance with the idea that dehydrochlorination of DDT analogues is a major resistance mechanism, deutero-DDT is a considerably better larvicide than DDT to the resistant strains (although it does not quite restore susceptibility), and Dilan, which cannot be dehydrochlorinated, is as toxic to resistant as to susceptible larvae.

### c. Other Species

As far as other insects are concerned, the information available is virtually limited to that listed in Table 1, except for fruit flies (*Drosophila* spp.) and the German cockroach (*Blattella germanica*) for which morphological markers are available and permit exploration of the genetical origins of resistance. Up to 12-fold increases in resistance have been found in *D. melanogaster* selected with DDT. Genetical studies of this resistance show that reciprocal crosses between the sexes are identical, giving no evidence of sex linkage or maternal effects. Crossing experiments between the naturally DDT-tolerant Fukuoka strain and a mutant marker strain have shown that DDT-tolerance is dominant and is associated with chromosome 2, the site being near the marker *vestigial* at locus 67. Experiments with a strain from Hikone, which had become resistant through exposure to DDT, also indicated the resistance to be associated with chromosome 2 and located near locus 66.7. This resistance extended to HCH, proved to be completely dominant and was associated with locus 66 on chromosome 2; accordingly, it seems that the same gene is responsible for both DDT-and HCH-resistance in the Hikone strain and for the DDT-tolerance in the Fukuoka strain.[43]

German cockroaches are not very susceptible to DDT and have been widely controlled with chlordane so that cyclodiene-resistance is of most concern. Some of the early work on the genetic basis of chlordane-resistance indicated it to be polygenic and probably involving little dominance. Later, when aldrin was used as the test cyclodiene, it became apparent that the resistance in two different strains was due to a single gene of intermediate dominance, as is usual with cyclo-

diene-resistance. *Blattella germanica* has 11 pairs of autosomes and crossing experiments between these two resistant strains and susceptible ones carrying morphological marker genes on chromosomes 2 to 7 showed that aldrin resistance is allelic in the two resistant strains, the resistance gene being located on chromosome 7.[409] Earlier work shows that DDT-resistance is associated with chromosome 2, so that the resistances to the two classes of organochlorines are located on different chromosomes, as with *M. domestica* or *C. fatigans*, and in contrast to the situation with *Ae. aegypti*.

While examining the mechanism of dieldrin-resistance in three strains of *B. germanica*, namely the Fort Rucker, London (Ontario) and Virginia Polytechnic Institute (VPI) strains, Matsumura showed that resistance in a derivative ($LG_3$) of the London strain, obtained by selection at the LD90 level for three generations, was monofactorial and intermediate in character.[410] Resistance (about twofold based on LT50) in the Fort Rucker strain was abolished by sesamex, indicating a possible oxidative detoxication as a causative mechanism, and the dosage-response regression line was accordingly moved into virtually the same position as that occupied by the line for the susceptible strain. Some degree of synergism (about twofold) with the VPI and $LG_3$ strains (each 20- to 30-fold more resistant than the CSMA susceptible strain according to LT50), resulted in corresponding shifts of the regression lines for these strains, while the resistant London strain gave a heterogeneous response to challenge with a combination of sesamex and dieldrin, resulting in a "stepped" regression line. The regression line for dieldrin response of hybrids ($F_1$) of the cross of $LG_3$ x CSMA (S) was steep (indicating eightfold resistance) and the corresponding line obtained with the sesamex-dieldrin combination indicated approximately twofold synergism with sesamex, as in the $LG_3$ parent. This $F_1$ colony was backcrossed to the susceptible parent for three successive generations, with mild dieldrin selection at each stage to remove susceptible individuals, and each cross gave a 1:1 ratio of susceptible and $F_1$ phenotypes.

In these experiments, the synergistic effect of sesamex toward the $F_1$ phenotypes carried through, in spite of the successive dilution of the original London genes by susceptible genes. This is a definite indication of a sesamex suppressible resistance factor in the original $LG_3$ strain. Thus, the Fort Rucker strain appears to have a sesamex-suppressible mechanism conferring a low order of resistance and this is also present in the VPI strain, which has another mechanism unaffected by sesamex and conferring a higher resistance level. The heterogeneous response to the sesamex-dieldrin combination shown by the London strain indicates that some specimens have both mechanisms, while others have only the sesamex-sensitive one, or the other one in a heterozygous form. The results of the $F_1$ (LG$_3$ x S) x S back-cross, showing 1:1 segregation of the progeny rather than four different phenotypes, indicates that the factors are genetically related. Another possibility is that the sesamex-sensitive mechanism is a modifier which is closely associated with the major resistance factor. Selection of the Fort Rucker strain with dieldrin for four generations produced a strain having both resistance mechanisms, but attempts to obtain a strain having only the sesamex-sensitive mechanism from crossing experiments with the London strain was not successful. One interpretation of these phenomena is that the Fort Rucker and the London strains represent successive stages in the transformation from complete susceptibility to the combined, full resistance as seen in the VPI and LG$_3$ strains, but in the absence of studies with morphological markers, which present difficulties in *B. germanica*, no full explanation is available. The synergistic effect with sesamex in these strains is indicative of an oxidative detoxication mechanism and there is now evidence that 9-hydroxydieldrin (see Figure 12, structure 10) is a metabolic product of dieldrin in this insect (see reference 695).

The amount of information available about the genetic origins of organochlorine resistance in insects of agricultural importance is rather limited. Dieldrin resistance in the cabbage maggot, (*Hylemya brassicae*) has been found in both Canada and the United Kingdom to be due to a single major gene of intermediate dominance,[411] as is the case for both dieldrin- and DDT-resistance in the spotted root maggot (*Euxesta notata*).[412] In both of these species, peculiar effects were found to be associated with the dieldrin-resistance. When pupae of *H. brassicae* from samples of dieldrin-resistant populations on Prince Edward Island were allowed to emerge in a greenhouse, the adults lived twice as long as a normal, dieldrin-susceptible strain, and produced twice as many eggs. This phenomenon has not been seen in the field, although it was noticed that after onion flies (*Hylemya antiqua*) became resistant to dieldrin in Ontario they persisted in the field for a much longer period than was observed previously. The increased adult longevity declined on continuous rearing, to a basal level which was still 70% greater than that of the susceptible strain. The effect is not a pleiotropic expression of the dieldrin-resistance gene allele since selection of a strain lacking this allele with an organophosphorus compound produced a similar effect. The initial investigations which showed that DDT- and dieldrin-resistance in *E. notata* are monofactorial, of intermediate dominance, and inherited separately, were conducted with impure strains, but the genetical analysis succeeded because the three genotypes are readily distinguishable, as in the case of some other insects already considered. Attempts to select the heterogeneous populations further, so as to produce pure strains uniformly homozygous for the resistance factor, resulted in a decrease in the resistance in each case. Determinations of the genotypical composition of successive generations indicated that selection caused disappearance of the resistant homozygotes (RR) at about the sixth generation ($F_6$) and the susceptible homozygotes died out at about $F_{13}$. This happened with either dieldrin or DDT, so that 16 generations of intermittent selection of the dieldrin-R strain with dieldrin produced a strain composed entirely of heterozygotes for dieldrin-resistance and almost entirely of heterozygotes for DDT-resistance; conversely, the DDT-resistant strain, selected for 16 generations with DDT, produced a strain consisting entirely of DDT-R heterozygotes and almost entirely of dieldrin-R heterozygotes. The two resistance genes are indicated to be on different chromosomes, as shown by the results of back-crosses of the dihybrids to the susceptible strains, and also by the fact that DDT selection initially increased the DDT-resistance and not the dieldrin-resistance, and vice versa. The final emergence of populations in which heterozygotes bred to produce heterozygotes (instead of the normal rr:Rr:RR of 1:2:1) was assumed to be due to the development of two balanced lethal systems which rendered the susceptible and resistant homozygotes nonviable in these two strains.

### 3. Postadaptation and Resistance

There is evidence both for and against the

ability of sublethal doses of toxicant to bring about postadaptive changes in insects. The laboratory exposure of houseflies to sublethal doses of DDT through successive generations caused disturbances in oogenesis in nearly 40% of the females, a great many of which contained degenerating ova for the first four generations and even showed a slight increase in susceptibility to DDT. Many of the follicles which actually developed were much delayed so that oviposition lacked synchrony. In the next four generations, however, the flies became more DDT-resistant and the disturbances in oogenesis disappeared. In field operations, disturbances in oogenesis following DDT applications appeared to be a prelude to the development of resistance; some observations during control operations in Russia indicate that during several years of field treatment, a lapse towards susceptibility occurred in the third year, followed by disturbances in oogenesis in the fourth year, and the development of resistance in the fifth and sixth years when the disturbances in oogenesis had subsided. These effects were attributed to a gradual adaptation of the reproductive process of the adult to the action of the insecticide and the observers found it difficult to accept that resistance arises solely through the selection of individuals already carrying genes for resistance.[413] However, subsequent work has not confirmed these phenomena in houseflies.

Some interesting experiments were carried out by Zaghloul and Brown,[414] in which they tested the effect of sublethal doses of DDT on the adults of a susceptible strain (from Guelph, Canada), a DDT-resistant strain and a slightly DDT-tolerant strain of *Culex pipiens* for six to seven generations. The levels of DDT applied were just insufficient to cause mortality in each strain. The treatment initially caused 25% of the ovaries to degenerate and reduced the proportion of females that fed and oviposited. In successive generations of the DDT-resistant and DDT-tolerant strains (both originally from Rangoon), the reduction in biotic potential became more severe, and there was no significant increase in resistance level. In contrast, the biotic potential of the susceptible strain increased and resistance developed (40-fold in adults and 240-fold in the larvae) so that this strain was eventually more resistant to DDT than the other two strains. The resistance so produced was stable with regard to reversion in the absence of DDT, and was accompanied by DDE production from DDT, a characteristic of DDT-resistance produced by genetic selection.

The amounts of DDT incorporated into the ovaries following various treatments were about equal in all strains and generations, and treatment at the larval stage resulted in about double the incorporation effected by treatment at the adult stage. Treatment of the adults was followed by the appearance of DDE in the ovaries of tolerant and resistant strains, but not in those of the susceptible strain until the fifth generation ($F_5$), by which time it had begun to develop tolerance. In this originally susceptible strain, the sublethal dose was raised as tolerance increased, but there was no increase in follicular-degeneration, which showed signs of decreasing by $F_5$. Furthermore, a stimulated oviposition rate was maintained, in contrast to the nonsusceptible strains for which the biotic potential decreased. Neither of the nonsusceptible strains increased their resistance level to DDT by more than twofold as a result of this continued sublethal treatment and the twofold increase in the tolerant strain disappeared when treatment ceased, results which argue against post-adaptational development of resistance.

In contrast, a Rangoon strain similar to the tolerant strain underwent a 250-fold decrease in larval susceptibility in five generations when subjected to selecting levels of DDT. Elements of selection were undoubtedly present in this work, as revealed by underdeveloped follicles, the failure of some females to feed and oviposit, and the failure of ova to develop, but these phenomena were common to all the strains. Nevertheless, the uniquely low proportion of developing ova which successfully produced larvae and pupae would suggest that there was in fact effective selection at this stage in the susceptible Guelph strain.

One consequence of treating *C. pipiens* larvae with sublethal levels of DDT appears to be an increase in the number of basal follicles by up to 20% in the adults. A 9% increase has been observed with *Ae. aegypti* following sublethal exposure of larvae to DDT, and it is tempting to suppose that this may be due to some effect of the toxicant on hormonal control. Despite this increase in ovarioles, the number of progeny per brood was decreased by about 50% when the two Rangoon strains of *C. pipiens* were treated with DDT at the larval stage, which was rather more severe than the effect of sublethal treatments of adults. These effects of sublethal exposure to toxicants clearly merit further study, and there is already a fairly extensive literature on the subject.[716]

In contrast to some of the findings mentioned previously, there is a good deal of evidence to support the conclusion that resistance is a preadaptive phenomenon. Luers and Bochnig, for example, reared *Drosophila melanogaster* in the presence of sublethal concentrations of DDT for 150 generations without any DDT-resistance developing.[43] Experiments involving the repeated topical application of toxicants to adult insects have shown that such treatment frequently results in increased susceptibility after a number of treatments, presumably because of the cumulative effect of the toxicant. Thus, the LC50 for normal houseflies given daily sublethal doses of $\gamma$-HCH for six days was 0.075 $\mu$g per fly at the end of the treatment, as compared with 0.094 $\mu$g for untreated flies.[416] In another experiment with DDT, the LC50 was more than halved by pretreatment with sublethal doses of it. Many other experiments have given similar results. Sublethal doses of $\gamma$-HCH given to houseflies in the larval medium for eight generations gave no mortality and produced no resistance, although resistance could be produced by applying doses sufficient to kill most of the adults. The evidence seems particularly convincing in those cases where all the immature stages were exposed, as must occur when toxicant is incorporated in the larval medium; in such cases, the eggs must be exposed and so the lack of development of resistance must indicate that no selection is occurring at that level. There is no information from these experiments about possible effects of the treatments in increasing the levels of enzymes metabolizing DDT or HCH, which is the phenomenon termed "enzyme induction." At any rate, the weight of evidence supports a requirement for mortality in order that Darwinian selection for resistance can occur.

During the last few years there has been much discussion concerning the possible contribution of enzyme induction to postadaptive resistance in insects. It is well known to mammalian pharmacologists that individuals can become tolerant to any drugs which they may be given, so that increasing doses of the drug are required to produce the same effect. In many cases, this effect arises because an administered drug will accelerate its own destruction in the body by increasing the level of liver enzymes whose function is to metabolize foreign compounds (Vol. II, Chap. 3 B.3c). In many cases, though not in all, these enzymes, which are located in the smooth endoplasmic reticulum of liver cells (also called the liver microsomes), convert bioactive or toxic compounds into less bioactive or less toxic compounds that can be eventually eliminated by excretion. It is now well known that insect tissues such as gut, fat body, malpighian tubules, etc. contain enzymes performing similar functions, and moreover, that they can be induced by foreign compounds such as applied toxicants. In other words, it is possible that a toxicant applied to an insect may accelerate its own detoxication and elimination just as occurs with mammals. This is an adaptive change which occurs in the lifetime of a single individual, be it mammal or insect, as a result of a challenge by a toxicant or bioactive compound. When the challenge is withdrawn, the enhanced enzyme levels usually, although not always, return to normal fairly quickly. Accordingly, the change is transient and not heritable although the capacity for response to the challenge varies between individuals and their offspring may inherit the same capacity, whatever it happens to be. Therefore, it is clear that an applied toxicant may cause an adaptive change in an insect which will tend to nullify the effect of the toxicant, and the important question is — what are the levels of toxicant required to bring this change about?

Enzyme induction in insects has been increasingly explored in the last few years and it seems to require rather large doses of inducing agents. Thus, the drugs chlorcyclizine, aminopyrine or phenobarbital, when incorporated into rearing medium containing sixth instar larvae of the wax moth (*Galleria mellonella*) at 5,000 ppm protected the larvae against parathion poisoning when they were subsequently placed in a parathion treated medium.[417] The protective effect was greater the longer the period of prior exposure to the drugs, each of which is known to stimulate enzyme induction in mammals. Thus, 2 days of pretreatment with chlorcyclizine in this manner reduced mortality due to parathion by about 30% of the control figure (obtained for larvae exposed to parathion without pretreatment) during the first 24 hr of exposure, but the effect diminished with longer exposure to parathion. Four days of pretreatment reduced 24-hr mortality by 60% of the control figure and 48-hr mortality by about 25% of the control; there was still a 16% reduction in mortality from the control figure after 72-hr exposure to parathion. After 8 *days* pretreatment with this drug, 24-hr mortality was reduced by 90% of that for the controls, and there was 50% reduction even in the 72-hr mortality.

In these experiments there were some difficulties due to the short duration of the larval instars in relation to the pretreatment times, larvae of earlier instars having to be chosen for longer pretreatments so that pretreated sixth instars were always taken for bioassay with parathion. There may also be a change in susceptibility to parathion depending on age within the instar, and seventh instar larvae are much more resistant to this toxicant than those in the sixth instar. Nevertheless, the trends in reduced toxicity of parathion seem consistent, and although no information was provided on parathion metabolism, it seems likely that this is a valid observation of the induction of metabolizing enzymes in an insect. In the mid-1960's the aromatic hydrocarbon 3-methylcholanthrene, a recognized inducer of microsomal enzyme activity in mammalian liver, was found to increase DDT metabolism in nymphs of the South American reduviid bug (*Triatoma infestans*) which thereby became less susceptible to this toxicant, and about the same time Agosin showed that pretreatment of these nymphs with DDT itself increased the incorporation of $^{14}$C-DL-leucine into microsomal proteins, as would be expected of a compound stimulating enzyme induction.[418]

For some time it has been clear that a number of organochlorine compounds, including DDT, HCH, and the cyclodienes, are potent inducers of microsomal enzymes in mammals, the inducing effect of chlordane having first become evident when anomalous results in some experiments on microsomal drug metabolism in rats were traced to contamination of the animal house with this compound. The enzymes induced by organochlorines and other foreign compounds are mainly, but not entirely, those called microsomal mixed function oxidases which effect biological oxidations requiring nicotinamide adenine dinucleotide phosphate and oxygen (Vol. II, Chap. 3B.3a,c).

The levels of organochlorines required to effect enzyme induction in the insects so far examined are high so that inductive effects can only be demonstrated with insects which are already resistant to the inducing agent. Thus, in a feeding experiment conducted by Plapp and Casida,[419] they found that inductive effects could be demonstrated by giving houseflies already resistant to these toxicants diets containing 1,000 ppm of technical DDT or 100 ppm of 90% dieldrin for at least 4 days. One of the strains used was the Orlando DDT- and dieldrin-resistant strain which can survive both diets, while the other was only dieldrin-resistant, and so could not be given the DDT containing diet. The effect of feeding these diets on the metabolism of aldrin, DDT and some other insecticides is seen in Table 4, which shows that the increase in metabolism ranged from

TABLE 4

**Effect of High Dietary Levels of DDT or Dieldrin on the NADPH-Dependent Metabolism of Some Insecticides by Homogenates of Housefly Abdomens**

Substrate metabolized (%) in 30 min (mean +S.E.)

| Substrate added to homogenate (Concentration in mμmol) | DDT-resistant strain (Orlando): diet | | | Dieldrin-resistant strain ("Curlywing"): diet | |
|---|---|---|---|---|---|
| | Control (no addition) | 100 ppm dieldrin | 1000 ppm DDT | Control (no addition) | 100 ppm dieldrin |
| Aldrin (0.14) | 37 ± 5 | 71·± 5 | 54 ± 1 | 31 ± 4 | 46 ± 2 |
| DDT (8.8) | 1.8 ± 1.2 | 2.5 ± 1.1 | 3.4 ± 1.2 | 1.3 ± 1.2 | 1.8 ± 0.5 |
| Diazinon (3) | 11 ± 1 | 17 ± 2 | 16 ± 2 | 9 ± 3 | 23 ± 5 |
| Propoxur (3) | 14 ± 3 | 21 ± 3 | 17 ± 4 | 12 ± 2 | 18 ± 3 |
| Allethrin (1) | 40 ± 9 | 77 ± 1 | 57 ± 2 | 39 ± 3 | 68 ± 3 |

(Data adapted from Plapp and Casida.[419])

marginal with propoxur for flies fed the DDT containing diet and 35% with DDT when dieldrin was fed, to as high as 150% with diazinon for houseflies fed the dieldrin-diet. The results obtained with the doubly resistant Orlando strain show that compared with DDT a lower concentration of dieldrin in the diet produced a greater average increase in metabolism. Thus, dieldrin appears to be a better inducer, as was shown in other experiments in which a tenfold increase in dieldrin concentration (that is, to 1,000 ppm) did not increase the inductive effect, whereas dietary DDT at 100 ppm was less active than at the higher concentration used in Table 4.

At Oregon State University, additional work with dieldrin has confirmed that it is a better inducer of housefly microsomal mixed function oxidases than DDT, and it is also evident that the inductive effect is greater in insect strains already possessing a high basic level of microsomal mixed function oxidase activity.[420] In these experiments the enzymic activities measured were those for two well-known and well-characterized oxidation reactions, namely, the oxidation of naphthalene, which proceeds via epoxide formation to give several water soluble products, and the epoxidation of a cyclodiene compound, in this case heptachlor. The same strains of housefly were used as in the experiments in Table 4, with the addition of a strain resistant to the carbamate insecticide "Isolan," the basic levels of microsomal mixed function oxidase activity being low for dieldrin-R "curly wing" mutant flies, intermediate for Orlando-flies and high for the Isolan-R strain. Microsomal naphthalene hydroxylation was enhanced by pretreating the Orlando-flies with dieldrin, which also afforded the flies pronounced protection against the toxic effects of carbaryl. However, both effects were abolished by pre-injecting the flies with a nontoxic dose (1 $\mu$g) of cycloheximide, an inhibitor of protein synthesis, as would be expected if the changes observed arose from increased enzyme synthesis (induction).

The activity of the enzyme that converts heptachlor into its epoxide (heptachlor epoxidase) was enhanced in all three housefly strains by pretreatment with dieldrin, the increase being in the same order as the original basal microsomal oxidase levels; on the basis of the maximum increases observed in epoxidase activity, induction in the Isolan, Orlando, and "curly wing" strains was approximately in the ratio 3:2:1. The $F_1$

hybrids produced by interbreeding the Isolan and "curly wing" strains have epoxidase activity intermediate between the activities of the two parent strains when induced with dieldrin. Thus, the strain having the higher microsomal oxidase activity has a greater potential for induction and appears to differ from the "curly wing" strain in the regulation of its enzyme levels, possibly by possessing more genes or gene sequences for the increased production of detoxifying enzymes. Following topical application of dieldrin (0.5 $\mu$g per male or female fly) to houseflies of each of the three strains, the epoxidase activity frequently began to rise within 3 hr and continued to increase for some 24 hr; other results involving pre-exposure of houseflies to dieldrin films in jars showed that the inductive effect of a 24-hr exposure persisted for some days.

To examine the inductive effect of DDT, the Orlando strain and another DDT-R strain called the $F_c$ strain were topically treated with various doses of DDT between 0.1 and 1.0 $\mu$g per fly, and the inductive effects of these doses were examined after a 3- or a 24-hr interval, using heptachlor epoxidase activity as the criterion, as before. The effect seen in the Orlando strain was a 16% increase in epoxidase activity 3 hr after the 0.1 $\mu$g DDT treatment. With the $F_c$ strain, the maximun increase (31%) in epoxidase activity was seen 3 hr after treatment with 0.25 $\mu$g of DDT, an effect which disappeared within 24 hr after the treatment, and higher topically applied doses appeared to depress epoxidase activity. The poor inducing capability of DDT appears to extend to rather similar aromatic structures, since chlorobenzilate, "Sulphenone," tetradifon, and ovex were also ineffective as inducers.[421]

It appears, therefore, that in order to effect an increase in heptachlor epoxidase in the $F_c$ or Orlando strains, a topically applied dose of at least 0.1 $\mu$g of DDT is required, which is about half the LD50 of DDT for the W.H.O. standard susceptible strain of M. domestica; the 0.25 $\mu$g dose which produced a 31% increase in heptachlor epoxidase in the $F_c$ strain exceeds the LD50 for the W.H.O. strain and is near to the LD50 for the $S_{NAIDM}$ strain, which is also widely used as a standard. Unfortunately, the minimum amount of dieldrin that will produce an effect when topically applied is unknown, since this series of experiments exposed houseflies to films of dieldrin on glass. The 0.5 $\mu$g topical application of dieldrin clearly produced a significant and lasting induction of epoxidase in all of the three strains treated, and this is about 30 times the topical LD50 of dieldrin for susceptible houseflies, but recent work of the same group indicates that a 0.1 $\mu$g (still about five times the LD50) can produce a significant effect. Generally, it appears that the doses required to produce significant enzyme induction are near to or higher than the LD50 levels, and are therefore more likely to kill susceptible individuals and to select populations for classical resistance than to induce their detoxifying enzymes; the situation may be different with endrin, which appears to be about 100 times more effective than dieldrin under similar conditions.[421] Nevertheless, it might be interesting to look for enzyme induction effects in what are possible marginal cases, for example, in the survivors of a treatment of houseflies with dieldrin at the LD50. Of course, there is a chance that some of the survivors of such treatment may possess the genes for resistance, although this is improbable because their frequency is normally so low that they can only be detected by treating large numbers of insects with very high doses of toxicant.

The foregoing observations show that enzyme induction can be demonstrated in insects already resistant to the inducing agent, and this adaptive response might add to the existing resistance level in some cases; perhaps, for example, in the $F_c$ strain, which is resistant to DDT because it has enhanced ability to degrade this toxicant oxidatively (see section B.2a). In this case, enhanced microsomal oxidation induced by DDT itself, or by some other nontoxic inducing agent, might add to the level of DDT-resistance, although the effect might not be easy to measure. On the other hand, dieldrin induces microsomal oxidases but there is no evidence that this phenomenon is related to dieldrin resistance. It seems that selection for classical dieldrin resistance can produce enhanced microsomal oxidase activity at the same time, but the distinction between the two is shown by Plapp's "curly wing" dieldrin-resistant strain, which has microsomal oxidase activity below that of several susceptible reference strains.

From a practical point of view, the real interest in induction concerns its possible contribution to cross-resistance to other compounds which are normally metabolized by the microsomal enzymes. This possibility is illustrated by the cross-resistance of the Orlando flies to carbaryl after they had been pretreated with dieldrin. On this basis, it can be seen that a dieldrin-resistant insect which comes into regular contact with dieldrin so that it always has considerable amounts of this toxicant in its tissues may exhibit enhanced tolerance to other compounds. The situation is most likely to occur with immature stages, as for example when dipterous larvae are selected for resistance in soil treatments with organochlorines and the resistant population remains in constant contact with high soil residues of the toxicant. Lepidopterous larvae feeding on foliage might similarly become resistant and also be in a permanent condition of microsomal enzyme elevation through regularly ingesting large amounts of residual toxicants with the foliage they feed on. One can imagine that these situations may confer some degree of initial immunity to insecticides such as carbamates or organophosphorus compounds which might be tried as alternatives.

## C. DEVELOPMENT OF RESISTANCE IN THE FIELD

### 1. Background

The first instance of insect resistance appears to have occurred in the state of Washington in 1908, when it was found that lime-sulfur doses ten times greater than those previously used successfully failed to control infestations of San Jose scale (*Quadraspidiotus perniciosus*) in apple orchards.[422] Until 1944, only 12 insect species were involved, resistance having developed in three scale insects to hydrogen cyanide, in ticks and four species of orchard caterpillars to arsenicals, and in two species of thrips to tartar emetic, with single cases of resistance to the fluorine derivative cryolite and to selenium.[373]

In view of the extremely widespread use of DDT during World War II, it does not now seem surprising that resistance to this compound soon appeared, with the housefly leading the way in Sweden and Denmark in 1946. In 1947, DDT-resistant houseflies and mosquitos (*Culex molestus*) were reported in Italy; in 1951 DDT-resistance was reported in the body louse of Korea and in mosquitoes (*Anopheles sacharovi*) in Greece. Each year since then, the incidence of resistance in the public health and veterinary fields has increased and about 57 species are currently resistant to the DDT group (Table 5).[43,373] When

## TABLE 5

Number of Insect and Mite Species Having the Main Types of Insecticide Resistance

| Insecticide group | Public Health Species[a] | Agricultural Species[b] |
|---|---|---|
| DDT | 57 | 29 |
| Cyclodiene (dieldrin-HCH) | 84 | 53 |
| Organophosphorus | 16 | 32 |
| Others | 2 | 27 |
| Total[c] | 105 | 119 |

(Data adapted from Brown and Pal[43] and Brown.[373])

[a]Includes species of veterinary importance.
[b]Includes stored product species.
[c]Less than the column totals, because some species are resistant to several groups.

insects are resistant to DDT, the resistance extends (cross resistance) to structural analogues such as DDD, Perthane®, and methoxychlor, but not to organochlorine insecticides of unrelated structural groups such as the cyclodienes or HCH. Dieldrin-resistance was reported in Californian houseflies in 1949, in the mosquito *Anopheles sacharovi* in Greece in 1952, in *A. quadrimaculatus* in 1953, and in *A. gambiae* in Nigeria in 1955. This type of resistance extends to other members of the cyclodiene group and also to toxaphene and HCH, and is quite distinct from resistance to the DDT group (see section B). The body louse (*Pediculus humanus corporis*) was also found to be resistant to the cyclodiene group in France and Japan in 1955, and in various parts of Africa, India, and Korea in successive years between 1955 and 1961. As in the case of DDT, cyclodiene-group resistance (henceforth called cyclodiene resistance and defined as indicated above) spread both geographically and in terms of the number of species involved, until, by 1970, some 84 species of insects of public health or veterinary importance were known to be resistant to this group of compounds.[43]

About 5 years after it first came into use for agricultural purposes, resistance to DDT appeared in the codling moth (*Carpocapsa pomonella*) in the United States (1951), and was soon reported in Australia (1955), South Africa (1956), Canada (1958), and the eastern Mediterranean (1966). DDD is an important agent for the control of certain fruit pests, and resistance to it was found in the red-banded leafroller (*Argyrotaenia velutinana*) in North America and in the light brown apple moth (*Epiphyas postvittana*) in Tasmania and New Zealand between 1954 and 1960. As time went on, DDT-resistance spread to pests of cabbage, potatoes, cotton, tobacco, and other crops, the current total being some 29 species of agricultural insect pests, including those that attack stored products.[373]

Insecticides of the cyclodiene group are widely used to control insect pests in the soil and resistance to chlordane was observed in the southern potato wireworm (*Conoderus fallii*) in 1955; in subsequent years resistance to aldrin and dieldrin has been reported in a variety of soil insects including two more wireworms, four species of corn rootworms, and nine different root maggots. Cyclodiene type resistance in cotton boll weevil (*Anthonomus grandis*) was first noticed in 1955 when toxaphene and endrin failed to control it satisfactorily. In 1968, the number of insect species of agricultural importance showing resistance to cyclodienes was 53, giving a grand total of 137 species, including both agricultural and public health pests that were resistant to this group of compounds. The corresponding total for the DDT-group was 86 species at this time. In addition, 32 species of agricultural insects were resistant to organophosphates and 27 to other insecticides, while the corresponding figures for public health species (including those of veterinary importance) were 16 and 2, respectively. Allowing for the fact that some species are resistant to two or three groups of insecticides, the total number of species affected by resistance at the present time must be more than 224, the estimate arrived at by Brown in 1968.[373,423]

Undoubtedly the phenomenon of insect resistance has been the greatest problem encountered in relation to the use of the organochlorine insecticides. Using the housefly as an example, resistance to DDT commonly appears in semi-tropical or Mediterranean type climates within 2 years of its first use as a control agent; a change to the cyclodienes or HCH results in dieldrin resistance after a further year, and 5 years use of malathion or diazinon will produce resistance to organophosphates. Of the 20 or so different species of flies that had become resistant to insecticides by 1968, five had become resistant to all three main groups.

The remarkable ability of organochlorine

insecticides to select for resistance is demonstrated by the increase in the number of resistant species from about 16 in 1951 to 25 in 1954, 76 in 1957, and 157 in 1963. Therefore, the 1968 figure of 224 given by Brown represents a 40% increase in the numbers of species found to be resistant in 1963, but the greatest rate of increase in these numbers corresponds with the period of maximum use of organochlorine insecticides in the 1950's. There was evidently some slackening in the rate of resistant species in the 1960's, but individual species tended to become resistant to several groups of insecticides, and the geographical distribution of resistance became wider. It may be that the reduced rate of increase in the incidence of resistance is due to the fact that the major pests have now been selected and also to the increasing use as substitutes for the organochlorines of organophosphates which are more easily degradable and therefore less persistent in the enviornment.[373,385] Resistance to organochlorines is not accompanied by resistance to organophosphates, but resistance resulting from exposure to the latter may be accompanied by cross-resistance to organochlorines, and several examples of this type of cross-resistance are known.

With the postwar prospect that DDT would be the panacea for all insect pest problems, it will be appreciated that the impact of resistance was somewhat painful, and it was at first difficult for field operators to accept the fact that failure to control a particular pest might be due not to some defect in the quality of the insecticide or in the application technique, but to an alteration in the pest itself. An understandable response in these circumstances is to apply more pesticide at more frequent intervals, and this technique will sometimes work, for example, with DDT-resistance in anopheline mosquitoes. With DDT-resistance, there is usually a lag-period in which the resistance builds up slowly and is unstable, whereas in the case of dieldrin, the development of resistance is rapid and complete.[384] With this chemical, therefore, more frequent application only results in intensification of the resistance. Thus, the natural tendency to overdose in order to "make sure" of control, coupled with increased application rates which seem at first to be an obvious measure against control failure, usually results in exactly that intense insecticide selection pressure which will result in high resistance. Besides encouraging

resistance these practices enhance any deleterious effects of the applied toxicant on natural pest predators and parasites.

Nowadays, it is understood that the complete eradication of a pest species is a difficult proposition which should only be attempted if adequate sampling of the population reveals the absence of naturally resistant individuals. Even with this ideal situation as a basis, eradication may be impossible if the species is not geographically isolated so that contamination with other populations can be prevented. Thus, it has to be realized that in most cases some economic damage by pests is inevitable and sufficient chemical should be used in combination with other methods if possible, to maintain this at an acceptable level.

## 2. Species of Public Health or Veterinary Importance
### a. Houseflies

In 1946, about a year after it was first used in the district of Arnas, and 2 years after it had been introduced into northern Sweden with excellent results, DDT failed to control houseflies in the Arnas area. DDT had been applied to manure heaps for the control of housefly larvae, and it was noticed that the knock-down time of the adult flies was up to ten times longer than normal. When Wiesmann compared them with a normal strain in Basle, he found that the median knock-down time (KD 50) was 93 min for Arnas flies, compared with 16 min for normal flies, while the lethal dose on tarsal application to the former was between 100 and 200 times higher than that required to kill the latter (0.025 $\mu$g).[43] The Arnas flies were more resistant to cold and heat than the normal Basle strain, and there were also some morphological differences such as heavier pigmentation of the tarsi and greater thickness of the cuticle, which suggested that resistance might be due to reduced cuticular penetration. Later surveys showed that DDT-resistance was prevalent in a large area of northern Sweden, with pockets of resistance in the center and the south.

In Denmark also, DDT was used for fly control on certain farms from 1944, and isolated control failures observed in 1946 became widespread in 1947.[424] Some specimens collected in 1948 survived for several days when treated with 33% DDT dust and there was cross-resistance to methoxychlor, DDD, and the $p,p'$-difluoro-analogue of DDT (DFDT), but no cross-resistance to the cyclodienes or to HCH. By 1949, DDT-resistance was general on Danish farms and HCH failed within about 3 months of its first use on one particular farm.[43] Chlordane was introduced for use in alternation with HCH, and resistance to the cyclodiene appeared in 1950 on the island of Lolland where HCH-resistance (1949) had already been added to the DDT-resistance that had appeared there in 1947. Specimens collected from areas in which all three compounds had failed were found to be resistant to toxaphene and dieldrin, and by 1951 resistance to all organo-chlorines was widespread in the southern islands, although chlordane remained effective in some places where it had been the sole control agent. Spraying with organochlorines for housefly control was discontinued in Denmark after 1952 when organophosphorus compounds came into use.

In West Germany, DFDT had been introduced for general use in 1945 and resistance to it appeared 3 years later; DDT-resistance developed in 1949, 2 years after its introduction. In order to supplement the failing control with DDT, HCH was added, but resistance to the mixture developed within 18 months. Finally, diazinon was introduced for housefly control about 1954. Although DDT had been available for housefly control in Switzerland since 1943, and had been extensively used against houseflies in canton Valais, a 1950 survey of housefly samples from various regions indicated only a low incidence of resistance, highest levels being found in canton Valais.[42] Chlordane was introduced at the same time as HCH and dieldrin (1950), and only one sample of flies out of 83 examined showed chlordane resistance. However, cyclodiene resistance usually appears much more rapidly than DDT-resistance, as is now well known, and widespread failure to control houseflies with either HCH or the cyclodienes was evident in 1951 and confirmed by 1952. Diazinon was again the main replacement.

The outlook was no better in the rest of Western Europe. Widespread pockets of DDT-resistance were evident in France by 1955; in Holland resistance was scattered, but severe in some places where compost heaps had been treated with DDT. In Britain, tolerance to DDT-aerosols was noticed in restaurants in 1949, and 1953, houseflies from a refuse tip in Hertford-shire, upon which DDT and HCH had been used

with progressively poorer effect for several years, were found to have tenfold resistance to DDT, 53-fold resistance to HCH, and 266-fold resistance to dieldrin. Also in 1953, Busvine examined flies from another refuse tip in Surrey and found them to be 12-fold resistant to DDT and $\gamma$-HCH, and 28-fold resistant to dieldrin, when compared with a normal housefly strain. In this case, the refuse tip had received successive 2-year treatments with DDT (1947—49) and HCH (1951—53). Organo-chlorine resistance was general in Spain by 1956, DDT-resistance having first been observed there in the late 1940's, and was present in North Africa (Morocco) within 2 years of the first applications of DDT and HCH. No instances of resistance had been reported in Czechoslovakia up to 1957, by which time organochlorine compounds had been used for some years, but resistance was said to be widespread in Yugoslavia by 1955 and had also been reported in East Germany in 1950.

DDT-resistance was reported in a Russian publi-cation of 1950 and later publications discussed DDT-resistance in Moscow and the Tashkent area. Resistance to both DDT and HCH was present in the Crimea, HCH resistance having appeared in Sevastopol by 1953 after its successful use for some years.[43] Although houseflies have always been a nuisance and have undoubtedly contributed to the spread of gastroenteritic infections in the temperate regions, the impact of resistance following a period of successful control is obviously likely to be much more serious in warmer or tropical climates, especially when the sanitation is not good. The appearance of housefly resistance to DDT in these regions was even more spectacular than in the temperate areas because in many cases there had been highly successful housefly control as a bonus from the antimalaria programs. In Italy, domestic spraying with DDT at 1 $g/m^2$, begun in March 1946 for malaria control, eliminated the houseflies as well, but in the early summer of 1947, after the second round of spraying, it was noticed that houseflies were surviving in sprayed houses near Anzio. After that, it was frequently seen that fresh treatments with DDT left a few survivors which later reproduced to give large numbers of individuals immune to any subsequent treatment. It seems that immunity to DDT in houseflies was quite general by 1948 and chlordane was used at 2 $g/m^2$ in Latina province, the coastal area to the south and east of Anzio. At this time, there was complete susceptibility to both chlordane and HCH since these had not been widely used, and chlordane was used successfully to control houseflies until late in 1950, when resistance began to appear and was associated with a great increase in the fly population in the treated areas.[42,43]

In Sardinia, excellent housefly control in 1946–47 followed the first antimalaria spraying with DDT but DDT-resistance appeared in 1948, and was followed by chlordane resistance in 1949, shortly after the first application of the latter compound. As would now be expected, the final result of these applications was cross-resistance to methoxychlor in the DDT series, and to heptachlor, aldrin, dieldrin, toxaphene, and HCH as a result of the chlordane applications. Tests performed by Busvine, in which the insecticides were topically applied in mineral oil to resistant flies from Sardinia, showed that resistance to HCH (8.5-fold) was about half that to DDT (16.5-fold) and similar to that for toxaphene (tenfold). The resistance to cyclodienes, chlordane (70-fold), dieldrin (40-fold), and aldrin (33-fold) showed the familiar cyclodiene pattern (Vol. II, Chap. 3B.3b) with least resistance to HCH.[425]

A characteristic of an Italian strain of houseflies from Torre in Pietra, which proved to be resistant to DDT but not to other organochlorine compounds, was that DDT took a long time to knock the flies down, in fact, seven times longer than the corresponding time for a normal strain.[426] In contrast, DDT knock-down with the Sardinian strain, having resistance to DDT and also to chlordane, was much more rapid. Flies in the Latina province also recovered from knock-down by pyrethrins, and the Torre in Pietra strain was found to be fivefold more resistant to this toxicant in terms of LD50 than the normal strain, although the Sardinian strain was nearly normal in this respect. Thus, there appeared to be an association between knock-down resistance to DDT and resistance to knock-down with pyrethrins, an early example of a phenomenon that is now well recognized in some insects. At about the same time, investigations of DDT-failure in Orlando, Florida, revealed that resistance was associated with knock-down times from 10 to 20 times greater than normal. Thereafter, knock-down resistance came to be associated with Orlando, which gave its name to the kdr-O (knock-down-resistance-Orlando) strain of DDT-resistant

houseflies (see section B.2a), still widely used as a reference strain.

To complete the story of field resistance to organochlorines in Italy, the introduction of dieldrin in the malarial areas in 1952 failed because of the pre-existing cross-resistance to cyclodienes. Mixtures of DDT with other compounds such as the Geigy carbamate insecticide "Pyrolan," or the DDT-synergist DMC ($p,p'$-dichlordiphenyl methylcarbinol), were effective for a time but had begun to fail by 1953. Diazinon was used effectively, both alone (from 1952–53) and in combination with DDT (from 1954), but resistance to it appeared late in 1956. These histories are typical of experiences throughout the Mediterranean area and the Middle East: failure of DDT after 2 years of effective control, with declining success towards the end of this period, and failure of the HCH and/or chlordane introduced as replacements after 1 to 2 years of use.

The public health consequences of these failures were soon apparent; the failure of first DDT and then chlordane to control houseflies in the Latina province between 1947 and 1952 was accompanied by an increase in the infant mortality rate from gastroenteritic disease, which had shown an encouraging decline following the first use of DDT. The United Nations refugee camps in Syria, Lebanon, Jordan, Palestine, and the Gaza Strip were first treated with DDT in 1949 to control the spread of dysentery, trachoma, and conjuctivitis by houseflies. The use of DDT in these circumstances was initially successful, but for some reason failed rapidly, the failure being accompanied by a great proliferation of the flies. This increase in numbers was presumably responsible for the high incidence of the above diseases during the following 10 months, since with such a rapid failure of control, it might be expected that had the numbers of carrier insects remained unchanged, the incidence of these diseases would remain unaffected. The major vector of trachoma and conjunctivitis in North Africa and the Middle East is believed to be the smaller fly (*Musca sorbens*), which breeds in human faeces. It does not normally enter houses and so it is not controlled by residual spraying inside them. The treatment of latrines and the lower outside walls with DDT and HCH in Morocco and the Middle East in 1948–1950 greatly reduced the occurrence of ophthalmic disease before insecticide-resistance developed, but the extent to which *M. sorbens* was involved in the general resistance to *M. domestica vicina*, the Mediterranean and Middle East type of *M. domestica*, is unknown; likewise the extent of its actual involvement in the incidences of disease mentioned here. In Egypt, treatment of fly-breeding places near Cairo with HCH dust for larval control failed late in 1949 after one successful year. In the following year, chlordane successfully maintained housefly control in two villages and continued to be used in one of them through 1951, when resistance appeared in this village. In the meantime, the other village was treated again with HCH so that HCH-resistance, having initially declined somewhat, soon reappeared. The correlation between reduced housefly numbers and reduced ophthalmic disease was very marked in 1949–50, but the improvement could not be maintained beyond 1950.[43] Similar phenomena were seen in the state of Georgia, where effective housefly control using DDT then chlordane or dieldrin in succession produced a marked reduction in gastroenteritic infections which unfortunately lasted little more than 1 year, due to the development of multiresistance to organochlorines.

The different geographical patterns of resistance that often follow the apparently similar uses of pesticides in various regions are difficult to understand. In India, the predominant houseflies are *Musca domestica vicina* and *M. domestica nebulo*, the latter increasing in proportion in the southeast. In spite of the extensive use of DDT in malaria control programs from about 1946, no serious housefly resistance problem was apparent by 1957. One authority attributed this to the short spraying season in India, while another suggested that since the flies in these hot climates stay at ground level during the day and rest on vegetation during the night, they might thereby reduce their time of contact with deposits of toxicant; either factor would tend to reduce the level of selection pressure. However, instances of HCH-resistance in houseflies (*M. domestica nebulo*) had been observed, by 1956, as a result of fly-control operations in Bombay and other places. In Japan, where the prevalent housefly is *M. domestica vicina*, there was also satisfactory control from the introduction of DDT in 1945 until 1951, again an unusually long time by other standards. However, DDT-resistance appeared in 1953 and was soon followed by tolerance to the cyclodiene substitutes. Resistance also seems to have been slow to develop in Australia, where the use of DDT on a large scale was avoided; there were a few scattered control failures by about 1950 and these appeared to involve *M. domestica domestica*.

An examination of these diverse reports suggests that in some areas there is a correlation between the slow development of DDT-resistance and the major presence of substrains of housefly other than *M.d. domestica*, although behavioral differences in warm climates may contribute. Housefly resistance to DDT developed in the early 1950's in Argentina, Uruguay, Chile, Brazil, Peru, Bolivia, and Ecuador, following malaria control campaigns, but there was general lack of resistance to this compound after several years of use in Venezuela, British Guiana, and French Guiana, the subspecies of housefly being *vicina* in these last countries and *domestica* in the others. On the other hand, resistance seems to have developed with normal speed in Central American areas such as Panama, Puerto Rico, and Trinidad, where houseflies are presumed to be *M. domestica vicina*. DDT was used for housefly control in Mexico from 1945 and flies resistant to DDT were common by 1951; subsequent use of HCH resulted in generalized organochlorine resistance by about 1956.

In the United States, the first reports of the failure of DDT for housefly control appear to have arisen in the area of Orlando, Florida, where the U.S. Department of Agriculture Laboratory had been heavily involved in the early wartime development of DDT. Flies collected from several dairy barns in the area in 1948 had extended knock-down times with DDT (up to 21-fold compared with a normal, DDT-susceptible strain), but could be controlled with cyclodienes or HCH. In another place where dieldrin had been used in 1949, resistance to it, other cyclodienes, and $\gamma$-HCH was present in 1950; the use of chlordane (from 1948) or toxaphene (from 1949) as DDT replacements in some dairies also produced cyclodiene resistance in that year.[427] By 1951, resistance to all organochlorines was widespread, although some of the reported levels seemed to be quite low compared with others reported with the same insecticides in different regions. Subsequent tests with organochlorines combined with synergists, for example DDT-DMC, were disappointing, the mixtures being either ineffective or

TABLE 6

Median Lethal Doses (LD50; μg/female fly) of Various Organochlorine
Insecticides to Normal and Resistant Houseflies in California

| | Housefly Strain | | |
| --- | --- | --- | --- |
| Toxicant | Bellflower (1948) | Pollard (1949) | Riverside normal (laboratory) |
| DDT | 11 | >100 | 0.033 |
| DDD | 60 | >100 | 0.13 |
| DFDT | 4.0 | 1.2 | 0.10 |
| Methoxychlor | 0.96 | 1.4 | 0.068 |
| Heptachlor | 0.06 | 1.5 | 0.03 |
| Dieldrin | 0.05 | 0.86 | 0.03 |
| Lindane | 0.08 | 0.25 | 0.01 |
| Toxaphene | 0.62 | 3.4 | 0.22 |

(Data adapted from March and Metcalf.[428,429])

effective for a short time only,[42] and residual spraying with organochlorines was abandoned in about 1952 in this area. By this time, DDT-resistance was present across the North American continent and is extensively documented.[43] Since the introduction of organochlorines occurred at about the same time and in the same order in most places, the pattern of development of organochlorine resistance was fairly uniform. Some of the resistant strains collected in the field in those days became well known to insect toxicologists in later years. Good examples are the strains collected and maintained by the Riverside group. For example, the Bellflower strain proved to be resistant to knock-down by DDT and had 330-fold resistance to it with cross-resistance to other DDT-analogues (Table 6), but was susceptible to γ-HCH and the cyclodienes. On the other hand, the Pollard strain, produced at the Pollard poultry ranch as a result of exposure first to DDT and later to HCH, was highly resistant to DDT and rather resistant to the cyclodienes as well.[428,429] Compared with the situation in the Bellflower strain, the extreme difference in the Pollard strain between the toxicities of DDT or DDD, and methoxychlor or DFDT is interesting (Table 6). However, the Bellflower strain seems not to have reached its maximum potential for DDT-resistance by 1948, since subsequent exposure to HCH in the field not only increased its resistance to HCH, but also substantially increased its DDT-resistance (by at least tenfold), an interesting observation, since the two types are normally quite distinct.

In their fairly recent survey of resistance in disease vectors for the World Health Organization, Busvine and Pal[48] comment that most reports of housefly resistance, except those from Hungary, India, and Canada, indicate no associated increase in fly-borne diseases. The impact of DDT on the incidence of enteric disease was established in Italy and the southern United States by carefully comparing the situation in treated and untreated areas in the early days before there were complications due to widespread resistance. Although fly-borne diseases may be transmitted in other ways, there is little doubt that failure to control flies due to resistance must result in a return of such sickness to the levels existing before pesticides were used.

The foregoing account demonstrates clearly the depressing versatility of the housefly in regard to its ability to withstand the challenge of man-made poisons. After the early and quite remarkable promise of DDT and other organochlorine compounds, it is not difficult to imagine the great disappointment that followed the rapid and widespread development of DDT resistance in the housefly.

b. Other Noxious Diptera

Sandflies (Phlebotomus spp.) of which P. papatasii is an example, are vectors of kala-azar and various forms of sandfly fever and Leishmaniasis. As a result of DDT applications in the home, this species practically disappeared from Italian towns in the early 1950's, but appeared again later. However, DDT still seems to be quite effective for control and cutaneous Leishmaniasis has virtually disappeared from Romania, Italy, and Greece. In the Eastern Mediterranean area also, cutaneous Leishmaniasis disappeared along with *Phlebotomus* as a result of house spraying against malaria. The rural or sylvatic form of *Phlebotomus* frequents rodent burrows and has therefore been unaffected by DDT spraying, so that oriental sore is found in certain areas. House spraying in central Iran ceased after malaria was eradicated, and there oriental sore has returned with *Phlebotomus*. Sandflies have also been controlled in the course of the malaria control program in West Pakistan, since house spraying with DDT is highly effective against them. However, abundant evidence of their survival indicates the likelihood of reinfestation when anti-malarial spraying ceases.[48]

Blackflies (*Simulium* spp.) are vectors of onchocerciasis in certain tropical areas and are a great nuisance in others since some species may enter the orifices of the body in great numbers, inflicting bites which may cause inflammatory fever and death. DDT and HCH have been used as larvicides for the control of *S. aokii* near Tokyo since 1954 and *Simulium* has been shown to be resistant in this area. There is some local resistance in Canada, where DDT is used to control blackflies in Quebec. DDT is used for the control of an onchocerciasis focus in Mexico and is occasionally added to the Victoria Nile and the Murchison Falls in Uganda for the control of this vector. In West Africa, one zone at Sikasso in Mali has been treated with DDT regularly since 1962, and areas in Ivory Coast and Upper Volta also receive treatment. There have been some trials in Ghana and Nigeria, and suggestions of DDT-resistance in *S. damnosum* in southern Ghana.[43,48]

Dieldrin is used extensively for the control of tsetse flies (*Glossina* spp.) which are vectors of the various forms of African sleeping sickness and so far there are no records of resistance in these species.[48] Of a number of other muscid flies that are of pest status, some have developed resistance to organochlorines and some have not. The horn fly (*Haematobia irritans*) was reported to be resistant to cyclodienes in Texas in 1959, and the stable fly (*Stomoxys calcitrans*) developed resistance to DDT in Denmark and in Sweden about 1948, after some 2 years of exposure to DDT. However, 4 years extensive exposure of a strain in Illinois to DDT did not change the LD50 at all.

The little housefly (*Fannia canicularis*), a frequenter of latrines, was reported to be resistant to DDT in northern Spain in 1953, and resistance has been indicated since then to DDT (in England, Japan, and the United States) and to cyclodienes (United States).

Another latrine-frequenting fly, *Chrysomyia putoria*, was first reported to be resistant to HCH in 1949, within 4 months after the first use of the compound as a larvicide at what was then Leopoldville in the Belgian Congo, and cyclodiene resistance has been reported since in Malagasy and Zanzibar.[43] The black blowfly (*Phormia regina*) does not seem to have developed resistance to DDT, but has natural tolerance to HCH. However, a colony of this species was selected for 30 generations with $\gamma$-HCH in the Riverside laboratory, without any change in the level of tolerance.[430] The sheep blowflies (*Lucilia cuprina* and *L. sericata*) developed cyclodiene resistance in various places in the southern hemisphere in the late 1950's, but there appear to be no records of resistance to DDT; the greenbottle fly (*Lucilia sericata*), the secondary screw worm fly (*Callitroga macellaria*), and the bluebottle fly (*Calliphora erythrocephala*) are reported in various places to have retained their susceptibility to DDT when exposed in dairy barns in circumstances where houseflies became resistant. The rapid development of HCH resistance in the midge (*Glyptotendipes paripes*) was reported in Florida as long ago as 1953, and that the borborid fly, *Leptocera hirtula*, can become resistant to organochlorines is apparent from the rapid successive appearance of resistance to DDT and then HCH, when these were used to control a laboratory infestation in Malaya.

Resistance to DDT and cyclodienes has also occurred in filter flies (*Psychoda alternata*); to cyclodienes in two other midges (*Chironomus zealandicus* and *Culicoides furens*); to DDT in clear lake gnats (*Chaoborus astictopus*) and to cyclodienes in eye gnats (*Hippelates collusor*). Organochlorine resistance has also appeared in the pomace fly (*Drosophila melanogaster*) and in another species, *Drosophila virilis*. Certain Japanese strains of *D. melanogaster* were found to have natural tolerance towards DDT without any selection, but selection with DDT as a side effect of the malaria control program at Hikone had produced there a much higher level of DDT-resistance in both of these species by the middle 1950's.[42,43]

## c. Mosquitoes

The development of resistance to DDT and dieldrin in Greece in 1951 and 1952, respectively, by *Anopheles sacharovi*, vector of *Plasmodium vivax* malaria, provided the first examples of resistance to organochlorine insecticides in anopheline mosquitoes; dieldrin resistance ultimately became so serious that this compound had to be abandoned, but DDT continued to be used. The national antimalarial house-spraying campaign begun in 1946 effected almost complete arrest of malaria transmission by the end of that decade, but it was noticed in 1949 that although adults of *A. sacharovi* were absent from DDT-treated dwellings and outhouses in the northwestern corner of Epirus, they rested in great numbers under road bridges nearby. Behavioral studies had shown that this species is strongly irritated by DDT deposits, which provoke it to flee from them toward the light, and it seems likely that this reaction had provoked the outside resting. The resistance was initially patchy, since at the time when control had failed in coastal areas of southern Peloponnesus, it was still complete in other areas on the mainland to the north, even in places where DDT had been used since 1946 as a larvicide in ricefields.[43]

In 1952, chlordane and dieldrin gave good control in some villages in southern Peloponnesus and poor control in others. In one place, $\gamma$-HCH deposited in a box at 0.15 g/m$^2$ gave 90% mortality in mosquitoes exposed to it for 30 min; 1 month later the same test gave only 13% mortality.[431] By 1953, DDT-resistance was evident throughout the country, and it was estimated that DDT spraying at this time prevented malaria transmission for 1 month, instead of 6 months as at the beginning. Resistance to both DDT and the cyclodienes appears to have been fairly general by 1954, by which time an increase in vivax malaria was also evident. The level of DDT-resistance continued to rise during 1954 and in 1956 there were about 2,000 cases of vivax malaria. About 1955, a survey of mosquito behavior in dwellings showed that spraying reduced the numbers of *A. sacharovi* to about a third of the level found in untreated buildings. DDT continued to be effective as a larvicide, and chlordane was effective until 1956, when it failed in some places. Where chlordane failed there was even greater resistance to dieldrin.

In contrast to the situation in Greece, 7 years of use of both DDT and HCH in Romania before 1955 had not produced resistance to them. In Turkey, *A sacharovi* showed no resistance by 1956, the species having been virtually eliminated by the early DDT treatments between 1946 and 1948. DDT-resistance was observed at two places in rice-growing areas in northern Lebanon in 1953, the insecticide having been used there for various purposes since 1951; it was, however, still sufficiently toxic to permit the elimination of these populations in 1956. Failure to control this species with DDT also occurred after about 5 years in the Kazerun area of southern Iran in 1956, where the LC50 measured by the Busvine-Nash test with DDT impregnated papers was well over 4%, against 1% for *A. sacharovi* from an unsprayed area. In Israel, dieldrin resistance was found in a strain of *Anopheles sergenti* in a district near the Dead Sea, but no resistance to DDT of local anopheline mosquitoes has been reported, and *A. sacharovi* has been rarely seen in the country since 1960.[43,48]

According to a recent W.H.O. survey, *A. sacharovi* has now developed resistance to both DDT and dieldrin in the Adana region of Turkey, and there is an area of persistent malaria in southern Turkey, although the insects there appear to be normally susceptible. In spite of resistance, malaria eradication is well advanced. The same survey indicates that in 1968, DDT-resistant *A. sacharovi* were found in the Ghab area of Syria and in some parts of neighboring countries. Some tolerance of *A. sacharovi* and *A. maculipennis* has been seen recently in Yugoslavia, without any persistent transmission of malaria. There is only a low degree of resistance to DDT in *A. maculipennis* in Bulgaria and Romania, although there is cyclodiene resistance in the latter, and these countries have eradicated malaria by using DDT. The Malaria situation is said to be unchanged in the United Arab Republic although *A. pharoensis* has become resistant to both DDT and dieldrin; control is mainly with DDT, since dieldrin is now only poorly effective.[48]

DDT-resistance appeared over the entire range of *Anopheles stephensi* in the Persian Gulf in 1957, after earlier reports of resistance at two oases in 1955; the LC50 for adult females from the oasis of Al Hasa was 5% by the Busvine-Nash test, compared with less than 0.5% for insects from an untreated area.[42] At Abadan, in Iran, less than 50% of adult *A. stephensi* were killed by 4% DDT in 1957, although they were still susceptible to γ-HCH and dieldrin. A consequential outbreak of malaria at Abadan was therefore controlled by using dieldrin. Control was at first much more effective with this compound, but resistance to it appeared after 2 years and rapidly became so serious that malaria recurred. In 1963, DDT was reintroduced and remained effective with no apparent change in the resistance level until 1966, when resistance began to appear again.[48] The main effects are in southern Iran and in the oil fields of Saudi Arabia, where the double resistance to DDT and dieldrin is seriously affecting malaria control and may necessitate a change to alternative compounds. In several places in West Pakistan, DDT-resistance has been confirmed in *A. stephensi* and *A. culicifacies,* without apparent effect on malaria control, and the latter is also resistant in some areas in India to the extent that replacement with HCH has become necessary.

For a number of years *Anopheles gambiae* has been resistant to HCH and dieldrin in large areas of Africa, this resistance having been noticed about 1955 in several places simultaneously. In November of that year, survivors of house spraying in northern Nigeria, where a campaign had been conducted for about 16 months, had an LC50 for dieldrin of 2% according to the Busvine-Nash test, this being about eight times the value found for a normal strain from another region. A colony of *A. gambiae* was subsequently reared in London from eggs collected at the village of Ambursa in this region, and the females were found to have LC50's for aldrin and dieldrin of more than 4%, the LC50 for γ-HCH being 0.18%, and for DDT 0.6%.[432] The colony was therefore not resistant to DDT and there was only slightly greater tolerance to malathion and pyrethrins. Resistance levels for the cyclodienes were obtained by extending the time of exposure to the impregnated papers from the standard one hour up to 18 hr, when LC50's for dieldrin, aldrin, isodrin, endrin, *trans*-chlordane, and *cis*-chlordane were found to be, respectively, 800, 400, 140, 90, > 1,000, and > 500 times the corresponding values for a normal strain from Lagos.[432] However, these figures are higher than they would be if converted to the standard exposure time for comparison with the 26-fold higher LC50 for γ-HCH obtained by the normal exposure. As is usually found, the cross-resistance to HCH was lower than that to the cyclodienes.

It turned out that the single female from which the colony had been founded and its mate must have been homozygous for dieldrin resistance, since Davidson (see section B.2b) was able to show that this resistance was due to the existence of a single gene allele.[379,380] A later survey of the original area (Western Sokoto) revealed that the majority of *A. gambiae* were homozygous for the resistant gene in both dieldrin- and HCH-treated areas; further east, in Kano, where HCH had been applied, the incidence of homozygotes was only 8%, there being 22% of heterozygotes having intermediate resistance. There was also evidence of up to 6% of resistant heterozygotes in the populations of untreated towns in the area, although surveys further south near Ibadan and Lagos revealed no resistant phenotypes where HCH had been used. Similar resistance was apparent in 1957 in Upper Volta, and a very high incidence of monozygotes for dieldrin resistance was encountered about the same time in inland Liberia, following two years of dieldrin treatment.[43]

DDT may substitute for the cyclodienes in forest areas, but it has not proved effective in preventing malaria transmission in the savanna areas. *Anopheles gambiae* like *Anopheles sacharovi* is one of a number of species said to exhibit "normal behavioristic avoidance"; it appears to be so irritated by exposure to DDT on surfaces within houses that it leaves after feeding but before receiving a lethal dose of the toxicant (see section A.) Such insects are quite susceptible to DDT when brought into normal contact with it; resistance to this kind of exposure seems not to have been encountered until 1967, when it was discovered in Upper Volta and Senegal. Because there is currently no satisfactory agent for the prevention of malaria transmission in the savanna areas, a large part of the African malaria eradication program faces an uncertain future.[48,433,434]

The history of malaria control in the Americas as a whole is one of mixed success. At one time, this debilitating sickness was present in a large area of the United States; in the 19th century, cases were not unknown as far north as southern Michigan. Improved living standards, better housing, better nutrition, and other factors forced the disease mainly into the southeastern states in the period before World War II. It reached its climax in the south in 1936 as a result of the years of depression, during which time malaria control could not be financed. In subsequent years, control measures centered on the destruction of larvae in breeding places, oil or Paris green being favored control agents. As a residual insecticide, DDT revolutionized this slowly proceeding control program; its use was vital at the end of the war, with the return of thousands of troops from war zones, many of them being carriers of malaria at the time. An antimalaria program was developed in 13 southern states in which DDT was applied to the walls of hundreds of thousands of homes. This program, together with a general rise in the standard of living after the war resulted in the virtual eradication of malaria in the United States.

*Anopheles quadrimaculatus* is common in the southern United States and instances of resistance to both DDT and dieldrin have been reported since the early 1950's, but mainly after malaria had been successfully eradicated. At least until 1958, experience with the control of *A. quadrimaculatus* in the Tennessee Valley, where DDT was extensively used as a larvicide, indicated that although larvae developed DDT-tolerance, the adults showed little tendency to do so, and any resistance that did appear tended to revert between treatments. Thus, despite the use of 90,000 pounds of DDT between 1945 and 1948, field populations of the mosquito failed to show any measurable resistance. It may be significant that this quantity of DDT was only applied to about one-fourth of the potential breeding areas during this entire period, so that any resistance selected by the treatments may have become diluted by influx of susceptible insects from nearby untreated areas. Also, spraying always began when the mosquito population reached a certain critical level and ceased when it began to decline in the autumn, so that several generations received lower exposure to DDT.[435]

Central and South America have generally been less fortunate than North America. *Anopheles pseudopunctipennis* and *Anopheles aztecus* have been under selection pressure with DDT in Mexico and other countries since the late 1940's, but there appears to be little evidence of any resistance to organochlorines before about 1958, when Brown cites observations of cyclodiene-resistance in Mexico, Nicaragua, Peru, Venezuela, and Ecuador.[373] The recent W.H.O. survey indicates that in the late 1960's, double resistance of *A. pseudopunctipennis* to DDT and dieldrin was widespread in Mexico.[48] In that country, the continued transmission of malaria certainly results from resistance in one eradication area, but in other zones it appears to be only a secondary

cause of persistant malaria. In Panama, systematic house-spraying with DDT from 1944 effected great reductions in the domestic populations of *Anopheles albimanus*. The potential malaria area of that country covers about 92% of the territory, and is inhabited by about 96% of the population. The total treatment of such an area presents great difficulties and was achieved for the first time in 1958 when dieldrin was applied to 163,000 houses. Widespread resistance of this species to actual contact with both DDT and dieldrin was observed throughout Central America in 1958, but there had been numerous reports of behavioristic resistance after 1952 when it became evident that engorged females no longer rested on DDT-sprayed house walls after feeding. There is a little evidence here of a changed reaction due to selection which may provide an example of true behavioristic resistance, in contrast to the situation with *A. gambiae*. Whereas in the early days of the program, a large proportion of the engorged females caught in houses died within 24 hr because they had rested on a DDT-treated surface, this proportion declined in successive cycles of treatment, showing that an increasing number of these insects were avoiding contact with DDT. When placed in contact with the insecticide, they were just as susceptible as a normal strain from an untreated area.

The problem of resistant *A. albimanus* also extends to other Central American countries;[48] in British Honduras and Costa Rica, serious dieldrin resistance arose recently and resulted in renewed malaria transmission, although in British Honduras, DDT still controlled the situation. In Nicaragua on the other hand, this species is resistant to both DDT and dieldrin in the cotton growing areas, where nearly 50% of the people live. These people would be fully exposed to the threat of malaria were not alternative insecticides available, with consequent economic loss for the country. In Venezuela, *Anopheles aquasalis* and *Anopheles albitarsis* developed dieldrin-resistance in the late 1950's, but remained susceptible to DDT, so that malaria eradication was still possible and a reappearance of malaria attributed to *A. albitarsis* was suppressed by DDT in 1965. In a limited area of western Venezuela, *Anopheles nuneztovari* recently showed some resistance to dieldrin; it remains largely susceptible to DDT, but its outdoor rather than indoor habits account for continued malaria transmission in this area. In the early days of the malaria control program (up to 1949), *Anopheles darlingi* was a major malaria vector in Venezuela; its incidence, and that of malaria, was remarkably reduced during this period by house-spraying with DDT. Since the other preponderant species of the time, namely, *A. albimanus* and *A. aquasalis*, were unaffected by this treatment, it was concluded that *A. darlingi*, which is strongly attracted to man (anthropophilic) and his dwellings (endophilic), was the most important carrier. The resistance to contact with DDT appeared to be greater for *A. albimanus* and least for *A. darlingi*, and *A. albimanus* was the mosquito most often found resting in houses. After 1949, an avoidance reaction appeared to develop in *A. darlingi*, since it was to be found resting on the outer walls of houses more frequently than before.[436] This situation appears to contrast somewhat with that in Panama in the early 1950's where *A. albimanus* apparently developed behavioristic resistance but not resistance to actual contact.

The main malaria vectors of Colombia are *Anopheles punctimacula, A. nuneztovari, A. darlingi*, and *A. albimanus*. All are susceptible to DDT, but *A. albimanus* is resistant to dieldrin. *A. albitarsis* has developed resistance to both DDT and dieldrin in rice-growing areas, but malaria eradication is proceeding satisfactorily using DDT.[48] In Brazil, the antimalaria campaign became nationwide in 1947–48, when it first became possible to use residual insecticides throughout the entire country. DDT was used, or DDT and HCH where both malaria and Chagas' disease were prevalent. These measures, together with the use of antimalarials such as chloroquin, are said to have reduced the incidence of malaria by 90% in the areas treated. By 1958, the control program promised to result in malaria eradication. In Brazil also, indications of avoidance reactions were seen in *A. darlingi, A. cruzii*, and *Culex fatigans* in the early 1950's. In 1966, *A. aquasalis* was reported to be resistant to DDT in the Para area of Brazil, where control otherwise remains good; in Bolivia there is no evidence of DDT-resistance in *A. pseudopunctipennis* or *A. darlingi*. Brown has listed 37 species of Anopheles that are currently resistant to dieldrin and 15 species that are resistant to DDT, some species being resistant to both.[43]

Resistance of culicine mosquitoes to DDT was already evident in the United States in 1948. Control difficulties were experienced with *Culex tarsalis, Aedes melanimon*, and *Ae. nigromaculis* in California, and with *Ae. sollicitans* and *Ae. taeniorhynchus* in Florida. DDT was first used to control larvae of *C. tarsalis*, the vector of encephalitis, in the San Joaquin valley in 1945, and lack of control was noticed in 1947–48, when DDT was followed successively by HCH (1947), toxaphene (1949), and aldrin (1951). A comparison of the susceptibility of larvae collected from treated and untreated areas in the valley in 1951 revealed that in one sample resistance was present to DDT and the cyclodienes to the extent of tenfold to DDT, 30-fold to toxaphene, 200-fold to aldrin, 1300-fold to heptachlor and 11-fold to $\gamma$-HCH;[437] neither heptachlor nor dieldrin had been used in the area, and a heavy application of dieldrin subsequently proved to be without effect. This is an example of the familiar cyclodiene cross-resistance spectrum expected to result from separate exposure to individual members of the cyclodiene group (Vol. II, Chap. 3B.3b). The irrigation water mosquitoes (*Ae. nigromaculis* and *Ae. melanimon*) were controlled by DDT from 1945 to 1948,[437] then with $\gamma$-HCH, toxaphene and aldrin until 1951, and finally with organophosphorus compounds such as parathion and malathion. Since that time *Ae. nigromaculis* has developed resistance to several insecticides of the latter type. DDT-resistance in the salt-marsh mosquitoes (*Ae. sollicitans* and *Ae. taeniorhynchus*) was first encountered in Florida in 1947 when DDT failed to control a flight of the two species. *Ae. sollicitans* became predominant over *Ae. taeniorhynchus* in 1949 and had to be controlled with HCH. Adults from a DDT-treated area at Cocoa Beach were found to be six to eight times, and larvae about ten times, as resistant as normal. There was also resistance to DDD, but not to $\gamma$-HCH, toxaphene or the cyclodienes, and $\gamma$-HCH appeared to be the best substitute for DDT at that time, although both chlordane and dieldrin gave good control. Some coastal salt marshes were extensively treated with HCH during 1950–1952, and larvae from these areas appeared to develop resistance to HCH and increase their resistance to DDT during this period. By 1952, the Cocoa Beach larvae were resistant to all organochlorines and even showed some resistance to replacement OP compounds.[42,43]

Among the culicine mosquitoes which are pests in forests, marsh lands and irrigated areas, *Culex fatigans* and *Ae. aegypti* are well known as vectors of the diseases Bancroftian filariasis and yellow fever, respectively. Dieldrin and DDT-resistance of *C. fatigans* was first observed in the United States and in India, respectively, in the early 1950's and of *Ae. aegypti* in Central America in 1953.[43] Adults and larvae of *C. fatigans* are evidently naturally tolerant to DDT; the LC50 for larvae of a Malayan strain was found to be about 20 times greater than the LC50 for larvae of *Ae. aegypti* or *Anopheles maculatus*. However, they are quite susceptible to HCH or dieldrin, but examples of very rapid development of resistance are on record. In one case, 3 years of treatment of the larvae of a Malayan strain with HCH produced only a tenfold level of resistance to both HCH and dieldrin without affecting the susceptibility to DDT. In two other instances from the same area, dieldrin failed to give satisfactory control of *C. fatigans* larvae within five months of its substitution for DDT.[42]

In the United States, where *C. fatigans* is frequently referred to as *C. quinquefasciatus*, larval resistance to HCH was first noticed about 1951, and it became apparent elsewhere that larvae from areas extensively treated with HCH were more resistant than those from untreated zones. Experience in a number of areas in the Americas indicated that *C. fatigans* larvae exposed to DDT were soon selected for DDT-resistance in addition to the natural tolerance they have for this compound. On the other hand, the Californian larvae that had become resistant to HCH in 1951 do not appear to have acquired much resistance to DDT by that time. Nevertheless, it soon became clear that only poor control could be obtained with DDT; for example, the poor control of *C. fatigans* in British Guiana and the Virgin Islands contrasted strongly with the complete elimination of *Ae. aegypti* in those places.

In Belém, Pará, Brazil, it was decided in 1952 to replace DDT by HCH because *C. fatigans* was resisting the DDT applied in houses there to control malaria. HCH was applied in quarterly cycles and was highly successful for the first two cycles, but only poorly effective in the third and ineffective in the fourth. Between 1952 and 1955, DDT was again used inside houses until dieldrin became available for testing in 1955. Not only was dieldrin ineffective from the beginning, undoubtedly due to the previous use of HCH, but a change appeared to occur in the mosquitoes in regard to their resting places in the houses.[438] Whereas previously 80% of specimens rested on walls and ceilings (which were subsequently sprayed) and 20% rested on furniture, clothing, etc., this ratio was reversed after treatment, the effect lasting for at least 30 days. Moreover, the original distribution had not been achieved 3 years later, during which period DDT spraying of houses had been resumed. In Africa, poor control of this species was experienced with DDT in the early 1950's due to natural tolerance, although wall deposits of HCH or dieldrin were highly effective; the latter compounds gave high mortalities for 1 month and 7 months, respectively, after spraying. The main effect of DDT deposits was to irritate the mosquitoes so that they left the huts without acquiring a lethal dose of the toxicant.[439]

Resistance has frustrated the satisfactory control of *C. fatigans* in many parts of South East Asia, and very soon after the start of the Indian National Filaria Control Programme in 1955, DDT proved to be unsuitable as an adulticide for this species. HCH was tried in 1957 and dieldrin in 1958–59, but resistance rendered them rapidly ineffective. The situation is currently somewhat similar in the western pacific region (Malaysia, Hong Kong, and parts of Australia) and in both areas recourse has been made to the use of oils and organophosphorus compounds as larvicides since 1960.[48]

*Culex pipiens* are considered to be temperate zone forms of *C. fatigans*, and the related subspecies *C. pipiens molestus* is distributed throughout Europe, North America, North Africa, and Australia. It was *C. pipiens molestus* which provided the first example of DDT-resistance in mosquitoes in 1947,[440] when a 1 year field exposure of an Italian strain to DDT produced a resistant strain which survived 3 days of contact with DDT deposits that killed a normal strain in 3 to 5 hr. Chlordane and HCH remained effective against this strain until 1950 when it was found to have become resistant to all organochlorine insecticides. *Culex pipiens* species in the United Arab Republic have shown resistance to both DDT and dieldrin, which has had some effect on control with DDT and has restricted the use of dieldrin, fortunately without any increase in filariasis.

The yellow fever mosquito (*Aedes aegypti*) is normally particularly susceptible to control by DDT; the LC50 for adults lies between that for normal *Culex* and normal *Anopheles*, and for larvae is usually less than 0.01 ppm (parts per million).[43] An eradication program has been in progress in the Americas since about 1953 and programs of house spraying with DDT in British Guiana, French Guiana, Mauritius, and St. Croix virtually eliminated the insect. By 1956, *Ae. aegypti* was claimed to have been eliminated from Paraguay, Uruguay, Chile, Brazil, Ecuador, Peru, Bolivia, British and French Guiana, and the Central American states, and in 1963, eradication was claimed for 17 countries.[441]

However, resistance has always lurked in the background. In 1954, for example, an outbreak of yellow fever in Port-of-Spain, Trinidad coincided with failure to control *Ae. aegypti* on the island due to DDT-resistance. Examination of larvae on that occasion revealed remarkably high LC50 values for DDT, applied both in emulsion form and as wettable powder, and it transpired that complete kills of larvae had never been achieved, even in treatments as far back as 1950. An eradication program was set in hand as a result of this outbreak, but the treatments always left a few highly resistant larvae; emulsions were more effective than wettable powders, but even then it took 48 hr at 30 ppm to kill these remaining larvae. A detailed examination of larvae of the Trinidad strain revealed that they were 1000 times less susceptible to DDT and 200 times less susceptible to DDD than a normal laboratory strain, but had nearly normal susceptibility to dieldrin, HCH or strobane.[442] Adults of the Trinidad strain were also resistant to DDT, and the resistance of neither adults nor larvae was overcome by admixture of DMC with DDT, although the larvae at least produced large amounts of DDE from DDT in vivo.

In the mid to late fifties DDT-resistance appeared in the Dominican Republic, in northeastern Colombia, in Haiti and in various parts of Venezuela, and around 1963, double resistance to DDT and the cyclodienes (see section B.2b) began to reverse the trend towards elimination of *Ae. aegypti* in the Caribbean Islands. Since then, the further spread of double resistance and reappearance of this mosquito on the mainland in Brazil, El Salvador, Honduras, and Guatemala constitute a serious threat to the hitherto successful programme.[48]

Besides being the yellow fever vector, *Ae. aegypti* is the transmitter of viruses responsible for haemorrhagic fever. Following an outbreak of this in the Phillipines in 1964, it has spread into Malaysia, Cambodia, Vietnam, Thailand, Burma, and India. The prospects for immunization against this pestilence are poor, and it is best contained by controlling the vector. Unfortunately, the chances of control by DDT are not good because of the widespread incidence of resistance to it, and previous experience suggests that the cyclodienes will not be effective for long. According to Brown, 22 species of culicines have developed resistance to organochlorines, of which some exhibit double resistance to DDT and the cyclodienes, as indicated, and a few have progressed to the stage of resistance to organophosphorus compounds as well.

In spite of all these setbacks, and although the deleterious effects of resistance on mosquito control are likely to increase, there are some indications that the rate of increase will diminish, if only because the major instances of resistance are already with us.[434] Greatest danger lies in the spread of multiple resistance to several groups of insecticides (see section B.2b). The malaria eradication campaign has been outstandingly successful in the temperate and semitropical areas, accounting for nearly three-quarters of the people needing protection, but permanent protection of the remaining quarter will be much more difficult (Vol. I, Chap. 2A.3d) The problem has hardly been touched in South and Central America, tropical Africa, and some areas of the Middle East, South East Asia, and Australasia. In the temperate regions such as Europe, resistance has not prevented the attainment of eradication, but it was usually DDT that completed the campaigns, dieldrin resistance being usually more intense and insurmountable. Where malaria is more firmly established, as in the Persian Gulf area and in Central America, simultaneous dieldrin and DDT-resistance, with the latter less intense, makes final eradication extremely difficult.

On the African continent there are great difficulties, and insecticides giving a rapid and high kill are required to arrest malaria transmission. HCH has been widely favored for this purpose, but dieldrin-resistance renders it ineffective in many areas and recourse has been had to compounds such as malathion and propoxur.[433,434] At this point the difference should be noted between eradication of a disease vector and of the disease that it carries. Thus, if the number of malaria vectors can be reduced, without eradication, to a level such that malaria transmission is arrested for, say, 3 years, then the disease itself can be *effectively* eradicated. Of course, even a low incidence of malaria vectors is always a potential transmission danger, but the fact remains that eradication of the insect is not entirely essential.

Because of environmental factors including increased urbanization and the growth of slums, and since *C. fatigans* is naturally refractory to insecticides and has fully developed resistance to organochlorines in many areas, there is a tendency for the spread of filariasis in many tropical countries. The eradication of this disease is a difficult proposition and would require about 10 years of uninterrupted mosquito control.[433] Similarly, there is a distinct possibility that the double resistance of *Ae. aegypti* to DDT and the cyclodienes may jeopardize the eradication of yellow fever, although the prospects for eradication of the insect itself seemed quite good at one time. Possible future resistance problems apart, some residual organophosphorus and carbamate compounds have shown promise for control of this vector.

### d. Lice, Fleas, Bedbugs, Reduviid Bugs, and Cockroaches.

The value of DDT as a control agent for the body louse (*Pediculus humanus corporis*) was remarkably demonstrated when, applied as a 10% powder at the rate of 28 g per person, it completely controlled lice and arrested a typhus epidemic in Naples in 1943.[12] By 1946, body lice had disappeared from Latina Province, and in Korea and Japan, two million people were successfully deloused in the winter of 1945—46. On the basis of laboratory experiments, there was some reason to hope that lice would not become resistant because the exposure of a colony to DDT at the LD75 level had not resulted in any loss in susceptibility to the toxicant after eight generations.

However the optimism was short-lived and in the winter of 1950—51, resistance appeared among lice infesting Korean military personnel.[42] When tested in sleeves treated with 5% DDT powder, these lice showed no increase in their natural mortality rate, a remarkable observation since a normal strain was completely killed by a 0.25% powder. The Korean lice were also resistant to methoxychlor, but completely susceptible to pyrethrins, γ-HCH, aldrin, dieldrin, chlordane, and

toxaphene. After this episode, investigations in Japan uncovered other instances of the failure of 10% DDT powders to give adequate control. Also in 1951, reports from refugee camps in the Near East indicated that mortality was delayed in some cases, and that reinfestation occurred rapidly after treatment. DDT dust had been used in Cairo from 1946 and resistance had arrived by 1950; in the Nile delta area, DDT became practically ineffective and was superseded by HCH. Most of the samples of lice from various parts of Europe that were examined in the early 1950's showed normal susceptibility, but a number of samples taken from individuals in the Marseilles area showed varying degrees of resistance ranging from greatly delayed mortality when exposed to 0.1% DDT dust to complete immunity. One of these samples was also resistant to γ-HCH.

A W.H.O. survey initiated in 1953 and published in 1957 involved 177 samples from 27 countries and revealed that lice from European countries and the untreated parts of Iran, as well as from Afghanistan, India, and Pakistan were normally susceptible to DDT.[443] However, high resistance was present in samples from DDT-treated areas of Iran, and the Korean type of resistance was shown to be present in Hong Kong and widespread in Japan; moderate resistance occurred in Cairo, in some parts of Turkey and in refugee camps in Syria, Jordan and the Gaza Strip. HCH-resistance was also observed in several places where this compound had been used, and was widespread in Japan.

Some laboratory experiments conducted with the Korean resistant lice provide an interesting example of the way in which toxicant pressure and the particular circumstances of toxicant use may influence the resistance level. Although these insects were rather resistant to 10% DDT dust, high mortality was obtained by exposing them to the same dust on cloth already impregnated with 0.01% DDT. However, rearing them in these circumstances of extreme selection with DDT resulted in complete immunity to 10% DDT dust in 15 generations. The eggs of this highly resistant strain were normally susceptible to dinitroanisole and γ-HCH but could not be killed with chlorophenyl chloromethyl sulfone, which is usually more effective than DDT. Susceptibility to organophosphorus compounds was retained and the "wild type" Korean lice lost their resistance when reared for eight generations in the absence of

DDT; those selected to the level of extreme resistance lost 75% of it after 15 generations without DDT and still retained a degree of residual tolerance after 45 generations without DDT pressure. Subsequent extended selection of these reverted strains on cloths impregnated with γ-HCH, the above mentioned sulfone or pyrethrins, produced no significant increase in tolerance to these compounds.[42]

On the basis of the use of 5% DDT dust as a discriminating dose for the detection of resistance in body-lice, the 1965 WHO survey[444] showed that varying degrees of resistance were present in India, and low levels in Hong Kong and Mexico. Examples of high resistance were found in France, Chile, the United Arab Republic, the Sudan, Jordan, Syria, the Gaza Strip, Afghanistan, and South Africa. In South Africa there is some suggestion that a recently increased incidence of typhus in eastern Cape Province may be due to the failure of DDT and HCH dust, and carbaryl (sevin; 1-naphthyl-N-methylcarbamate) is now being used as a replacement.

The patterns of distribution of resistance described suggest strongly that the degree of resistance is a measure of the extensive use of DDT dust for louse control since this dust has been used for a long time in the United Arab Republic, the Gaza Strip, and Afghanistan. It has also been used extensively for lice control in eastern Europe and more recent information indicates the presence of resistance there. Up to the present time, dusting with DDT has proved effective in arresting the spread or onset of disease due to lice, but there is now some degree of doubt regarding the future of this method of control. DDT and HCH are also widely used to control the head-louse, *Pediculus humanus capitis*, but there is now evidence for resistance to both and malathion is a favored control chemical.

Many instances of the failure of DDT to control domestic fleas (*Pulex irritans*) were reported in South America in the early 1950's, after several years of satisfactory control. Flea populations had been greatly reduced in Italy and Greece by this time, but in 1951 a resurgence of *P. irritans* was claimed to be occurring in Greece, and was assumed to be due to DDT-resistance, although the incidence of fleas was still much lower than before the use of DDT. By 1952, resistance was also present in the flea populations of the Palestine refugee camps and one or two instances of

resistance to 10% DDT dust in Lebanon were effectively controlled by the use of DDT emulsions instead. DDT-resistance to cat fleas (*Ctenocephalides felis*) and dog fleas (*Ctenocephalides canis*) was widespread in the Americas by the early 1950's and by 1956 the organophosphorus compound malathion was being used to control them, since the use of chlordane or dieldrin soon produced characteristic cyclodiene resistance.

The oriental rat flea (*Xenopsylla cheopis*) is an important carrier of the plague, and can be satisfactorily controlled by application of DDT dusts. No clear evidence for the incidence of resistance in this species had appeared by 1957, although indications were available of its presence in Ecuador and in the state of Georgia. Attempts to produce DDT-resistance by selecting fleas in the laboratory had only doubled the tolerance. However, by 1963, extensive resistance to DDT had developed in India, and since the use of HCH also produced a rather low level of resistance to this toxicant, there have been control failures in Southern India during the last few years and a number of cases of the plague. In Vietnam, there have been severe outbreaks of plague in recent years and high resistance to organochlorine insecticides has necessitated their replacement by the organophosphorus insecticide diazinon. Since strains of *X. cheopis* resistant to DDT and dieldrin are also present in Thailand, the situation in Vietnam is a source of some anxiety. Recent outbreaks of plague in places such as Indonesia and Tanzania show that the situation requires constant vigilance. Resistance in fleas may be a side consequence of the use of insecticides in the antimalaria campaigns and may hamper the elimination of isolated foci of the plague. *X. cheopis* is also resistant to DDT in Israel where it is a vector of murine typhus.

The bedbug (*Cimex lectularius*) and the tropical bedbug (*Cimex hemipterus*) are a widespread nuisance pest and unfortunately resistance to their control by insecticides is now very common in warm and tropical countries. As pointed out by Busvine and Pal,[48] the impact of their resistance goes far beyond the loss of a simple method of removing a nuisance pest; the failure of measures to combat bedbugs or houseflies causes suspicion and antagonism among householders, who may not thereafter be amenable to the application of measures for malaria control. Thus, an apparently simple situation can jeopardize the success of an antimalarial campaign.

In 1947, 3 years after the first application of DDT to control *C. lectularius* at Pearl Harbor, some failure of control was noticed. In the next few years, resistance appeared in Korea (1951), Japan (1951), Belgian Congo (1952), Greece (1951), Italy (1953), the Americas (1953–54), Iran (1956), and Israel (1956). The Italian occurrence involved a refugee camp from which *C. lectularius* had previously been eradicated. Detailed examination showed that the strain was completely resistant to DDT, methoxychlor and dieldrin, showed partial resistance to chlordane, and was quite susceptible to $\gamma$-HCH, diazinon, and malathion. Many instances of resistance to DDT were observed in the United States in the early 1950's, and one instance in French Guiana in South America (1955). This last episode illustrates the ease with which hitherto successful control programs can be thwarted. There were no reports of resistant *C. lectularius* elsewhere in South America and the pest was completely controlled by DDT in French Guiana between 1950 and 1955, but significantly, DDT had not proved effective in eliminating bedbugs from St. Lucia in the Windward Islands. The appearance of DDT-resistant bedbugs in Cayenne was linked with three houses occupied by recent immigrants from St. Lucia. The infestation subsequently spread to a barracks, where it was found to be controllable with HCH.[42] The North American cases of DDT-resistance were also usually controllable by HCH or chlordane, unless there had been previous extensive exposure to these compounds. DDT-resistance in Israel was soon followed by the failure of the alternative use of HCH. Spray tests indicated resistance levels to HCH of 20 times and 160 times normal, respectively, in two strains from different places; the higher level reverted to a 40-fold resistance when the strain was removed from HCH pressure for 6 weeks. Although DDT has been successfully used for 20 years to control bedbugs in northern Europe, resistance does not appear to be a problem there.

In 1952, after 2 years of DDT spraying with a wettable powder at an army camp in southern Taiwan, *C. hemipterus* became highly resistant to DDT but could still be controlled with sprays of HCH wettable powders. In the mid-1950's, DDT-resistance in this species was also found in Kowloon, Hong Kong, where DDT had been rarely used, and control failures were also reported in Singapore. By 1955, control of *C. hemipterus* with DDT dusts began to fail in Bombay State, and although DDT emulsions were sometimes used successfully in place of the dusts, satisfactory control could only be obtained with dieldrin or HCH emulsions. Inevitably, resistance to these two compounds was evident by 1956. In Africa, dieldrin resistance developed in some places in consequence of the use of this compound for malaria eradication but in these cases, *C. hemipterus* retained its normal susceptibility to DDT. The present-day situation seems to be one of widespread global resistance (except in northern Europe as indicated previously) which very frequently extends to both DDT and the cyclodienes.

Reduviid bugs (Rhodnius, Triatoma, and Panstrongylus) are vectors of Chagas' disease, a fatal form of human trypanosomiasis found in South America and caused by *Trypanosoma cruzi*. Therefore, they are of considerable public health importance in southern and central America up to Mexico. In Brazil and Venezuela, Chagas' disease rivals malaria in importance, and since there are no prophylactic drugs, vector control campaigns are the only effective recourse in the present circumstances. The nymphal forms of these species are naturally tolerant to DDT, and although the adults are more susceptible, DDT is not a very satisfactory control agent.[44] HCH and dieldrin are therefore the favored control agents for these species and one million houses in Venezuela and several million in Brazil are sprayed annually with one or other of these agents, which have so far given satisfactory control. However, a recent report[48] that there may be resistance to these chemicals in both countries emphasizes an urgent need to have replacement insecticides available, since any spread of organochlorine resistance would quickly have a serious impact on these control campaigns.

The German cockroach (*Blattella germanica*) is naturally rather tolerant to DDT, but the early DDT spraying programs reduced its numbers dramatically at first because of its abundance. There are indications that this cockroach became resistant to DDT during the DDT spraying program in Greece in the early 1950's, and in the same year in the United States, laboratory tests confirmed that a case of control failure by DDT treatments was due to the development of a resistant strain. Resistant strains were produced by selection with DDT in the laboratory (for example, 22-fold and 198-fold resistance for males and females, respectively),[445] but the resistance level declined rapidly in the absence of selection pressure. Because of the situation with DDT, chlordane became the control agent of choice, but resistance to it became a serious problem in the United States by 1952. According to a laboratory test, female cockroaches surviving treatments with a 2% chlordane solution at Corpus Christi, Texas were 12 times as resistant as the males and 276 times as resistant as females of a normal strain.[446] After nine generations without chlordane pressure, the level of resistance in the females had fallen to 108 times; after 2 years, it was 14 times and after 3 years six times the normal level. By 1957 the strain had reverted to normal susceptibility. The strain showed cross-resistance to γ-HCH, aldrin, dieldrin, and heptachlor, but little increase in tolerance to DDT, pyrethrins, and diazinon.

In subsequent years, up to 20-fold levels of chlordane resistance became common in the United States, especially in the south and the resistance extended to all organochlorines. Mixtures of malathion with dieldrin, malathion with two parts of Perthane, or various pyrethrin formulations became popular for control. Later, the organophosphorus compound diazinon was recommended as an additional control agent. Soon afterward, moderate levels of pyrethrin resistance were noted and about 1956 there was some indication of malathion resistance, although failure of the mixture of this compound with Perthane (called malrin) was attributed by some to the development of resistance to Perthane rather than to malathion.[42]

The response to a recent W.H.O. questionnaire indicates that HCH-dieldrin resistance was widespread by the late 1960's, instances being reported from the United Kingdom, Germany, Poland, Czechoslovakia, Hong Kong, Japan, British Honduras, and Canada.[48] Malathion is a good choice as an alternative control agent because of its low mammalian toxicity, but resistance has now definitely been reported in some places including the United States. Because the insect occurs in domestic situations, the question of toxic hazard becomes important and the choice of alternatives is limited, although several carbamates and organophosphorus compounds are under scrutiny. One advantage of the stability of dieldrin is that it can be applied to surfaces as a persistent lacquer and is a very effective control agent when used in this form, before the onset of resistance, that is.

### e. Acarina (Ticks and Mites)

Ticks and mites can be controlled by spraying their habitat while they are waiting for a host, or by treating the infested host. Habitat treatment has been recommended for various hard ticks of medical importance and for the scrub typhus mite (Trombicula spp., including Leptotrombidium), while soft ticks of the Ornithodoros family, vectors of relapsing fever, have been controlled by limited treatments of dwellings. The eradication of scrub typhus is not feasible since it is endemic in such large areas, but excellent and sustained control can be achieved by the treatment of wide areas with lindane, aldrin, or dieldrin sprays.[44] Dieldrin is uniquely valuable for this purpose; DDT is not very satisfactory and the organophosphorus compounds are not sufficiently persistent. Ornithodoros moubata and some other Ornithodoros spp. are naturally tolerant to DDT, and γ-HCH has been used as a dust in African dwellings against O. moubata and against O. tholozani and O. coniceps in the Middle East. These treatments do not appear to have produced any definite resistance so far. In eastern Europe, Ixodes persulcatus has been controlled for some years by area-spraying with DDT, and no incidence of resistance has been recorded so far.

In the veterinary area, control by treating the infested host is more applicable and the extensive use of this method may be the reason for some of the severe resistance problems encountered with cattle ticks. The first recorded instance of organochlorine resistance in the brown dog tick, Rhipicephalus sanguineus, a vector of Q-fever in man, occurred in New Jersey in 1955 after 6 years of successful control by 2% chlordane sprays applied four times annually. This infestation was completely controlled by a 0.25% emulsion of γ-HCH. By 1956, other resistant populations had appeared at several points in the United States, and in the meantime, the lone star tick (Amblyomma americanum) had allegedly become resistant to cyclodienes in Oklahoma. Cyclodiene resistance is now so widespread that organophosphorus compounds have to be used to control this pest.[44]

The most remarkable story of insect resistance is that involving the cattle ticks, Boophilus *decoloratus* (the South African blue tick) and *Boophilus microplus* (the Australian cattle tick). *B. decoloratus*, the vector of Texas cattle fever, was controlled in South Africa after about 1910 by the use of cattle dips containing sodium arsenite. Resistance to this compound was first recorded near East London in Cape Province. Resistance to arsenic had spread along the coast by 1940 and by 1945 was found as far north as Zululand.[447] When the resistant ticks were dipped in a concentration of arsenic that allowed only 1% of normal ticks to survive and lay eggs, 63% of these survived to reproduce. Initially, HCH-dips afforded a most promising substitute since an HCH concentration of 0.003% gave good control, and these dips were introduced in 1946. Three cases of failure to control with HCH-dips arose in the East London district 18 months later, and tests showed that only 4% control could be achieved with HCH concentrations that had previously been fully effective.

When these ticks were compared with ticks within a neighboring arsenic-resistant area but outside the HCH-resistance area,[448] both strains still had arsenic resistance (although the use of these dips had ceased 4 years previously), but the ticks from outside the area of HCH-resistance were still susceptible to γ-HCH and toxaphene. The East London strain could still be controlled by toxaphene, although the tests showed it to be somewhat resistant to this organochlorine. The HCH-resistance was accompanied by resistance to dieldrin and neither strain was particularly susceptible to DDT in any case. In 1956, DDT resistance was found on a farm in the East London area, and diazinon was substituted for cattle tick control.

*Boophilus microplus*, a vector of piroplasmosis, became resistant to arsenic in central and northern Queensland about 1937 after many years of satisfactory use of this kind of dip. DDT was found to be an effective substitute for sodium arsenite, and after 1948, dips containing 0.05% γ-HCH were also used. Failure to control with HCH was noticed in 1952 at Rockhampton, within the arsenic-resistance area, and in terms of the concentration of γ-HCH that would prevent 50% of adult females from laying or the LC50 for the larvae, both stages were about 180-fold resistant to the toxicant. The adult females were also resistant to dieldrin and toxaphene, and had some tolerance for chlordane and DDT. In southern Queensland, arsenic resistance appeared in 1949. HCH was

introduced in 1950 and failed in 1952. DDT was successfully used in 1953, but is less toxic than HCH, toxaphene or the cyclodienes and toxaphene was introduced in 1954, only to fail immediately. This failure presumably occurred because of cross-resistance conferred by the selection with HCH. DDT still controlled these ticks which were also fairly susceptible to sodium arsenite; the toxaphene resistance was 20-fold and they were also resistant to aldrin and dieldrin.

At about the same time, larvae from an area near Brisbane, where dieldrin had failed to control the ticks, were found to have a 2,000-fold resistance to this toxicant and most of the adults could still oviposit after the usual dieldrin treatment.[449] In central Queensland where 0.5% DDT sprays had given fairly good control since before 1950, DDT-resistance appeared in 1954—55, and in Rockhampton where dips had been used since 1950, it appeared in 1956—57. According to LC50 determinations, the Rockhampton larvae had a 22-fold resistance to DDT. Fortunately, DDT-resistance developed slowly and satisfactory control was maintained on many properties in spite of it until the chemical was finally withdrawn along with the other organochlorines in 1962. *B. microplus* is also present in South America and it appears that ticks resistant to all the organochlorines are quite common there. What is particularly noticeable about the history of arsenic and organochlorine resistance in these pests is the close parallel in its development in *B. microplus* and *B. decoloratus* in different countries. In both Australia and South Africa HCH resistance was found about 18 months after its introduction for tick control, and in both countries DDT-resistance appeared in 1956.[450]

Apart from the resistance problem which was evidently more severe with the cyclodienes and HCH than DDT, it was the need to reduce organochlorine residues in meat that led to abandonment of organochlorines in 1962. At that time, there appeared to be no problem in replacing them with a number of effective organophosphorus or carbamate chemicals. However, a year later, the failure of the organophosphorus compound dioxathion to control ticks in the field, followed by the discovery in central Queensland of the Ridgelands strain of *B. microplus* having some resistance to all the new acaricides in use, heralded the creation of a most serious problem for cattle owners and control authorities in Australia. The remarkable story of the emergence of the Biarra and Mackay strains, having multiple resistance to all the toxicants currently in use, is outside the scope of this account but has been referred to frequently elsewhere.[450] When the older remedies, arsenic and DDT, were tried on the Biarra strain, they were as effective on this strain as on susceptible ticks. However, in comparison with modern chemicals, arsenic is a less effective acaricide, even on arsenic susceptible strains, and DDT cannot be used because of the residue problem. Failure to control the cattle tick means an increasing incidence of tick fever and other diseases. One way to combat the tick problem is to introduce tick-resistant cattle such as those of the Zebu type, and there are also indications that such cattle are less susceptible to tick fever.

### 3. Species of Agricultural Importance
#### a. *Impact of Resistance in Agricultural Insects*

The advanced methods of communication available today ensure that people in the developed countries are rapidly made aware of fresh outbreaks of disease or of sudden food shortages due to freak climatic conditions such as flood or drought. To the uninvolved onlooker, the impact of disease has always been greater than that of malnutrition, and the background condition of chronic food shortage that usually underlies actual famine is taken largely for granted. For these humanistic reasons and also because the modern pesticides were developed in wartime, when the emphasis was on disease control and hygiene, it is not surprising that the possible impact of resistance in agricultural pests was at first disregarded by all except a few experts. A third reason was undoubtedly that the consequences of the use of compounds such as DDT for disease control took a little time to be appreciated; lives saved from disease, or generally made healthier and more vigorous, produce an increasing demand for food, and food production must rise to meet that demand.

Thus, it happened that although the W.H.O. had been fully aware since the middle 1950's of the possibly serious impact of resistance on the malaria eradication and other disease vector control programs, warnings given in 1959 by the Food and Agriculture Organization (FAO) Panel of Experts on Pesticides in Agriculture of the comparable significance of resistance as a threat to crop protection provoked no formal action until 1962. At the FAO Pesticides Conference for Member Countries in 1962, the Director General of FAO was recommended to establish a Working Party on Pest Resistance to Pesticides, and this recommendation was formalized in 1963. The main objectives of the working party were to study (a) the development and application of standardized methods for the detection and evaluation of resistance (of which about 12 covering the major pests are now available[44]); (b) the collection and interpretation of information on the occurrence of resistance; and (c) the evaluation and use of alternative existing pesticides with particular reference to resistant strains. The investigation under (a) is especially important in relation to (b), since the great diversity of the pests, climatic conditions, and methods of toxicity measurement, to mention only three of the many variables, makes the significance of reports of resistance very difficult to assess. The first world survey of the status of resistance in agricultural pests was undertaken in 1965, and it soon became apparent that resistance was increasing wherever pesticides were in regular use, so that the relative effectiveness of existing chemicals against agricultural pests was slowly but continually changing. In some areas, resistance was evidently becoming so serious as to render the chemicals concerned useless for crop protection.

There are a number of economic consequences of resistance in agricultural pests that differentiate this problem from that of the public health pests. Increasing resistance results in progressively poorer pest control with corresponding increases in crop losses and in crop production costs. In cases in which a crop is a vital element in the national economy, the effect may be disastrous as the cost of evaluating alternative pesticides, applying greater quantities of the same pesticide, etc. make increasing inroads on the financial benefits previously gained from increased crop production. As in the public health area, the necessity for frequent changes in the chemicals used means that the materials cannot be stockpiled in advance, nor is there much incentive for manufacturers of agricultural chemicals to face the great odds, currently stated to be at best 3,000:1 and more usually 7,000 to 10,000:1, against a chemical reaching the market, only to see their product rendered useless by resistance.[4,5] Apart from the problems of environmental contamination that may be a consequence of the higher dosage rates used in attempts to control a resistant pest, increases in residue levels in the crops and in the foodstuffs produced from them can have serious direct economic consequences for trade between countries; following the advent of resistance, higher pesticide residues may occur which are above the legally acceptable level in a country formerly importing the foodstuff. Although they incorporate wide safety margins, such legal tolerances may not be easily changed to allow for the new situation. As an example, DDT was still somewhat effective in controlling *Boophilus microplus* in Australia at the time of its withdrawal in spite of widespread resistance to the compound; the residue tolerances imposed by countries importing Australian beef finally forced its abandonment in 1962.

From the 1965 resistance survey and from information continually accumulating afterwards, it was soon evident that the number (119) of resistant arthropods of agricultural (including stored product) importance has now overtaken the number (105) resistant in the public health or veterinary area (Table 7). In the public health area, the Diptera (80 species) strikingly outweigh the other orders (25 species); in the agricultural area, the main pests are the *Hemiptera*, *Lepidoptera*, *Acarina*, and *Coleoptera*, totaling 101 species, against only 11 species of Dipteran pests.[373]

When using pesticides such as the organochlorines which have long residual lives, considera-

tion has to be given to possible toxic effects on beneficial organisms and on natural parasites and predators that control the pests whose destruction is sought. Therefore, some encouragement was available from reports that the honey bee (*Apis mellifera*) in California and Louisiana, the parasitic wasp (*Macrocentrus ancylivorus*) in Ontario, mayfly nymphs (*Heptagenia hebe* and *Stenonema fuscum*) in New Brunswick, and the boll weevil parasite (*Bracon mellitor*) had become resistant to DDT.[423] At the same time, fishes such as the golden shiner (*Notemigonus crysoleucas*), green sunfish (*Lepomis cyanellus*), bluegill sunfish (*Lepomis machrochirus*), and the mosquito-fish (*Gambusia affinis*) have proved themselves able to develop resistance to organochlorines, even to endrin in the case of mosquito-fish exposed to it in waters draining from endrin-treated cotton growing areas in the southern United States. In these areas also, cyclodiene resistance has been observed in the northern cricket-frog (*Acris crepitans*) and the southern cricket-frog (*Acris gryllus*). White mice (*Mus musculus*) selected with DDT in the laboratory have been shown to develop resistance to it, and less desirably, the pine mouse (*Pitymys pinetorum*) has become resistant to the endrin used to control it in Virginia.[423]

Tables 7 to 11 are adapted from the valuable summaries of Brown,[373,423] and illustrate the wide incidence of organochlorine resistance among agricultural pest species. As indicated previously,

TABLE 7

Number of Resistant Species in Each Arthropod Group

|  | Public Health area. | Agricultural area. | Total. |
|---|---|---|---|
| Diptera (flies, mosquitoes) | 80 | 11 | 91 |
| Lepidoptera (caterpillars) | 0 | 29 | 29 |
| Hemiptera (aphids, leafhoppers, scale insects) | 2 | 33 | 35 |
| Acarina (mites, ticks) | 8 | 16 | 24 |
| Coleoptera (beetles, wireworms) | 0 | 23 | 23 |
| Siphonaptera (fleas) | 5 | 0 | 5 |
| Orthoptera (cockroaches) | 3 | 0 | 3 |
| Phthiraptera (lice) | 7 | 0 | 7 |
| Ephemeroptera and Hymenoptera (mayflies, bees) | 0 | 3 | 3 |
| Thysanoptera (thrips) | 0 | 4 | 4 |
| Total | 105 | 119 | 224 |

(Data adapted from Brown.[373,423])

## TABLE 8

### Agricultural Species Resistant to DDT, DDD, and Related Compounds[a]

Insects Attacking:

Apple and peach:
Codling moth (*Carpocapsa pomonella*)
Oriental fruit moth (*Grapholitha molesta*)
Red banded leafroller (*Argyrotaenia velutinana*) (DDD)
Light brown apple moth (*Epiphyas postvittana*) (DDD)
Apple leafhopper (*Erythroneura lawsoniana*)
White apple leafhopper (*Typhlocyba pomaria*)
Peach aphid (*Myzus persicae*)

Cabbage:
Imported cabbage worm (*Pieris rapae*)
Cabbage looper (*Trichoplusia ni*)

Potato:
Colorado potato beetle (*Leptinotarsa decemlineata*)
Potato flea beetle (*Epitrix cucumeris*)
Potato tuber moth (*Phthorimaea operculella*)

Cotton:
Pink bollworm (*Pectinophora gossypiella*)
Cotton bollworm (*Heliothis zea*)
Egyptian cotton leafworm (*Spodoptera littoralis*)

Tobacco:
Tobacco budworm (*Heliothis virescens*)
Tobacco hornworm (*Protoparce quinquemaculata*)
Tuber flea beetle (*Epitrix tuberis*)

Grape:
Grape leafhopper (*Erythroneura variabilis; E. elegantula*)

Miscellaneous

Spruce budworm (*Choristoneura fumiferana*)
Diamond-back moth (*Plutella maculipennis*)
Beet armyworm (*Laphygma exigua*)
Citrus thrips (*Scirtothrips citri*)
Onion thrips (*Thrips tabaci*)
Coffee thrips (*Diarthrothrips coffeae*)
Alfalfa plant bug (*Lygus hesperus*)
Rice weevil (*Sitophilus oryzae*)
Grass grub (*Costelytra zealandica*)
Spotted root maggot (*Euxesta notata*)
Spider mites (*Tetranychus urticae*)

> (Ovex, genite, fenson,
> chlorobenzilate, chlorobenside,
> dicofol, tetradifon)[b]

European red mite (*Panonychus ulmi*)

[a] Resistance is to DDT unless otherwise indicated.
[b] Acaricides

DDT- and cyclodiene-resistance are distinct; when an insect appears in both types of tables, this may not necessarily mean that both resistances are present together, since the insects referred to may have been samples from different localities. Nevertheless, it is frequently the case that the two main types of organochlorine compounds have in fact been used successively so that both resistances are present in the same insect. The original work should be consulted for details. In the cyclodiene tables, the compound listed is the one which was in use when resistance was first noticed; cross resistance to all other cyclodienes and to HCH is normal.

Resistance may arise at many points simultaneously, and then coalesce to cover a large area as happens with houseflies, mosquitoes, and body-lice. In the case of plant-feeding insects, it may appear in certain fields or orchards and either spread from them or remain relatively localized for some time; resistance of the western corn rootworm to aldrin spread from a single point in Nebraska into seven states in 5 years, while the incidence of DDT-resistant codling moth has sometimes been confined to a few orchards for several years.[384] Resistance is most likely to occur if a high level of kill (high insecticidal pressure) is combined with use of the toxicant over a wide area so that few susceptible survivors remain to dilute the population of resistant individuals. Reproduction rates are usually higher in subtropical climates than in temperate ones so that the selection process is speeded up on this account.

The history of organochlorine use for boll weevil control provides a good example of the way in which the principal crop of a whole region or country can be threatened by resistance (Table 11). The treatment of cotton with toxaphene and endrin led to resistance in the boll weevil, cotton leafworm, cotton fleahopper, and cotton aphid in the southeastern United States and to resistance in the leaf-perforator and salt marsh caterpillar in the southwest. This resistance was first recognized in 1955 in Louisiana and it later spread over the main range of the boll weevil in Mexico and Venezuela as well as in the southeastern United States.[373] The boll weevil is naturally tolerant to DDT, which was successfully used to control the cotton bollworm (*Heliothis zea*) and the pink bollworm (*Pectinophora gossypiella*)[24] until these pests became resistant to it in Mexico (1959), Texas, Louisiana, Arkansas, and Georgia (1962—64).

## TABLE 9

### Agricultural Species Resistant to Cyclodiene Insecticides and HCH

| Insects Attacking: | Insecticide first selecting resistance |
|---|---|
| **COTTON** | |
| Boll weevil (*Anthonomus grandis*) | Toxaphene |
| Cotton leafworm (*Alabama argillacea*) | Toxaphene |
| (*Spodoptera littoralis*) | Toxaphene |
| Salt marsh caterpillar (*Estigmene acraea*) | Organochlorines |
| Cotton leaf perforator (*Bucculatrix thurberiella*) | Organochlorines |
| Cotton fleahopper (*Psallus seriatus*) | Organochlorines |
| Cotton aphid (*Aphis gossypii*) | HCH |
| Spiny bollworm (*Earias insulana*) | Endrin |
| Cotton stainer (*Dysdercus peruvianus*) | HCH |
| | |
| **SUGAR CANE** | |
| Sugar cane frog hopper (*Aeneolamia varia*) | HCH |
| Sugar cane borer (*Diatraea saccharalis*) | Endrin |
| | |
| **TOBACCO** | |
| Tobacco hornworm (*Protoparce sexta*) | Endrin |
| Tobacco budworm (*Heliothis virescens*) | Endrin |
| Dark-sided cutworm (*Euxoa messoria*) | Dieldrin |
| Sandhill cutworm (*Euxoa detersa*) | Aldrin |
| Potato tuber moth (*Phthorimaea operculella*) | Endrin |
| | |
| **RICE** | |
| Rice leaf beetle (*Lema oryzae*) | HCH |
| Rice stem borer (*Chilo suppressalis*) | HCH |
| | |
| Rice water weevil (*Lissorhoptrus oryzophilus*) | Aldrin |
| Smaller brown planthopper (*Delphacodes striatella*) | HCH |
| Rice paddy bug (*Leptocorisa varicornis*) | HCH, endrin |
| Black rice bug (*Scotinophora lurida*) | HCH |
| Green rice leafhopper (*Nephotettix cincticeps*) | HCH |
| | |
| **STORED PRODUCTS** | |
| Rust red flour beetle (*Triboleum castaneum*) | HCH |
| Rice weevil (*Sitophilus oryzae*) | HCH, dieldrin |
| Granary weevil (*Sitophilus granarius*) | HCH |
| Maize weevil (*Sitophilus zeamais*) | HCH |
| Clothes moth (*Tineola biselliella*) | Dieldrin, endrin, chlordane |
| | |
| **MISCELLANEOUS** | |
| Black cutworm (*Agrotis ypsilon*) | Aldrin |
| Singhara beetle (*Galerucella birmanica*) | HCH |
| Chinch bug (*Blissus pulchellus*) | HCH |
| Cocoa capsid (*Distantiella theobroma*) | HCH |
| Wooly apple aphid (*Eriosoma lanigerum*) | HCH |
| Pear psylla (*Psylla pyricola*) | Dieldrin |
| Serpentine leaf miner (*Liriomyza archboldi*) | Aldrin |

(Data adapted from Brown.[373,423]

## TABLE 10

### Soil Insects Resistant to Cyclodiene Insecticides

| Insect | Insecticide selecting resistance |
|---|---|
| Southern potato wireworm (*Conodorus fallii*) | Chlordane |
| Tobacco wireworm (*Conoderus vespertinus*) | Dieldrin |
| Sugarbeet wireworm (*Limonius californicus*) | Aldrin |
| Western corn rootworm (*Diabrotica virgifera*) | Aldrin |
| Northern corn rootworm (*Diabrotica longicornis*) | Aldrin |
| Southern corn rootworm (*Diabrotica undecimpunctata*) | Aldrin |
| Banded cucumber beetle (*Diabrotica balteata*) | Aldrin |
| Alfalfa weevil (*Hypera postica*) | Heptachlor |
| White fringed beetle (*Graphognathus leucoloma*) | Dieldrin |
| Onion maggot (*Hylemya antiqua*) | Dieldrin |
| Bean seed maggot (*Hylemya liturata*) | Dieldrin |
| Cabbage maggot (*Hylemya brassicae*) | Dieldrin |
| Seed corn maggot (*Hylemya platura*) | Dieldrin |
| Turnip maggot (*Hylemya floralis*) | Heptachlor |
| Barley fly (*Hylemya arambourgi*) | Dieldrin |
| Spotted root maggot (*Euxesta notata*) | Dieldrin |
| Large bulb fly (*Merodon equestris*) | Aldrin |
| Carrot rust fly (*Psila rosae*) | Dieldrin |

(Data adapted from Brown.[373],[423])

## TABLE 11

### Countries in Which Principal Crops are Threatened by Insect Resistance

| Country | Crop | Pest | Toxicant |
|---|---|---|---|
| U.S.A.: | | | |
| Southern | Cotton | Boll weevil (*Anthonomus grandis*) | Endrin |
| Washington State | Apple | Codling moth (*Carpocapsa pomonella*) | DDT |
| New Brunswick | Spruce | Spruce budworm (*Choristoneura fumiferana*) | DDT |
| Ghana | Cocoa | Cocoa capsid (*Distantiella theobroma*) | Lindane, aldrin |
| United Arab Republic | Cotton | Cotton leafworm (*Spodoptera littoralis*) | Toxaphene |
| Japan | Rice | Asiatic rice stem borer (*Chilo suppressalis*) | HCH |
| Trinidad | Sugar-cane | Sugar-cane froghopper (*Aeneolamia varia*) | Lindane |
| Australia (Queensland) | Cattle | Cattle tick (*Boophilus microplus*) | Cyclodienes, DDT, OP |

(Data adapted from Brown.[373],[423])

Apart from their resistance to DDT, the bollworm and the tobacco budworm (*Heliothis virescens*; DDT-resistant since 1961) have assumed greater significance as pests in recent years because insecticide control programs directed at other insect pests have reduced their natural enemies. Both *Heliothis* species soon developed resistance to cyclodienes[451] and for a time there was no effective means of chemical control; high dosages of DDT, or DDT with toxaphene were effective, but the danger of contaminating nearby forage crops with organochlorine residues prevented such usage in many areas. Since that time, control has been achieved with carbamates and organophosphorus compounds, but recent work shows that *H. virescens* is resistant to methyl parathion in Mexico and Texas and that cross-resistance to other organophosphorus insecticides is widespread;[452] fortunately wild populations of *H. zea* show no widespread resistance to these compounds so far. The widespread use of toxaphene for the control of the cotton leafworm (*Spodoptera littoralis*) in Egypt is another case in which severe selection pressure finally produced high resistance in 1961, with consequent failure to control the pest. In this case, as in many others, the necessary change to organophosphorus compounds has resulted in resistance to these also. DDT-resistance in the codling moth and in the spruce budworm threatened equally serious consequences for crop production in the two states involved (Table 11), and a period of satisfactory and valuable control of the sugar cane borer in Trinidad was brought to an end with the advent of resistance to HCH.

The problems experienced with cattle ticks in South Africa and Australia were discussed in the previous section, but the two remaining situations listed in Table 11, namely, the resistance of the rice stem borer and other rice pests in Japan, and of the cocoa capsid in Ghana, are worth further discussion.

According to the Food and Agriculture Organization, rice is the principal food of half of the world's population. At the end of World War II, the production level in Japan was only 5.9 million tons annually and food had to be donated from abroad. For the next 10 years, production fluctuated around 9 million tons until the effect of modern production methods involving intensive cultivation, combined with heavy fertilization and pesticide usage, boosted it to 12 million tons in 1956, 13 million tons in 1962, and 14.5 million

tons in 1967;[453] by 1970 there was overproduction of rice. The efficient production of rice requires its careful protection against a number of pests and the situation is complicated because of the need for irrigation and the consequent interaction of aquatic soil and plant systems. The main rice pest is rice stem borer (*Chilo suppressalis*), but the green rice leafhopper (*Nephotettix cincticeps*), the smaller brown plant hopper (*Laodelphax striatellus*), brown plant hopper (*Nilaparvata lugens*), rice stem maggot (*Chlorops oryzae*), paddy stem maggot (*Hydrellia sasakii*), and some others are also of importance.

As its name implies, the rice stem borer attacks the rice plant directly, while the leafhoppers and plant hoppers cause serious damage by transmitting pathogenic infections such as rice dwarf, yellow dwarf, stripe, and black stripe diseases. Control of these insects requires several applications of insecticides annually in some areas. In 1951, the organophosphorus insecticide parathion was found to be effective against rice stem borer and was followed by a number of other compounds of this type; parathion resistance developed about 1960 and, together with the toxic hazards associated with its use, led to its final abandonment in 1970.[323] HCH dust applied to the surface of paddy field soil was effective against *Chilo* larvae (first observed in 1956) and later, HCH granules were found to be effective when applied to soil or to paddy field water.[323] Because of its stability and persistence HCH is superior to other insecticides for this type of application. It is ineffective against the green rice leafhopper and has to be combined with carbamates or organophosphates for the simultaneous control of *Chilo* with *Nephotettix* and some of the other pests. The plant hoppers are widely distributed in the rice growing areas of Southeast Asia and Australasia and have great potential for mass outbreak, so that the discovery of HCH and cyclodiene-resistance in them in the early 1960's in Japan, Taiwan, and Fiji posed a serious problem for rice growers.

About 1966, HCH-resistance in *Chilo* was observed in several regions of western Japan, a fivefold decrease in susceptibility of the pest being sufficient to produce complete control failure. Concurrent with the appearance of HCH resistance in a number of rice pests was a marked increase in the populations of *Nephotettix* (unaffected by HCH) associated with the destruction, by increasing HCH treatments, of natural predators like the ricefield spider;[323] this predator is much more susceptible to HCH than to several of the organophosphorus and carbamate compounds currently in use. For economic reasons, technical HCH has been widely used in Japan, the peak year being 1968. The residues arising from the use of technical HCH consist mainly of the beta-isomer, and in 1969, the levels of this compound in meat and milk, arising mainly from the use of rice straw as cattle feed, were found to be unacceptable. The combination of these and some other considerations involving the environment led to a halt in DDT (introduced for rice pest control in 1947) and HCH manufacture for domestic consumption at the end of 1969 and the use of HCH ceased almost entirely in Japan early in 1971. Since the application of aldrin, dieldrin, and endrin has been severely restricted since 1970, the use of organochlorine insecticides in that country is now very small.

A number of the organophosphorus compounds that followed parathion proved to be valuable for the simultaneous control of several of the rice pests, but there are signs of resistance in *Chilo* to compounds such as fenthion and fenitrothion. Malathion found early success in controlling the green rice leafhopper but was replaced by the carbamate insecticide carbaryl after resistance to the former appeared as long ago as 1955. In the past, there has been talk of negative correlation in resistance to toxicants, the idea being that resistance to one chemical may be accompanied by enhanced susceptibility to another. Since not many examples of this have been found, it is interesting that some strains of *Nephotettix* selected in the laboratory for high malathion resistance appear to be about twice as susceptible to DDT and to γ-HCH as the normal malathion susceptible strain.[454]

The Japanese situation is a particularly interesting one since the intensive use of pesticides appears to provide ideal conditions for the rapid development of resistance. Accordingly, constant vigilance is necessary to recognize resistance at an early stage so that the chemical involved can be quickly replaced. There has been an intense pressure on the Japanese chemical industry to produce new chemicals in order to keep ahead of these requirements. This situation may be changed somewhat by the current overproduction of rice.

An equally remarkable example of the potential impact of resistance on crop production is provided by the progress of cocoa production in West Africa in the 1960's. Cocoa is a principal crop produced for export in Ghana and Nigeria and is therefore of vital importance to the economy of those countries. Among the pests attacking cocoa, the cocoa capsid (*Distantiella theobroma*) provides the greatest threat to production. There was no extensive spraying of farm cocoa for the control of this pest before 1956 in Ghana or before 1958 in Nigeria, and cocoa production showed roughly the same trends in both countries up to the time when chemical control was introduced.

National production of cocoa in Ghana in the first quinquennium (1957–62) of chemical control exceeded that for 1952–57 by 33%, and for Nigeria the corresponding increase was 40%. Although some contribution to these changes undoubtedly came from improvements in growing techniques, etc., there is no doubt that these remarkable improvements resulted largely from the introduction of intensive spraying programs for capsid control employing aldrin (Aldrex-40®) and γ-HCH (Gammalin-20®) formulations. Trials conducted in Ghana, Nigeria, and Sierra Leone indicated a remarkable reduction in damage due to capsids within months after spraying commenced and spectacular improvements in tree condition were apparent in subsequent seasons. Sustained treatment sometimes resulted in cocoa pod yields up to twofold or more greater than the yields from untreated trees.[455,456] The similar and parallel increases in production in both Ghana and Nigeria, despite the 2-year lag before spraying began in the latter country, suggested the involvement of some additional factors, but the value of the spraying program was never in doubt. Moreover, since an increase of only 10% in the yield of raw cocoa was of considerable economic value to these countries, there was ground for considerable optimism regarding the future of the industry.

γ-HCH resistance was first found in 1963 in Ghana[457] and suspected in Nigeria the same year. Although a number of reported control failures were found to be due to causes other than resistance, laboratory tests confirmed that γ-HCH resistance had appeared in widely separated areas in Ghana. Since there appeared to be normal cross-resistance to other cyclodienes a discriminating dose of dieldrin could be used to detect γ-HCH resistance in the laboratory; in the normal cyclodiene resistance pattern, resistance to γ-HCH is lower than to cyclodienes so that tests with cyclodienes may provide a more sensitive indication of cyclodiene-type resistance than is available by using γ-HCH itself. Similar surveys in Nigeria at the time indicated no significant incidence of resistance although it seemed likely that resistance potential was present but not yet fully evident due to the delayed development of the spraying program in that country. The interest in this case history of resistance lies in its marked contrast to the situation in Japan. This country has a highly developed pesticide chemical industry, together with all the machinery in University and Government organizations for the rapid evaluation of alternative chemicals under local conditions, and is readily able to change to a fresh control agent. Even if a promising chemical is available, it cannot be evaluated quickly under local conditions if no such machinery exists. Sufficient experience of cyclodiene resistance was available by this time to suggest that continued use of Gammalin or Aldrex would spread and intensify the resistance, so the decision had to be taken either to abandon the use of these compounds until a suitable alternative could be evaluated or to continue their use in the meantime. Clearly, such decisions have to be based on both political and economic considerations. Faced with a possible return to pre-pest control production levels, and providing that control failure has not become absolute, the only reasonable alternative would appear to be continued use of existing control agents while work proceeds on suitable alternatives. In a peasant farming situation like that found in West Africa, the toxic hazard of alternatives is a particularly important consideration.

### b. Resistance in Soil Insects

There is no doubt that the introduction of chlordane, heptachlor, aldrin, and dieldrin as soil insecticides in the late 1940's has made a remarkable contribution to world grain production. In the area of soil pests as in other areas, there has been a marked rise in the incidence of resistance to these compounds since the early 1950's. Many of the species listed in Table 10 are drawn from observations in North America so that at least three species of wireworms and four species of *Diabrotica* rootworms are now involved. The cabbage maggot, onion maggot, and carrot rust fly have become resistant to dieldrin in virtually all areas where they are exposed to cyclodienes, and

the turnip maggot, bean seed maggot, and seed corn maggot are not far behind.[373]

Several reports from the Chatham Laboratory of the Canada Department of Agriculture Research Branch provide typical histories of the development in various soil insects. Most of the tobacco growing land in southwestern Ontario was treated annually after 1954 with aldrin or heptachlor (1.5 lb active ingredient (a.i.) per acre) for the control of cutworms or wireworms. From 1958 onward, increasing numbers of bean seed maggots (*Hylemya liturata*), which had previously been of minor importance on tobacco, attacked tobacco transplants in the area. It became evident that the control measures adopted for cutworms and wireworms had not produced resistance in these but had favored the development of a new pest of tobacco and had produced resistance in it (255-fold compared with a susceptible strain) after about 16 generations of selection pressure. The resistance subsequently spread throughout the tobacco growing areas and through other agricultural areas of southwestern Ontario. Serious damage to crops was apparent in many areas between 1958 and 1964. The resistant insects showed cross-resistance to heptachlor (214-fold), heptachlor epoxide (174-fold), dieldrin (120-fold), endrin (tenfold), isobenzan (eightfold), lindane (sixfold), and endosulfan (twofold) but not to DDT or diazinon; the latter was subsequently used to control these infestations.[458]

Severe infestations of onion maggot (*Hylemya antiqua*) appeared in 1958 in areas of Ontario where it had been successfully controlled between 1953 and 1957 by a 3 to 5-lb per acre application of aldrin, dieldrin, or heptachlor.[459] In this outbreak, the crop was destroyed in fields that had been treated with aldrin at 2.5 to 10 lb per acre in preplanting broadcast applications. Other outbreaks occurred in British Columbia (1957), Manitoba (1958), and Quebec (1957—1958), and had been seen previously in the state of Washington (1953), Michigan (1956—1957), and Wisconsin (1957), in areas in which aldrin had been in use for 4 to 6 years. The aldrin-resistance level was nearly 600-fold, and there was cross-resistance to heptachlor (647-fold), dieldrin (357-fold), endrin (27-fold), and lindane (14-fold), but not to DDT or diazinon.

Failure of the excellent control of carrot rust fly (*Psila rosae*) provided by aldrin in the early 1950's was reported in 1957 in Washington where treatments with aldrin and other cyclodienes in the Lake Sammanish area were completely ineffective on carrots. Similar incidents occurred near Vancouver and at Bradford, Ontario, in 1961; at Bradford, preplanting treatments of 80 lb of 5% aldrin granules per acre, followed by aldrin sprays, were so ineffective that many fields of carrots and parsnips were too damaged by the pest to be worth harvesting. Between 1958 and 1961, the Chatham laboratory conducted tests on the susceptibility to cyclodienes of a number of soil pest species collected from different areas in Canada. These tests showed resistance in the cabbage maggot, the spotted root fly (*Euxesta notata*) (see section B.2c) and the dark sided cutworm (*Euxoa messoria*), as well as the spread of resistance in the species mentioned previously. *E. messoria* larvae reared from field collected adults were resistant (about tenfold) to dieldrin, DDT, and diazinon, but the cutworm species of economic importance in southwestern Ontario, namely, the black cutworm (*Agrotis ypsilon*) and the variegated cutworm (*Peridroma saucia*) were not resistant to cyclodienes.[460]

Failure to control cabbage maggots (*Hylemya brassicae*) was reported in Newfoundland (1959), British Columbia and Quebec (1960), Washington (1962), and Ontario, after 5 or more years of success using aldrin or heptachlor (1.5 to 5.0 lb of active ingredient per acre) as broadcast or furrow treatments. Resistance to aldrin and dieldrin was about 1000-fold and much lower to endrin (37-fold) and isobenzan (45-fold). The cross-resistance patterns to cyclodienes mentioned here appear to be fairly typical of those found with a number of species; notably, resistance is always lower to endrin and is generally least to γ-HCH.

The history of these cases shows that resistance developed in areas in which broadcast or band applications of cyclodienes had been made during several consecutive years. In some cases, several crops were grown annually (radishes, for example), and fresh insecticide was applied with each crop. Following control failures, retreatment with the same, or other cyclodienes was quite common; in one case, treatment with aldrin was followed by retreatment with heptachlor and finally with endrin, all without effective control! Therefore, in all these cases, the entire population of soil animals was subjected to intense selection pressure by cyclodienes.

Similar cases occurred in Britain. In 1963,

severe infestations of the cabbage root fly (*Erioischia brassicae*) occurred on a farm in Oxfordshire, where the numbers of first generation larvae were so great that about 140 acres of the early brussels sprout crop were destroyed and further areas severely damaged in spite of the extensive use of aldrin as a broadcast treatment to the soil before transplanting and also as a spot treatment after planting out.[461] A high degree of control is generally obtained by the recommended use of aldrin or dieldrin as a root dip for transplants, or for treatment of the soil around the plants after transplanting, in which case the amount of active ingredient applied is not usually more than 0.4 lb per acre. There was no evidence indicating that resistance had previously developed in Britain when these recommendations were followed. Most fields on the farm had been intensively cropped with brassicas during the preceding 8 years and both broadcast and spot treatments had been used on each crop. Analysis of soil samples from the infested fields indicated aldrin concentrations as high as 20 lb per acre around plant stems and up to 2 lb per acre between the plants, with additional amounts of dieldrin present in these positions. DDT had also been used occasionally for the control of foliage pests and was present at 1.8 lb per acre in the interplant spaces.

Thus, it became evident that the soil populations had been subjected to unusually intense selection pressure by organochlorine insecticides, and a comparison of adult cabbage root flies from this area with those from an area subjected to normal control showed that the former had developed a 35-fold resistance to dieldrin, with cross-resistance to aldrin and γ-HCH.[461] Since that time, resistance of the cabbage root fly has been found in several areas in which there has been exposure to high aldrin pressure. The infestations are frequently remarkably localized so that a field with a resistant infestation may be surrounded by an area in which only small numbers of resistant insects are present. In Britain, cyclodiene resistance has also been reported in the carrot rust fly (*Psila rosae*), in the bean seed fly (*Delia platura*), and in the onion fly (*Delia antiqua*), but not so far in the wheat bulb fly, (*Leptohylemya coarctata*) or the large bulb fly (*Merodon equestris*). As in the cases mentioned previously, diazinon was found to give effective control of the resistant insects.[462]

At the time of the Oxfordshire outbreak of cabbage root fly, it was noticed that, whereas pupae of a normally cyclodiene susceptible population of this insect were 40 to 60% parasitized by certain *Coleoptera* and *Hymenoptera*, no parasitism was observed in pupae of the resistant strain from Oxfordshire.[461] This finding indicates that besides selecting the cabbage root fly for resistance, the cyclodienes also killed off the natural enemies of this insect so that no natural constraints were placed upon the population once the resistant cabbage root flies had begun to multiply. Canadian accounts of the resistant seed maggot outbreaks indicate that these insects had not previously been a serious pest on tobacco, which suggests that their numbers are normally controlled by natural predators which may have been eliminated by the pesticide while the pest was being selected for resistance. Another possibility is that increased fecundity may have contributed to the population explosions. For example, in the case of houseflies there have been a number of observations of increased numbers following selection with cyclodienes. When houseflies in latrines in Georgia were treated with dieldrin, resistance developed within a few months and a concurrent 200-fold increase in their numbers was found to be associated not with increased fecundity, but with the unusually solid nature of human excrement in the privies due to destruction by the insecticide of another fly (*Hermetia illucens*) whose larvae normally maintain the excrement in a semiliquid state that is inhospitable to housefly larvae.[463] Chlordane and γ-HCH gave a similar result which was not apparently seen with DDT. Other investigations showed that when some strains of normal houseflies are selected with dieldrin at LD75 (the dose required to kill 75% of the population), the survivors of this treatment laid more eggs. The $F_1$ generation (their immediate offspring), but not the $F_2$ or $F_3$ generations, also produced 70% more eggs than usual.[464] Similar results have been noted with fruit flies (*Drosophila melanogaster*) and increased housefly production following treatment with cyclodienes has since been observed in many other areas besides those in the United States. Since the work of Edwards and others has shown that organochlorines do affect predatory species in soil,[28] sometimes selectively, it seems probable that this factor and increased fecundity are operating together in many cases.

## c. Resistance in Stored Product Insects

The development of insecticide resistance in stored product insects was reviewed by Parkin,[465] and by Lindgren and Vincent.[466] Recent developments, especially those regarding the rust-red flour beetle (*Tribolium castaneum*) have been discussed by Dyte.[377,467] Besides being of considerable practical importance, the effect of toxicants on these insects has much theoretical interest. The principal cases of resistance in stored product species are summarized in Table 12, which includes examples of resistance produced by laboratory selection, as well as those arising in the field. Thus, a strain of the granary weevil (*Sitophilus granarius*) selected in the laboratory for high resistance to pyrethrins (138-fold) is also resistant to DDT (29.5-fold) and to DDT-analogues.[468]

The case of *T. castaneum*, which is a major pest of stored food in the tropics, is particularly interesting. This is the species most frequently found on produce imported by temperate countries such as Britain, Canada, and the United States, and is therefore a suitable species in which to determine the occurrence of resistant strains that may have been distributed by means of international trade. Accordingly, Dyte and Blackman have conducted a survey of strains of *T. castaneum* obtained directly from various countries or from imported commodities being unloaded in British ports to determine their susceptibility to γ-HCH or malathion.[467] Using a discriminating dose technique (see section A.) involving filter papers impregnated with γ-HCH at a rate which knocks susceptible insects down in 24 hr, they tested 18 strains from 11 countries for lindane resistance and found it in 15 strains from Australia, England, Gambia, Kenya, Malawi, Malaysia, Senegal, the Seychelles, and Zambia.

The countries of origin, products infested, and response to the toxicity test are found in Table 13. Of 13 strains obtained from cargoes unloaded in British ports, 11 were γ-HCH-resistant, and the cargoes carrying the resistant strains had been loaded in India, Burma, Kenya, Nigeria, Tanzania, and the Sudan. Five strains of *T. castaneum* taken from ships or their cargoes in Australian ports were resistant to γ-HCH, having resistance factors of 13-fold to 417-fold. The discovery of resistant insects in these cargoes does not prove the occurrence of resistant strains in the countries exporting the commodities, since cross-infestation

TABLE 12

Insecticide Resistance in Pests of Stored Products

| Species | Compounds to which resistance is shown | | | | | | |
|---|---|---|---|---|---|---|---|
| | DDT | Lindane | Pyrethrins | Malathion | Other OP's | Carbamates | Other Insecticides |
| *Beetles* | | | | | | | |
| Tribolium castaneum | + | + | + | + | + | + | |
| Tribolium confusum | + | | | + | | | Methyl formate |
| Trogoderma granarium | | | | + | | | |
| Sitophilus granarius | + | | + | | + | + | (Perthane (Bulan (Methyl bromide |
| Sitophilus oryzae | + | + | | + | + | | Dieldrin |
| Sitophilus zeamais | | + | | | | | |
| Oryzaephilus surinamensis | | | | + | | + | |
| Oryzaephilus mercator | | | | + | | | |
| Caryedon serratus | | + | | | | | |
| Dermestes maculatus | | + | | | | | |
| *Moths* | | | | | | | |
| Plodia interpunctella | | | | + | | | |
| Ephestia cautella | | | + | + | + | | |
| Tineola bisselliella | | | | | | | (Dieldrin (Endrin and (Chlordane |

(Data adapted from Dyte.[377])

can arise on board ship, but it is interesting that some cargoes came from Burma, India, Tanzania, and the Sudan, where resistant *T. castaneum* has not been reported. Nearly all of the strains obtained from these cargoes were also resistant to malathion, the resistance being of the specific type involving hydrolysis by carboxyesterase attack. These investigations show that both γ-HCH and malathion resistance have become widespread in *T. castaneum* and are being distributed through international trade. A discriminating dose test involving topical application of DDT has shown that γ-HCH-resistant strains of *T. castaneum* from Kenya, Malawi, and Senegal are also somewhat resistant to DDT.

## 4. Stability of Resistance

A most important question from the point of view of practical pest control with chemicals concerns what happens to a species that has developed resistance when the selecting agent is removed from its environment. The usual situation is that a satisfactory and economical pesticide is used in a control program until resistance develops to a point at which further use becomes uneconomical either because control fails completely or because the cost of the larger amounts of toxicant required to maintain it is prohibitive. The use of larger amounts of toxicant for control may also be unacceptable because of legal requirements for residue tolerances. At this stage, the usual procedure is to employ another pesticide so that pressure on insect populations due to the first one is relaxed. What happens to the existing resistance then depends basically on the population genetics of the resistance factors and their combinations. In general, populations consisting entirely of highly resistant individuals (homozygous for resistance-RR) are unlikely to arise in the field. Environmental factors tend to select against the genes for resistance, which also becomes diluted through mixing with susceptible populations.[424]

Since in most cases the resistance genes are rare before selection, the individuals carrying them must be at a disadvantage compared with the remainder of the population; an apparent exception is the unusually high incidence of dieldrin-resistant phenotypes in *Anopheles*

TABLE 13

Occurrence of Lindane and Malathion Resistance in Adult *Tribolium Castaneum*
Collected From Imported Cargoes Being Unloaded in Various British Ports

| Country and port of origin | Commodity | Percent knock-down at doses giving complete response in susceptible insects | |
| --- | --- | --- | --- |
| | | Lindane | Malathion[a] |
| Australia, Sydney | Linseed meal | 100 | 100 |
| Burma | Goundnut cake | 99 | 63 |
| Egypt | Rice bran | 100 | 100 |
| India, Bombay | Rice bran | 74 | 83 |
| India, Bombay | Rice bran | 93 | 5 |
| India, Kakinada | Rice bran | 95 | 51 |
| India, Vizagapatnam | Rice bran | 92 | 4 |
| Nigeria, Apapa | Groundnut kernels | 97 | 22 |
| Nigeria, Apapa | Groundnut kernels | 97 | 27 |
| Kenya, Mombasa | Maize | 29 | 72 |
| Kenya, Mombasa | Groundnut cake | 87 | 94 |
| Sudan | Groundnut kernels | 75 | 43 |
| Tanzania, Dar-es-Salaam | Cottonseed expeller | 62 | 35 |

(Data adapted from Dyte and Blackman.[467]
[a]With the carboxyesterase inhibitor triphenylphosphate, 100% knock-down was obtained in all but one case, showing the metabolic nature of the resistance mechanism for malathion.

*gambiae* in parts of West Africa that have not been treated with the insecticide. Generally, an important factor in the development and persistence of resistance is the development of fitness for survival of resistant phenotypes and genotypes by natural selection concurrent with insecticide selection. This means in practice that the longer the selection with a pesticide has been maintained the fitter is the resistant population for survival.

Laboratory experiments with DDT illustrate the selection of fitness for survival. A somewhat naturally DDT-tolerant strain of *Aedes aegypti* from Malaysia increased its LC50 44-fold after selection with DDT for only two generations; this resistance regressed rapidly when the DDT selection pressure was removed, returned quickly immediately selection was resumed, and now regressed only slowly when selection pressure was relaxed once more. After a third spell of selection the resistance was high, and stable in the absence of DDT. The early regressions occurred because the resistant variants were initially handicapped by low oviposition and hatch rate, which was cor-

rected following further selection.[469] There are numerous other examples of regression in laboratory strains of resistant insects, either after removal of selection pressure or by artificial selection for differing characteristics among the population. When monofactorial cyclodiene-resistance developed spontaneously in a strain of houseflies undergoing selection for early emergence of adults to increase susceptibility to DDT ( a well-known method of selecting for DDT-tolerance is the selection of the late emerging adults), Georghiou eliminated the cyclodiene resistance by selecting and propagating that portion of the population which was susceptible to knock-down by $\gamma$-HCH.[470] In another experiment, dieldrin-resistant *Anopheles albimanus* and normal *Culex fatigans* acquired only low resistance after prolonged selection with *m*-isopropylphenyl *N*-methylcarbamate, but the dieldrin susceptibility was substantially restored in *A. albimanus* by this selection and a preexisting tenfold resistance to DDT also disappeared.[471] In the case of dieldrin, the population initially consisted of dieldrin-susceptible, hydrid, and resistant phenotypes in the ratio 10:44:46; the frequency of homzygous susceptibles increased from 10% to 83% in the course of selection with the carbamate. This appears to be a special case of negatively correlated cross-tolerance, which normally means that resistance to an insecticide A is accompanied by enhanced susceptibility to insecticide B. In this case, the resistance to the carbamate induced in *A. albimanus* by selection was only threefold, so that concurrent substantial restoration of dieldrin susceptibility represents an ideal situation. However, these are laboratory situations, in which the stability of resistance can be quite different from that found in field populations. According to the Hardy-Weinberg law of genetics, a large randomly-mating population in the absence of immigration, mutation, and selection is stable with respect to both gene and genotype frequencies. Laboratory populations are unusually small but protected from immigration or random selection; theoretically a colony that is completely homozygous for resistance will maintain this condition indefinitely without any recession.

What then, is the prospect of an organochlorine compound to which resistance has developed being used again after a period of use of an insecticide of a different type? This question is best considered by reference to field-experiments with a number of insects, housefly control in Denmark and California being good examples. High resistance to DDT, HCH, chlordane, and dieldrin developed after 2 or 3 years of their use in residual sprays on Danish farms.[424] When they were replaced by other insecticides, the organochlorine resistance quickly regressed into a heterogeneous state in which susceptible homozygotes, resistant heterozygotes, and resistant homozygotes were all present together. Therefore, it is not surprising that when DDT and chlordane were tested again after an interval of up to 8 years in some cases, high resistance reappeared after only a short period of effectiveness. Even now, more than 20 years after DDT was first used, anywhere between 10 and 90% of flies tested on farms where resistance has appeared in the past are highly resistant to this compound. The dosage-response regression lines for these populations are flat and extend between doses of 0.2 to 50 $\mu$g per fly, showing the range of phenotypes present between complete susceptibility and complete resistance. When considering this phenomenon it must be

remembered (a) that there is toxicant selection pressure due to the organophosphorus compounds that are currently in use, and (b) that an association exists between this type of selection and the selection for DDT-resistance due to *Deh* and in some cases DDT-*md* (see section B.2a).

After a lapse of more than 10 years in the use of HCH and dieldrin, resistance to them shows an apparently complete regression. On nine farms tested in 1967, dieldrin-tolerance was usually less than threefold, and intermediate resistance (8 to 32 times the normal LD95) was found in up to 7% of the flies on four of the farms; no highly resistant flies were found. This remarkable regression stands in contrast to the situation on farms in California where the same development of organochlorine resistance was followed by more intensive use of malathion, diazinon, etc. as alternatives.[472] In this area, a high level of DDT- and dieldrin-resistance remained 13 years after the use of organochlorines had ceased; the regression lines for the response of various samples to dieldrin-treatment were flat and occupied a region between 15 and 65% kill with a dose range of 0.02 to 100 $\mu$g per fly, while those for DDT also showed a heterogeneous response, the doses ranging from 0.5 to 100 $\mu$g per fly and the kills from 5 to 75%. It seems likely that the use of cyclodiene compounds for agricultural treatments may have helped to maintain the high dieldrin-resistance found in this area, while the continuing high level of DDT-resistance is likely to be related to the intensive use of organophosphorus insecticides, for the reasons indicated previously. The influence of organophosphate selection in the field on organochlorine resistance seems to be demonstrated in some cases by the reversion of the organochlorine resistance of field strains in the laboratory when the OP-infuence is removed.

In mosquito control campaigns, DDT-resistance in *Anopheles* mosquitoes has been found in some cases to revert when HCH or dieldrin is substituted for 2 or 3 years. However, return to DDT has been prevented by the rapid return of resistance when it is used again. The unusually high frequency of genes for dieldrin-resistance in West African populations of *A. gambiae* means that the onset of cyclodiene-resistance is rapid and complete, the resulting resistance being stable and persistent. For other *Anopheles,* the situation is better, partial, or complete regression having been reported for *A. culicifacies* in India and *A. sundaicus* in Java. In these cases, dilution of the resistant populations with populations from untreated areas was of importance for the regression. Among other insects of public health importance, DDT-resistant body lice seem to vary in their response to removal of insecticide pressure, both regression and prolonged persistence of the resistance having been reported, depending on the locality. Ten years after the use of DDT-treatments for body lice control had been superseded by HCH, lice from certain villages in the United Arab Republic still had high resistance to DDT.[424]

The Australian cattle tick (*Boophilus microplus*) has a high ability to develop resistance to insecticides of diverse structural types. Again, experiences of regression are variable, but a field populations of *B. microplus* described by Stone[473] that had been treated for 8 years with DDT showed no regression of DDT-resistance following 17 months of control with dioxathion as a replacement. The strain was apparently not homozygous for resistance since there was partial reversion in a laboratory colony. Metabolism experiments have shown that the larvae of this tick produce at least 17 metabolites from DDT, including both DDE and products of oxidative reactions.[474]

However, the Australian researchers conclude that there is no causal relationship between resistance to DDT and its metabolism.[474] High dieldrin-resistance developed by the sheep blowfly (*Lucilia cuprina*) in New South Wales had reverted to a heterogeneous condition 8 years after diazinon had replaced dieldrin, showing that a high frequency of resistant phenotypes still existed in the population.[424]

Among insect pests of agricultural importance, the codling moth (*Cydia pomonella*) slowly lost its resistance to DDT when the chemical was abandoned in Virginia after 9 years use. The regression in abandoned orchards was incomplete even after 12 to 16 generations (6 to 8 years) without the use of any insecticide. Other observations indicate that for some populations there is a reasonably rapid fall to a constant level of resistant individuals which is then maintained indefinitely. When methyl parathion replaced endrin for control of the cotton boll weevil (*Anthonomus grandis*) in the southern United States, considerable regression of endrin-resistance occurred within 10 to 20 generations (4 to 6

years), but this was not a sufficiently good reversion to permit effective reuse of endrin.

It is a regrettable fact that in those strains which show a partial reversion of organochlorine resistance (or even a complete reversion as with some examples of regression of cyclodiene-resistance) high resistance levels are rapidly restored after a few generations of renewed selection pressure with the appropriate organo-chlorine. One reason for this is that even if the frequency of resistant individuals has declined considerably, a background remains of nonspecific factors that assist resistance and were developed during the severe selection, so that the second selection starts from a more solid base. Accordingly, the prospects for successful reuse of organochlorine insecticides are not good once resistance has occurred. One exception is the use of DDT against some DDT-resistant anopheline mosquitoes, which is possible because the resistance is usually recessive and only of a moderate level. If there is to be any hope, early replacement of the failing compound is essential before resistance becomes fully established. This is a difficult proposition for cyclodienes because resistance occurs rapidly without the apparent need for remodeling of the genetic background that is seen with DDT. The use of mixtures of compounds, or the alternate use of compounds of completely unrelated types has been recommended. According to genetic calculations, resistance is likely to develop less quickly if two unrelated compounds are used successively than if they are used as a mixture and in some cases this is supported by laboratory experiments. Both alternation and combined use have met with success in field application; alternation of azinphos-methyl with endrin slowed the development of endrin resistance in boll weevil and a mixture of toxaphene with DDT successfully controlled this insect, although it selected for resistance to organophosphorus compounds and carbamates.[384]

The continuing fight against disease-carrying insects, together with ever increasing demands on world food resources, ensures that the use of insecticides is a way of no return for modern man. Therefore, once resistance appears, recourse must be had to a continual alternation of chemicals, each new development hopefully keeping a little ahead of the resistance. The only alternative, apart from nonchemical control measures, would be a return to the situation that existed before the advent of modern pesticide chemicals, which is rather difficult to expect of a civilization fully geared to accepting the benefits of this 20th Century technology in terms of increased health, food, and fiber.

Lest the picture presented in the foregoing sections should appear too dismal, it must be remembered that, in fact, only about 5% of the 5,000 or so species of insects and mites which are of economic significance as pests or disease vectors have developed resistance to organochlorine insecticides thus far. However, it is unfortunate, that this minority includes some of the most important disease transmitting species. On the positive side, there is the blessing that DDT-resistance is often of such a nature that disease vector control with it can still be effective, while in the meantime, alternative chemicals to combat resistant strains are gradually coming forward (Vol. I, Chap. 2A.3d).

Chapter 3

## ACTION OF CHLORINATED INSECTICIDES

### A. INTRODUCTION

The observed action or inaction of a toxic substance which has been received by a living organism is the resultant of a complex sequence of events occurring between the point of contact with the toxicant (for example, insect cuticle, or mammalian skin; the alimentary canal of insects or mammals) and the sensitive site of action (target) within the organism.[475] Modification of the toxic action may result from an alteration in the amount of poison reaching the target or in the response evoked there, and such phenomena as resistance or selective action of the toxicant towards different species can be attributed to these causes, although natural tolerance might result from the complete absence of a particular target in a nonsensitive organism. The fact that the principles of toxic action and toxicant modification are generally similar for insects and mammals makes truly selective toxicants difficult to devise; insects are usually killed selectively because their morphology and behavior makes them more vulnerable than mammals to contact with particular toxicants, which may often be equally toxic to both when LD50's from equivalent modes of application are expressed in comparable units. Accordingly, the first principle of selectivity is to ensure that only the target organism receives the toxicant.

The processes which may intervene between a toxicant and its site of action are shown in Figure 7. In what might be called "normal" routes of application, the insecticide first has to pass the walls of the alimentary canal or the integument (that is, the insect cuticle or mammalian skin). The difference between cuticle and skin is not so great as might be imagined; for example, O'Brien has presented evidence to show that DDT penetrates equally rapidly into rat skin and American cockroach cuticle.[2] The rate of penetration into insects can be considerably influenced by the vehicle used for application (see Table 2 in Vol.I). When applied directly to the cuticle in a volatile solvent so that there is direct entry into the epicuticular wax, hydrophilic compounds diffuse into the more polar underlying layers, whereas nonpolar ones have a greater affinity for the wax and therefore penetrate more slowly. Thus, phos-

phoric acid penetrates cockroach cuticle much more rapidly than DDT.[2] Having passed the integumental barrier, the toxicant enters the mammalian bloodstream or the insect haemolymph and may be transported to all parts of the organism in solution if water soluble, or bound to proteins or dissolved in lipid particles if lipophilic. The mode of penetration and transport of the highly lipophilic organochlorine insecticides in insects is not easy to understand. There is evidence that some insecticides applied to the cuticle travel laterally in the epicuticular wax and gain access to the insect central nervous system via the nearby tracheoles of the respiratory network, although direct passage through the cuticle is the commonly favored route of entry.[476,477] Since the cuticle might become changed so as to retard toxicant penetration or transport, it represents a possible first line of defense for the insect.

During the penetration and circulation processes in either insect or mammal some part of the toxicant may have been taken up and stored in inert tissue (such as adipose tissue), so that it is removed from active participation in bioaction. There will also have been opportunity for non-enzymatic chemical transformations as well as for enzyme-catalyzed biotransformations leading to toxication (bioactivation) or detoxication. Since the unaltered toxicant and any of its transformation products may be excreted, enhanced excretion also represents a possible protective mechanism against the toxicant. A change in metabolizing enzyme activity to increase detoxication or reduce bioactivation is another obvious defense mechanism and there may also be internal structural barriers such as the neural sheath (lamella) of insects, which prevent or retard final access to the target.

At the target itself, the toxicant produces an effect depending on its concentration and affinity for the target, and activity (intrinsic toxicity) is now simply a function of molecular structure, separated from all the other influences previously discussed. At this point, tolerance to the toxicant might be procured by an alteration in the structure of the target so as to reduce its affinity for the toxicant. For poisons such as the organophosphate or carbamate anticholinesterases, intrinsic toxicity

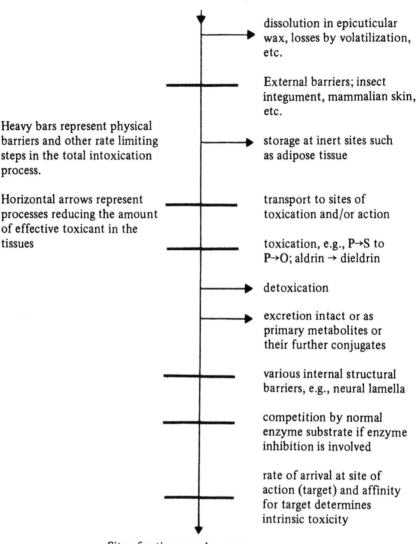

Toxicant or toxicant precursor

→ dissolution in epicuticular wax, losses by volatilization, etc.

External barriers; insect integument, mammalian skin, etc.

Heavy bars represent physical barriers and other rate limiting steps in the total intoxication process.

→ storage at inert sites such as adipose tissue

Horizontal arrows represent processes reducing the amount of effective toxicant in the tissues

transport to sites of toxication and/or action

toxication, e.g., P→S to P→O; aldrin → dieldrin

→ detoxication

→ excretion intact or as primary metabolites or their further conjugates

various internal structural barriers, e.g., neural lamella

competition by normal enzyme substrate if enzyme inhibition is involved

rate of arrival at site of action (target) and affinity for target determines intrinsic toxicity

Site of action may be on an enzyme involved in some vital function, such as synaptic transmission (acetylcholinesterase) or at a membrane surface as is thought to be the case with organochlorines

FIGURE 7.    Pharmacodynamics of a toxicant in an insect or other living organism.

can be judged by measuring their inhibition of cholinesterase preparations in vitro, and in this case it has been possible to show that certain examples of organophosphate resistance in insects are associated with altered cholinesterase.[450]

Since the precise mode of action of organochlorine insecticides is unknown, there is no correspondingly simple way to test their activity in vitro, although the indications are that some idea of the intrinsic activity can be obtained by examining their effects on the propagation of action potentials in isolated insect nerve cords and on synaptic transmission in insects.[478] The view has been prevalent for some years that the primary action involves a reversible physical interaction with nerve membrane (rather than a chemical one involving the formation of covalent bonds) having a number of secondary consequences which, when combined together, are probably the real cause of mortality.[479] This view has gained strength recently, as will be seen in the discussion of mode of action (see section D.).

When organochlorine insecticides are used as tools for investigations of structure-activity relationships and mode of action, it is important to recognize this difficulty of measuring activity at the site of action (intrinsic toxicity) and to remember that measurements of toxicity in vivo are the resultant of the whole process* just discussed; metabolic conversions are particularly significant and must be allowed for before valid conclusions can be reached about the relationships between structure and activity. Indeed, an examination of the information regarding biodegradable organochlorine insecticides that has become available as a result of the interest in environmental problems shows clearly, as in the case of DDT analogues for example, that much of the early work on the structure-activity correlations should be reexamined.[52] The total pharmacokinetic process discussed above results in particular distribution patterns in living organisms that have been much discussed in recent years. These patterns in insects and mammals are of special importance to those who wish to kill pest insects selectively, and are examples of the complex equilibria that result when toxic chemicals enter the environment.

## B. ENVIRONMENTAL BEHAVIOR AND METABOLISM

For the purpose of this discussion, the physical (abiotic) environment, consisting of air, water, and the inorganic parts of soil is considered along with the biotic environment, which comprises all living organisms. This is appropriate. since an organochlorine compound (or any other pesticide) introduced by any means into the total environment must eventually equilibrate between the physical and biotic elements of it. The situation is something like the result of shaking organic compounds with water and an immiscible solvent such as hexane; partition occurs between the aqueous and organic phases, the hexane taking almost all of the lipid soluble compounds while the water contains those which are more polar (water loving; hydrophilic). Living organisms are complexes of proteins, lipids and water moving around in the abiotic environment (here likened to the aqueous phase) and must, like the hexane phase, eventually equilibrate with the lipid soluble materials in it. However, the important distinction between the environment and the model system is that the "immiscible" phases of the former can each change the nature of the compounds partitioned into them:

```
                                            pesticide
                                               |
                                               v
chemical transformations<——abiotic environment ⇌ biotic environment
                                               |
                                               v
                                        metabolic products
```

The single arrows indicate that the transformations are usually irreversible. The major routes in both elements of the environment are oxidative, reductive, or hydrolytic and there is a tendency, although this is not invariable, for the products to be more polar (that is, more hydrophilic). This does not necessarily mean high water-solubility, but, as will be seen, living organisms have means of converting their initial metabolic products into water soluble compounds that can be excreted. Some conversions that occur or are likely to occur in the abiotic environment under atmospheric conditions (photochemical reactions, for example) were mentioned in the chemical sections and it is obviously not always easy to distinguish biological transformations from the chemical ones because of the close association between the two types in many situations. For example, the soil is a complex interface between the abiotic and biotic elements of the environment since it consists of a mixed inorganic and organic matrix populated by living organisms; pesticides in or on it may be degraded both chemically and by soil animals and there may be a complex combination of both activities. Pesticides on plant surfaces may be degraded by atmospheric exposure and the residues taken up by the plant tissues. These may also take up transformation products formed in

*The absorption, distribution, metabolism, and excretion characteristics of a drug or toxicant are called its *pharmacokinetics* and interactions with the target site its *pharmacodynamics*. In some texts of all these processes are classed together as *pharmacodynamics*.

the soil so that "metabolites" found in tissues may not necessarily have been produced in the plant.

The current status of organochlorine insecticide biodynamics and metabolism studies is discussed in a number of recent reviews, and the following sections cover the main features of the current situation, details being available from the references cited.[480-486]

## 1. Biodynamics

This term is used here to refer to the behavior patterns of organochlorine insecticides and their metabolites and transformation products in living organisms. In vertebrates, such patterns have been called *pharmacokinetics* and are one part of the whole complex of events, including action on the target, referred to in the introduction. They are determined by making balance studies of the amounts excreted and remaining in the tissues following administration of the chemical in various ways over short or long periods of time. Similar investigations conducted with insects, plants, and microorganisms and the patterns observed have often given evidence that metabolic detoxication must be occurring long before metabolites have actually been identified.

### a. Insects

Many investigations of the behavior of organochlorine toxicants have been conducted in insects in relation to the resistance problem. Typical experiments involve determination of the fate of a single dose of insecticide at various times following its application to groups of insects held with food and water in a suitable vessel. The doses used have usually been at or near the toxic level and whole body levels have often been measured without much regard for the distribution between different tissues and organs. However, the investigations of Sun[487] and of Gerolt[476] involved more careful examination of dieldrin distribution in houseflies and quite recent investigations of the resistance question have used combinations of autoradiographic detection of tritiated or [14]C-labeled dieldrin with electron microscopy to determine its distribution in the intact nerve cords of German cockroaches and in the thoracic ganglion of houseflies.[488,489] In studies of penetration, the technique of rinsing the intact insect with acetone to recover the removable "unpenetrated" toxicant is often employed. This method dissolves the epicuticular wax and the insecticide in it and is really a measure of toxicant loss into the lower layers of the cuticle.

There is often a linear relationship between the logarithm of the amount remaining on the "outside" and the time of penetration, showing that penetration is a first order phenomenon, the rate at any time being proportional to the amount of unpenetrated insecticide. In such cases, the time to 50% penetration ($t_{0.5}$) can be measured and the penetration rate constant (k) is calculated from the equation $k = 0.693/t_{0.5}$.[2] In other cases, there is extremely rapid penetration immediately after application, followed by a slower first order phase of penetration. Thus, the penetration of the major HCH isomers into cockroaches is biphasic and both the initial rapid phase and the following slow phase appear to be first order processes, the relative rates of penetration of the isomers being $\delta \gg \gamma > \alpha \gg \beta$. Whole body autoradiography shows that $\beta$-HCH penetrates the cockroach body much more slowly than the $\gamma$-isomer, the latter reaching all parts of the central nervous system, crop, and gizzard within 15 min, but surprisingly, not the inside of the central nervous system.[490] This situation differs from that found with dieldrin in houseflies, since in this insect dieldrin localization occurred along the membranes surrounding the axons (see section D.2b); a supposedly similar mode of action would presuppose similar localization.

Less useful but frequently used curves are obtained by plotting percent penetrated against time of penetration. Sun has used such data to derive rates of penetration which are plotted against time and compared with similarly derived rates of detoxication; the difference in the areas beneath the two curves then represents the actual total amount of toxicant in the tissues during the experiment.[491] This method highlights the effects of any agent, such as a synergist, which blocks metabolism; for example, in the case of some experiments conducted with dihydroaldrin by the author, the 1,3-benzodioxole synergist sesamex more than doubles the amount of toxicant in housefly tissues during the 6-hr period of the experiment.

If sufficient toxicant is applied and there is no metabolism or excretion, penetration will continue until the tissues are saturated and will then cease so that the internal level remains constant with time; this condition will also be achieved when the rate of any metabolic or other elimination process balances the rate of penetration, resulting in a steady state concentration of toxicant. In general, the externally applied toxicant in a single application cannot maintain the internal level of toxicant when metabolism is occurring, so the internal level passes through a maximum as metabolism proceeds and the external level is depleted. The situation has some resemblance to a series of consecutive reactions.[492] In equation (1), A, B and C are normally different products formed successively

$$A \xrightarrow{\quad k_1 \quad} B \xrightarrow{\quad k_2 \quad} C \qquad (1)$$

by the decomposition of A. If, however, A represents toxicant applied externally to an insect, B the internal toxicant and C the conversion products, then

$$\frac{d[B]}{dt} = k_1[A] - k_2[B] \qquad (2)$$

where [A] and [B] are concentrations at time t, $k_1$ is the penetration rate constant and $k_2$ the rate constant for conversion to C. In this simple view, the entry of A is a first order process. A steady state concentration of B (i.e., internal A) is attained when $k_1[A] = k_2[B]$, and [B] declines when metabolic rate, $k_2[B]$, overtakes entry rate, $k_1[A]$. C may be either a toxication or a detoxication product and modifications of the rate constants with consequent changes in the levels of B and C can be critical for the insect's survival; insecticide synergists, for example, affect $k_2$ and may also affect $k_1$, either directly or indirectly. This sort of situation is expected when a single dose of DDT is externally (topically) applied to a resistant housefly which can dehydrochlorinate it to DDE enzymatically. On the other hand, dieldrin and heptachlor epoxide are stable in both normal and susceptible houseflies and the excretion rate is not high, so that a virtually constant internal level is eventually obtained.

Hewlett[493] considered the equation

$$x = a(1-e^{-z/a})$$

to be the most useful relationship between the amount of insecticide (x) absorbed by houseflies from an external dose z in a given time, so that the level $a$ is approached asymptotically as z increases indefinitely. An appropriate value for $a$ is ten in 24-hr absorption experiments with topically applied DDT, and by using this value in the equation, it can be shown that the value of z required to produce a given internal concentration necessary for toxicity is increased to infinity by a combination of two mechanisms, each successively effecting a twofold reduction in the amount of toxicant that would otherwise reach the site of action. This illustrates the way in which just two mechanisms conferring modest levels of resistance by themselves can confer near immunity when present together in an insect. That such situations exist has been verified by recent work in which high levels of resistance have been produced in houseflies by combining modest resistance factors from genetically pure strains (Vol. II, Chap. 2B.2a). With the exception of lindane, which appears to be fairly readily degraded, the organochlorine insecticides used commercially are rather stable in many insects, which accounts for their efficiency as toxicants; high internal levels build up somewhat quickly after contact and are maintained until death ensues. When natural tolerance occurs (as for example that of the Mexican bean beetle to DDT, which is associated with dehydrochlorination) it is worth investigating, since these exceptions can illustrate hitherto unrecognized principles.

### b. In Soil, Plants, and Microorganisms

The dynamics of organochlorine behavior in plants and soils is closely associated with the subject of "insecticide residues," which are the levels of the toxicant and/or its transformation products remaining in or on these elements of the environment after all the processes of "weathering" (that is the combined effects of wind, rain, air, and light and interaction with other chemicals) and biotransformation have occurred. For edible plants, a time limit for these transformations is usually imposed by the interval between toxicant application and harvesting. A major difficulty in investigating the behavior of toxicants in plants and soil relates to the difference between experiments in the carefully controlled environmental conditions of the laboratory or greenhouse and those conducted under field conditions where changes in climatic influences and weathering may be frequent. The former undoubtedly serve to demonstrate the general principles of behavior, especially in regard to any metabolic changes occurring, but results in the field, as with toxicity evaluations, can be very different.

Each individual soil type is a complicated system with characteristics determined by its range of particle size, origin, organic matter content, water content, and fauna and flora, so the variable persistence of organochlorine insecticides is not surprising. However, certain generalizations are possible, and have been extensively discussed by Edwards, who calculated average half-lives of persistent insecticides from published work.[59] Such half-lives are necessarily approximate, since the disappearance of organochlorines is not truly exponential, it being the resultant of a number of processes which are influenced by the variables previously mentioned as well as by seasonal changes. According to some authorities, the general pattern of persistence in soil is that DDT and dieldrin remain longest, followed by lindane, chlordane, heptachlor, and aldrin in order of decreasing half-life. Half-lives recently listed by Menzie follow a slightly different order and are (time in years) DDT (3 to 10), dieldrin (1 to 7), aldrin (1 to 4), isodrin-endrin (4 to 8), chlordane (2 to 4), heptachlor (7 to 12), lindane (2), and toxaphene (10).[494] These are the values obtained when the compounds are worked into the soil. As might be expected, they are normally reduced to weeks if the toxicants are merely applied to the surface. If the approximate half-life of a toxicant is known, predictions can be made regarding the likelihood of its accumulation in soil after successive treatments.[495] Thus, if the breakdown is assumed to be roughly exponential, half-lives up to 1 year will result in residues of not more than twice the annual addition whether this is divided or added all at once. For a half-life of 4 years, the maximum accumulation is about six times the annual application, rising to not more than 15 times the annual treatment for a half-life of 10 years.

Decker has demonstrated the validity of such predictions for aldrin by comparing the calculated values for residues in Illinois cornfields after annual treatments for up to 10 years with the actual residual levels obtained by sampling; there is remarkably good agreement between the two sets of results.[496] Apart from direct evaporation of the parent compounds, it is likely that some chlorinated insecticides are dissipated from soil by volatilization of their degradation products. For example, $p,p'$-DDE and $\gamma$-1,3,4,5,6-pentachloro-cyclohexene, soil degradation products of $p,p'$-DDT and $\gamma$-HCH, respectively, are much more volatile than the parent compounds and their volatilization has been suggested as a major cause of loss of those insecticides from soil.[90a,90b] Much information regarding the behavior of organochlorine insecticides in soil has been provided by the extensive investigations of Lichtenstein,[497] and Harris.[498] Harris has presented a detailed investigation of the behavior of chlordane and dieldrin in soil in relation to their efficiency as soil insecticides.[499,500]

Organochlorine insecticides have no systemic activity although there is now evidence that they are taken up by plants in small amounts, the degree of translocation being influenced by plant nutritional status in some cases. Early reports indicated that extracts of crops treated with heptachlor and aldrin contained the corresponding epoxides, but epoxide formation may have occurred either on the crop, in it, or possibly in the soil followed by translocation into the crop. Later work showed that the epoxides were present in pea plants grown in sand that contained high levels of the same two toxicants but itself epoxidized little of them, and dieldrin has been shown to translocate in cereals and forage crops. Experiments conducted in 1963 indicated that the endosulfan isomers and some of their conversion products such as endosulfan-diol and endosulfan ether (Figure 11, structures 9 and 13, respectively) can be translocated into the aerial parts of plants from the soil, but as with other organochlorines, the amounts of the toxic isomers transported in this way are not effective against pests attacking the plants.[248] Interest in the dynamics of organochlorine behavior in plants therefore stems from the need for more detailed information about the nature and behavior of residues and research in this area is continually expanding. Plant tissues have recently been shown to effect metabolic conversions such as aldrin epoxidation in vitro,[504,505] and numerous recent investigations at the Institute for Ecological Chemistry at Schloss Birlinghoven in Germany have shown that metabolism of organochlorines occurs in plants.[111,486,486a]

Table 14 summarizes the results of experiments in which various [14]C-labeled organochlorine insecticides were applied to the leaves of white cabbage and the residues determined after various time intervals.[486] The persistence of total residues is seen to increase from left to right, dieldrin, photodieldrin, and the DDT-group compounds

# TABLE 14

Total Residues of DDT-group Compounds, Cyclodiene Insecticides, and Lindane, Labeled with Carbon-14, After Application to the Leaves of Growing Cabbage Plants[a]

| | γ-HCH | Technical Heptachlor | Aldrin | Endrin | trans-Chlordane | Isodrin | Dieldrin | Photo-dieldrin | DDD | DDT | DDE |
|---|---|---|---|---|---|---|---|---|---|---|---|
| Amount applied (mg/plant) | 1 | 1 | 1 | 1.3 | 1 | 1 | 1 | 2.5 | 2 | 2 | 0.5 |
| % of applied [14]C recovered | 8.3(4) | 15(4)[b] | 17(4) | 11(4) | 35(4)[b] | 30(10)[e] | 40(4) | 75(4)[d] | 63(14)[k] | 65(14) | 56(14) |
| | 4.7(6) | 11(12) | 12(6) | * | 20(10)[c] | | 29(6) | | | | |
| % of recovered [14]C as unchanged toxicant | 30–40(4)[i] | 57(4) | 4(4)[f] | 90(4)[h] | | 5(4)[e] | 75(4)[g] | | 93(14)[l] | | |
| | 30–40(6) | 54(12) | 6(6)[f] | | | Nil(10) | 79(6)[g] | | | | |

(Adapted from data of the Institute for Ecological Chemistry, Schloss Birlinghoven.[111,244,486])

*Tobacco plants treated with 2.08 or 1.04 mg of [14]C-labeled endrin/plant under various conditions contained a total residue to 32 to 47 per cent of the applied [14]C 6 weeks after application.

[a]Time (weeks) after application is given in parentheses.

[b]90 per cent of this in the leaves.

[c]80 per cent of the residues in leaves consisted of mainly three unidentified metabolites.

[d]30 per cent of this inside the leaves, of which 15 to 33 per cent consists of at least three metabolites.

[e]Remainder of [14]C at 4 weeks and whole of [14]C at 10 weeks is conversion products: mainly endrin and Δ-keto-endrin plus other, more hydrophilic, compounds.

[f]Remainder of recovered [14]C consists of six conversion products, including one very hydrophilic metabolite and small amounts of dieldrin and photodieldrin.

[g]Remainder of recovered [14]C consists of photodieldrin plus more hydrophilic metabolites.

[h]Remaining ten per cent consists of a very polar metabolite together with a small amount of Δ-keto-endrin.

[i]Remaining [14]C consists of five more polar metabolites.

[k]84 per cent of this in the treated leaf area.

[l]In the treated leaf area.

being the most residual. Volatility undoubtedly accounts for the low total residues of lindane, technical heptachlor (artificial mixture), endrin and aldrin, and the highest proportion of unchanged toxicant is present in residues when epoxidation is not a transformation process. Experiments with the simulated technical heptachlor (a mixture of 74% heptachlor, 22% trans-chlordane and 4% nonachlor, all labeled with [14]C) indicates that the material not lost by volatilization penetrates the leaf surface rather rapidly, as does trans-chlordane applied individually, but detailed evaluation is difficult with the technical mixture. trans-Chlordane appears to be considerably more persistent than technical heptachlor and experiments in which the latter's individual components, each labeled with [14]C, were applied to cabbage, gave radioisotope recoveries after 5 weeks of 75, 31, and 5%, respectively, following the separate nonachlor, trans-chlordane and heptachlor treatments. By this time, however, heptachlor and nonachlor were extensively transformed (66 to 74%) to other compounds in both plants and soil, whereas trans-chlordane proved to be rather stable. However, 80% of the residues of the latter in leaves after 10 weeks was in the form of transformation products.

When aldrin and isodrin are applied to plants, the conversions aldrin→ dieldrin → photodieldrin (Figure 12, structure 5), and isodrin → endrin → Δ-keto-endrin (Figure 14, structure 10) ensue. An interesting feature of the aldrin application (Table 14) is that the residues are mainly conversion products other than dieldrin and photodieldrin, whereas isodrin is converted mainly into an unidentified polar compound (on leaf surfaces), together with endrin and Δ-keto-endrin. When dieldrin and endrin are applied, the residues are mainly the unchanged toxicants. Photodieldrin residues are evidently more stable than either dieldrin or aldrin residues, and although Δ-keto-endrin transforms more rapidly than endrin itself, its residues (mainly one hydrophilic metabolite) appear to be more stable than endrin residues. Both the sequences mentioned are therefore in the direction of increasing persistence although aldrin and its transformation products are metabolized somewhat more readily than those from isodrin (endrin), and endrin residues are evidently translocated in the plant and appear in the soil, a process said to be important for the reduction of

endrin residues.[486] Open field trials conducted with [14]C-aldrin as a soil application (3 lb per acre) or seed dressing have been conducted under various climatic conditions with potatoes, sugar beet, wheat, and maize, and show that the residues contain a higher proportion of hydrophilic metabolites than of dieldrin. These experiments have shown that the unchlorinated bicycloheptene nucleus of aldrin can be cleaved to give the dicarboxylic acid (Figure 12, structure 7), which appears as a major component of the hydrophilic products present in plants, soil and the drainage water from treated soil.[486]

Lindane is more than half converted into five more polar products, not yet identified, within a few weeks of its application to cabbage plant foliage (Table 14), and a similar series of compounds appears in spinach or carrots grown in lindane treated soil, about half of the absorbed toxicant again being degraded within a few weeks. The structural difference between spinach and carrots results in a different residue distribution pattern in each species so that it is difficult to generalize about such patterns; each crop or crop type must be treated as an individual problem. Carrots are a good example; the fleshy roots pick up organochlorine insecticides readily and provide a sensitive indicator of soil residues, but the distribution within the root varies considerably among the varieties. In some, the residues are confined to the outer layers and can be entirely removed by peeling while in others residues occur within the root. Some roots convert aldrin into dieldrin, while others do not (a reflection of differences in epoxidase enzyme content).

The long residue life of the DDT-group compounds is illustrated by the foliage application experiments of Table 14. Almost all of the residue of the applied 2 mg of DDT after 14 weeks was found in the areas of application, about three quarters of it having penetrated the leaf tissue. With four successive applications (totaling 6 mg) at weekly intervals, the residue reached 66% of the total and there was some translocation to untreated parts of the plants and to the soil. In this experiment about half the recovered material was in the leaves and in either case about 1% of the external and 5 to 10% of the internal chemical consisted of conversion products. Only 1% of the DDE remaining 14 weeks after a single application (Table 14) was translocated, although most of it penetrated the tissues in the area of application.

Multiple applications of the same dose at weekly intervals resulted in a somewhat higher degree of translocation while degradation was lower than for DDT. Field experience with DDD has proved its ability to control insects that are not readily accessible to DDT and the experiments with cabbage revealed an interesting tendency to spread from the areas of application that was not found with either DDT or DDE, although the total residues of all three were similar 14 weeks after application. DDD and DDE were found to be conversion products of DDT on and in the leaves of treated cabbage and carrot plants and the more hydrophilic fractions contained DDMU, DDA, and DBH (Figure 8). Although the scope of some of the greenhouse experiments is somewhat limited, they add to evidence from other sources, including the field trials with [14]C-labeled material in the case of aldrin, that $\gamma$-HCH, heptachlor, aldrin, and to a less extent isodrin and endrin, are dissipated fairly rapidly following their application. Volatilization may of course contribute to the spread of the unchanged toxicants from the point of application. A most interesting point which will emerge in the later discussions on metabolism is that for aldrin and heptachlor there is accumulating evidence that the presence of the unsubstituted double bond confers "opportunity factors" for both chemical and biological detoxication in soil that are less readily available for their epoxides.

The majority of investigations relating to interactions between organochlorine insecticides and microorganisms are designed to determine the identity of the metabolites produced. For example, with soil microorganisms such investigations afford information about the dynamics of organochlorine behavior in the presence of the organism but usually say little about the dynamics in the organism itself, which may be critical for its survival. On the other hand, total metabolism of an added compound in a given time presumably implies that it has passed through the organism during that time. As one example of a study on biodynamics within a microorganism, Lyr and Ritter[506] examined the penetration of [14]C-HCH isomers into the yeast Saccharomyces cerevisiae and showed that the rate of uptake increased in the order $\beta < \delta < \gamma < \alpha$, which differs from the order of their solubility in organic solvents and also from the order of their toxicity to the organism, this being $\delta > \gamma > \beta > \alpha$. When the cells in culture are exposed to 3 x $10^{-5}$ M concentrations of the individual isomers in the medium, the intracellular concentrations of the $\alpha,\beta,\gamma$, and $\delta$-isomers reach constant levels corresponding to the uptake of 60, 18, 40, and 24%, respectively, of the compound in the medium, so that the most bioactive $\delta$-isomer is taken up only marginally more readily than the slightly active $\beta$-HCH, while the least toxic $\alpha$-isomer is absorbed most readily. For this reason, the stereochemical differences between isomers are considered to have overriding influence on the activity.

As an example of the other kind of investigation, Harris examined the ability of 92 pure cultures of soil microorganisms to metabolize aldrin and showed that fungi could be classified according to their ability to convert aldrin into dieldrin (a) rapidly after an initial lag phase, (b) linearly from the beginning of exposure, (c) rather poorly, or (d), as in (a) or (b) but with a peak in dieldrin formation followed by a decrease in its concentration, suggesting that it is further metabolized.[507] Similar classes are observed among aldrin converting actinomycetes and bacteria from soil and there are indications that aldrin can be degraded directly by some organisms in a process independent from dieldrin formation. This is certainly the situation with heptachlor, for which it is now clear that several other metabolic conversions are effected by soil microorganisms (Figure 10), apart from the well-known epoxidation reaction. In fact, it now seems that microorganisms can transform most cyclodienes, the conversion of dieldrin being the most difficult to demonstrate. Dieldrin at 0.15 ppm in cultures of Aspergillus and Penicillium spp. was not metabolized in a 2 to 3-week period, in which time aldrin suffered extensive change into other compounds besides dieldrin. On the other hand, [14]C-photodieldrin at 0.059 ppm in the culture medium was somewhat readily metabolized by A. flavus (38%) and P. notatum (23%) in 3 weeks. Matsumura and Boush examined 600 soil microbial isolates and found ten which had degradative ability towards dieldrin. One of the most active cultures, referred to as Shell 33, was found in dieldrin contaminated soil from the Shell manufacturing plant in Denver and has now been induced to grow in a medium in which dieldrin is the sole carbon source. All the soil isolates found to convert dieldrin also convert endrin and further investigation of soil samples has produced 25 other

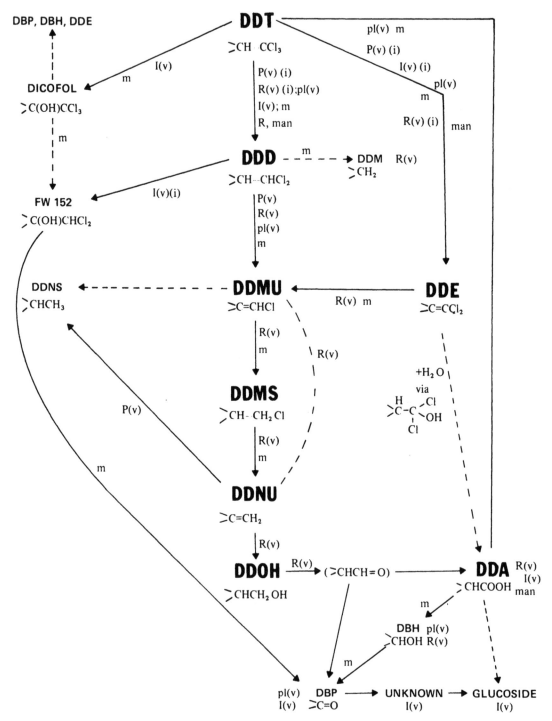

FIGURE 8.   Pathways and products of DDT metabolism in some living organisms. I = insects; m = microorganisms; P = pigeon; pl = plants; R = rat; i = in vitro; v = in vivo

cultures capable of converting it, sometimes very extensively.[485,508]

Concern about the effects of DDT on the aquatic environment has resulted in a number of investigations of its uptake by algae and other small food chain organisms. For example, six species of marine algae were found to differ significantly in their accumulation of DDT from the surrounding medium. Uptake apparently occurred by a passive process; concentrations attained were high compared with the surrounding medium and were usually greater for organisms having smaller cell size.[508a] Aquatic organisms vary greatly in their sensitivity to dieldrin. The photosynthesis of some species of algae is unaffected by 1,000 parts per billion (ppb) in the ambient medium, whereas in others it is inhibited by concentrations greater than 1 ppb. Considerable concentrations of dieldrin may accumulate in these organisms but the accumulation factors seem to be lower than those found for DDT. For *Cyclotella nana*, accumulation rapidly reaches a plateau level for a given ambient concentration and for some organisms, dieldrin accumulation increases linearly with increase in ambient concentration up to 1,000 ppb. This seems surprising unless the algae take up colloidal or particulate dieldrin, since the water solubility is considerably exceeded at such high "concentrations."[508b]

### c. In Vertebrates

The pharmacokinetics of organochlorine insecticides is principally the study of distribution patterns of the unchanged toxicants in vivo following their administration as a single dose or continuously over a long period of time. Investigations of this kind are of great importance and have been the subject of much controversy since they bear directly on the question as to whether living organisms exposed for a long time to small amounts of these chemicals in the environment will continue to accumulate them indefinitely. Background information and references to the literature on this subject are given in various reviews.[481,482,509,510]

As indicated previously, living organisms are subject to physical laws and therefore must eventually come to equilibrium with lipid soluble compounds in the environment. The final balance attained in the tissues depends critically on the availability of various elimination processes and the ability to metabolize such compounds to more polar derivatives that can be excreted is particularly important. In fact, there has been evidence from various balance studies in feeding experiments (cited in the reviews mentioned) conducted since the early 1950's that constant exposure to organochlorine insecticides results in steady-state tissue concentrations of these compounds, such results providing evidence that metabolism is occurring.

Robinson[511] has examined in detail the available pharmacokinetic data for cyclodiene and DDT behavior in man and experimental animals including farm livestock, rats, and monkeys from the standpoint of four postulates, namely, that (i) the concentration of an organochlorine insecticide in a specific tissue is a function of the daily intake, (ii) the tissue concentrations attained depend on the time of exposure, (iii) the concentrations in different tissues are functionally related, and (iv) tissue concentrations decline when exposure ceases, there being a functional relationship between the rates of decline at any time and the amounts in the tissues at that time. Organochlorine insecticides may enter the body by inhalation, percutaneous absorption, or ingestion, of which only the last route has been systematically examined. However, this is appropriate since ingestion is the main route by which humans and most wildlife acquire their organochlorine residues. With regard to postulate (i), the results indicate for DDT and HEOD a functional relationship between tissue concentrations and dietary exposure for man and experimental animals. Thus, the relationship in man between DDT concentration in subcutaneous fat and daily dose is given by the equation $\log_{10} C = 0.0142 + 0.6539 \log_{10}$ ($10^2$ x daily dose of DDT in mg), and the daily ingestion ($\mu g$) of HEOD is given by concentration of HEOD ($\mu g/ml$) in blood/0.000086, or, concentration of HEOD ($\mu g/g$) in fat/0.0185, the concentration in fat being some 200-fold higher than that in blood.[511]

Information available for rats, dogs, and men indicates that the concentrations of DDT and dieldrin in tissues are functionally related to time of exposure and the data are generally more consistent with an asymptotic approach to a dose dependent upper limit of tissue concentration than with other types of relationship that have been proposed in some cases. This is certainly the case for HEOD concentrations in the blood and adipose tissue of human volunteers who are fed dieldrin.

There is an evident correlation (see postulate iii) between the HEOD concentrations in blood and adipose tissue of man (as mentioned above), rats, and dogs; the situation is most explored for HEOD but there is evidence of similar relationships, in rats for example, for DDT. When exposure to DDT or HEOD ceases, the tissue levels decline in a manner depending on the compound and tissue considered (see postulate iv). Thus, the disappearance of HEOD from rat adipose tissue follows a simple first order relationship but the decline of DDT in the body fat of rats, monkeys, and steers is not simple and is probably best represented by a biphasic decay involving two exponential terms, although a log-log relationship (power function) between concentration and time has been suggested in the case of steers. A complementary approach to the measurement of tissue levels is the one involving the measurement of excretion rates. Thus, the concentrations of DDT and DDA in the excreta of Rhesus monkeys are related to the dietary intake and hence to the concentration in fat as is also the case in man (see postulate i). In rats, measurements of excretion rates using [14]C-cyclodienes have shown that after a time the excretion consisting mainly of hydrophilic metabolites balances the daily intake, indicating the attainment of plateau levels.[511]

Apart from $p,p'$-DDT and HEOD, human adipose tissue contains $p,p'$-DDD, $p,p'$-DDE, heptachlor epoxide, and HCH isomers (mainly $\beta$ and $\gamma$), and the limited information available suggests that their dynamics resemble those of DDT and HEOD. Endrin differs remarkably from both DDT and dieldrin in being eliminated very rapidly from vertebrate tissues; it shows little tendency to store and can only be detected in human tissues immediately following an acute overexposure.[303] Poultry continuously fed endrin have a greater tendency to store this compound in fat, but elimination is rapid when exposure ceases. The absence of measurable endrin levels in human subcutaneous fat or blood, even in areas in which it is extensively used (as in India and the Lower Mississippi area) leads to the conclusion that in spite of its high acute toxicity, endrin is a relatively nonpersistent insecticide in man. Both photodieldrin (Table 15) and the half-cage hexachloro-ketone from endrin ($\Delta$-keto-endrin) are reported to have relatively short half-lives in mammals, so that the persistence of these compounds is rather the reverse of that

found on plants. The short half-life of $\beta$-dihydro-heptachlor in the rat is in sharp contrast to the behavior of heptachlor and isobenzan and undoubtedly relates to its susceptibility to degradation; in contrast, heptachlor is converted into a stable epoxide and isobenzan is a rather stable ether resembling heptachlor epoxide, so that their persistence is not surprising. For endosulfan, the cyclic sulfite ester moiety provides opportunities for detoxication that are unique among insecticides derived from hexachlorocyclopentadiene. Following its dietary administration to sheep, pigs or cows, a little endosulfan sulfate is temporarily stored in the fat and may be excreted in milk but rapidly disappears when administration ceases.[248]

There have been many measurements of residue levels in the tissues of other vertebrates but few detailed investigations of the sort just discussed.[482,483] For dieldrin, DDT, DDD, DDE, DDMU, and DDMS, half-lives of 47, 28, 24, 250, 27, and 8 days, respectively, have been indicated in the pigeon and there are numerous indications in the literature that other birds as well as fish and other aquatic organisms accumulate steady-state levels of these compounds (which may be rather high in some cases) and that these levels decline when exposure ceases. While studying enzyme induction effects, Bitman has shown that 2,4'-isomers of DDT and DDE accumulate to a much lower extent than the 4,4'-isomers in the lipids of rats and quail when the compounds are fed at comparable levels.[512]

The results obtained following chronic exposure to HEOD are consistent with expectations if the animal is regarded as a two-compartment model of the mammillary type used to explain the pharmacodynamic properties of drugs; for acute exposures it appears that a more complex model involving three or more compartments may be required to describe the results.[513] Although the compartments are conceptual and do not necessarily equate with particular tissues, a useful model for interpretation of the available data consists of a central compartment (as might be liver plus blood) which is in contact with a second, otherwise isolated, inert storage compartment (such as adipose tissue), and receives insecticide which it can transform or eliminate:

TABLE 15

Pharmacokinetic Characteristics of Some Cyclodiene Insecticides and DDT

| Compound[a] | Dose/time[b] | Tissue | Time to Plateau Level (Days) | Biological Half-life (Days) | Reference |
|---|---|---|---|---|---|
| HEOD | 10 ppm [c]/56 | Rat liver[e] | | 1.3(m);10.2(m) | 513 |
| HEOD | 10 ppm/56 | Rat blood[e] | | 1.3(m);10.2(m) | 513 |
| HEOD | 10 ppm/56 | Rat adipose | | 10.3(m) | 513 |
| Photo-dieldrin | 10 ppm/26 | Rat adipose | | 1.7(m);2.6(f) | 514 |
| Telodrin | 5-25 ppm | Rat adipose | ~170(m) | 10.9(m) | 515 |
| | (Various times up to 224 days) | Rat adipose | ~200-250(f) | 16.6(f) | |
| HEOD | 211 μg daily/550 | Human blood | ~365 | 84-112(m) | 516 |
| Endrin[d] | 0.8 ppm/1 | Rat[f] | | 1-2(f) | 517 |
| | 1.6 ppm/1 | Rat[f] | | 6(f) | |
| Endrin | 0.4 ppm/12 | Rat[f] | 6(m);6(f) | 3(m);3(f) | 517 |
| HHDN | 0.2 ppm/84 | Rat[f] | 53(m);200(f) | 11(m);100(f) | 508 |
| β-DHC | 50 μg (single i.v.) | Rat[f] | | 1(m) | 518 |
| DDT | 5 ppm/168 | Rat adipose | 140-168(m) | | 511 |
| DDT | Various/7 years | Rhesus monkey adipose | 365-465 | | 511 |
| DDT | 35 mg daily/600 | Human adipose | 365(m) | 1-2 years | 511 |

(Data adapted from Brooks.[483])

[a]HEOD, the major component of the insecticide dieldrin, is 1,2,3,4,10,10-hexachloro-6,7-exo-epoxy-1,4,4a,5,6,7,8,8a-octahydro-1,4-endo,exo-5,8-dimethanonaphthalene. HHDN, the major component of the insecticide aldrin, is 1,2,3,4,10,10-hexachloro-1,4,4a,5,8,8a-hexahydro-1,4-endo,exo-5,8-dimethanonaphthalene.
[b]Times in days unless otherwise indicated.
[c]Concentration in the daily diet.
[d]Single administration at these concentrations in the diet.
[e]Half-lives are given for fast and slow phases of biphasic elimination.
[f]Data based on pattern of excretion of $^{14}$C-labeled compounds from intact rat; other data shown are based on measurement of toxicant in the tissues.
(m)=male; (f)=female.

---

insecticide enters at constant rate $(\alpha)$

$$\text{inert storage compartment } (C_s) \underset{k_1}{\overset{k_2}{\rightleftharpoons}} \text{central compartment } (C_t) \xrightarrow{k} \text{elimination}$$

(metabolism occurs)

This model is in some sense an elaboration of the consecutive reaction system discussed earlier in which continuous addition of insecticide (at a constant rate $\alpha$) and storage are allowed for. In this case,

$$\frac{dC_t}{dt} = \alpha - kC_t - k_1C_t + k_2C_s$$

In chronic feeding experiments, when a steady state is attained:

$$\frac{C_s}{C_t} = \frac{k_1}{k_2} = K \quad \text{(the partition coefficient between the compartments)}$$

so that,

$$\frac{dC_t}{dt} = \alpha - kC_t, \quad \text{whence} \quad C_t = \frac{\alpha}{k}(1 - e^{-kt}) + C_o e^{-kt}$$

or,

$$C_t = \frac{\alpha}{k}(1 - e^{-kt}), \quad \text{assuming, for simplicity, that } C_t = 0 \text{ when } t = 0.$$

The last equation requires that the concentration in the central compartment approaches a definite upper limit ($\frac{\alpha}{k}$) as the time of exposure becomes large, and represents a simplified version of the model, which is fully discussed by Robinson.[519] If the rate ($\alpha$) of intake is increased or reduced, the steady state (plateau) level will rise or fall correspondingly to a new value dictated once again by the balance between intake and elimination, the simplest form of elimination equation being $C_t = C_0 e^{-kt}$, in which $C_0$ is the concentration in the central compartment when exposure ceases. This simple equation applies to the elimination of HEOD from rat adipose tissue but the elimination from liver and blood is biphasic and best represented by an expression containing two such exponential terms.[513] The results for HEOD with this animal suggest that the early, rapid phase of elimination may relate to the depletion of HEOD already in the blood and liver, regarded as one compartment, and the slower phase to its gradual release from fat (the second compartment) into blood before elimination. The application of such models requires that the rate constants and sizes of the compartments do not change during the experiments, but both DDT and dieldrin are known to induce increased liver microsomal metabolizing enzyme activity and the overall value of the elimination rate constant might well be altered on this account if the level of toxicant administration exceeds the no-effect level for such induction.

There is increasing evidence that the principles just discussed are generally applicable in the environment. Important support comes from a consideration of the surveys of organochlorine residue levels in human adipose tissue, which have now been conducted regularly for many years. Although there are interesting geographical variations in some of the levels that have not been adequately explained, a significant observation is that they have now been constant for some years in the U.S. and the U.K., as is to be expected if a balance between intake and elimination has been attained, and have even tended to decline in some areas. For example, in the United Kingdom there is good evidence that the result of limitations placed on the use of organochlorine insecticides since the mid-60's is a decline in concentrations of

their residues in several elements of the biosphere such as fatty foodstuffs, human adipose tissue, the fat of sheep reared in the U.K., and the eggs of shags (*Phalacrocorax aristotelis*) inhabiting British coasts.[54] The 1969–71 values for residues in human adipose tissue indicate a reduction of about 30% in dieldrin content and a little over 20% in total DDT (*p,p'*-DDT plus *p,p'*-DDE) since 1965–67, showing a continuation of the decline that has been evident since 1963–64; total HCH residues (mainly $\beta$-HCH) have fallen by about 30% over the whole period.[520] In shags' eggs, which have been used to measure organochlorine insecticides in the marine environment, the maximum concentration of HEOD (dieldrin) appears to have occurred in 1966, and of *p,p'*-DDE about 1967–68, with declines since then of about 66% and 47%, respectively.[521]

According to a recent appraisal by Matsumura,[58] a similar situation exists in some sectors of the North American environment. The relatively rapid decline in organochlorine levels in certain ecological systems when use of these compounds is restricted suggests that the levels in such systems are not maintained by rapid equilibration with some large reservoir of organochlorine contamination, as some discussions of this topic seem to imply. Several attempts at systems analysis relating to DDT fate and effects in the environment have been made recently[522,523] and there is much to be said in favor of this approach in the long term. However, discrepancies between the predictions of such models and recent observations of the kind mentioned above suggest that current knowledge of organochlorine behavior is inadequate to permit fullest use of the technique at present.

Elaborations of the "laboratory ecosystem" approach recently developed by Metcalf seem likely to provide useful information about pesticide accumulation along food chains and the influence of other factors such as metabolism.[38,524] In these model systems, aquaria are established containing representative terrestrial and aquatic life, including plants, insects, and fish with suitable food chain relationships. Isotope-labeled insecticides applied to the plants are then consumed by plant eating insect larvae and are conveyed to the water through excretion and in other ways either unchanged or as metabolites, thence to be picked up by fish, mosquito larvae, and other species. Subsequent analysis of all elements of this limited but controllable environment indicates the nature of metabolites and the accumulation factors to be expected. In this way, Metcalf has shown that DDT and especially DDE accumulate to relatively high levels in fish, *Gambusia affinis* (about 100,000-fold compared with the concentrations in water). With *p,p'*-DDD, methoxychlor, ethoxychlor and methylchlor (see section C.1a) accumulation factors of 42,000, 1,500, 1,550, and 140, respectively, are found, demonstrating the influence of biodegradability in this series. However, the situation seems to be almost reversed with the snails (*Physa*) used in these experiments, since the last three compounds mentioned accumulate to higher levels than DDE, DDT, or DDD, the latter being least accumulative (13,000-fold) apart from methiochlor (300-fold). Methoxychlor (accumulation factor 120,000) appears to be rather stable in the snail, which illustrates the importance of species comparisons. No doubt the experience gained with this kind of experiment can be applied to the development of more elaborate models which may provide accurate information for use in systems analyses applied to the environment.

## 2. Nature of the Transformation Products

Stimulated by the controversy surrounding the organochlorine insecticides, studies of their transformation in both the abiotic and biotic environments have proliferated enormously in recent years, but many details of the pathways to these products remain to be elucidated.[483-486] In many situations such as on plant surfaces or in the soil, it is not always easy to determine whether the transformations observed are purely chemical, purely biological, or both, since the same end products frequently result. The abiotic conversions are mainly associated with photochemical processes and have been discussed already in the appropriate sections in Volume I. The products are mentioned in the following sections when appropriate but, except for photodieldrin (Figure 12, structure 5), which has probably received most publicity, little is known about their metabolism in living organisms. Biochemical aspects of organochlorine metabolism are discussed in section B.3.

### a. DDT Group

Some of the transformations definitely established or postulated for DDT are shown in Figure 8. Tentative pathways are indicated by dashed arrows. When the pathway to a product is completely unknown, the symbol for the organism is placed alongside the product. The three most common degradative routes for DDT are dehydrochlorination to DDE, oxidation to dicofol and other compounds not yet identified, and reductive dechlorination to DDD. DDE is a product of, or an intermediate in, the slow transformation of DDT by light and air, and can be further converted into DBP as well as DDMU and its relatives by the action of light. Casida[525] has recently demonstrated a sensitization of the photochemical conversion of cyclodienes on plant foliage by rotenone, and it may be that surface effects on foliage and interaction with plant pigments are involved in the conversions of DDT observed on plants. Any of the products may penetrate the plant surface and add to the biotransformation products or themselves be further converted and so the true origin of residues becomes obscured. At any rate, DDE is formed to some extent by most of the living organisms investigated and it is clear that the compound is long-lived in the environment, as shown, for example, by its half-life of 250 days in pigeons, which is much greater than that of the other DDT conversion products that have been characterized.[526] DDE appears to be less acutely toxic than DDT to a number of organisms and has long been regarded as a detoxication product, but this may not be true in all cases as indicated by recent feeding experiments with pigeons, quail, and blackbirds. DDT is converted primarily into DDE in the first two species, whereas DDD is the favored product in the blackbird (*Turdus merula*), which appears to eliminate DDE much faster than the other two can; DDE is actually more toxic than DDT to the pigeon.[527]

Chemically, the most logical elimination route for DDE in all species would involve direct hydration of the double bond to give DDA, via hydrolysis of the acid chloride which is the expected initial product; this possibility has been much spoken of, but its occurrence does not seem to have been conclusively demonstrated. Furthermore, Datta recently presented evidence that DDE is slowly converted in rats into DDMU then DDNU, without the intermediate formation of

DDMS. He also indicates that liver effects the conversion to DDNU and kidney the conversion DDNU → DDOH → DDA, both in vivo and in vitro.[528] Comparison with the Peterson and Robison pathway,[529] which is the backbone of Figure 8, therefore suggests that DDNU is formed either by reduction of DDMU to DDMS followed by dehydrochlorination, or by direct reductive dechlorination of DDMU in rats. In any case, DDE appears not to be converted directly into DDA in rats. In human males ingesting technical DDT, DDE formation was extremely limited, and there was no evidence for the conversion of directly ingested DDE into DDA, which observation appears to account for the high stability of DDE deposits in human adipose tissue. However, DDD was readily formed and ingested DDD was rapidly converted into DDA.[530]

Oxidative pathways for DDT seem to be the particular prerogative of insects, the benzylic hydroxylation to dicofol being first reported in 1959 (see B.3a). The discovery soon followed of an oxidative enzyme system in houseflies and cockroaches that requires NADPH and oxygen and is analogous to the mixed function oxidase complex found in mammalian liver. By this time DDT dehydrochlorination to DDE was well established as a resistance mechanism in houseflies and was known to require reduced glutathione (GSH) for its operation. When reporting the oxidizing enzyme, Agosin and colleagues[531] pointed out that since both processes require NADPH, their competition for it might be expected. Since that time, several instances of insect resistance to DDT have been discovered that rely on oxidative degradation rather than dehydrochlorination. Apart from dicofol formation from DDT, and the analogous conversion of DDD into FW 152 (Figure 8), the identities of some other compounds produced by various insects such as houseflies, mosquitoes, cockroaches, triatomid bugs, and grain weevils have yet to be determined.[484,485]

Surprisingly, aromatic ring hydroxylation has so far been demonstrated only in a resistant strain of the grain weevil (*Sitophilus granarius*) which excretes a complex glucoside containing labeled 4-hydroxy- and 3-hydroxy-4-chlorobenzoic acids when fed on wheat impregnated with ring-labeled [14]C-DDT, as well as producing DDD, dicofol, FW152, DDE, DBP, and DDA.[532] The loss of one aromatic ring of DDT to give benzoic acids raises the possibility that microbial action may have been involved in this conversion. Traces of 4-hydroxybenzoic acid are found in weevil cultures on untreated wheat and are found when DDT-susceptible weevils are reared on DDT-treated wheat, along with DDE and traces of DDD and DDA. The insect frass, which is a likely source of microbial activity, produced only a little DDE, DDD, and DDA, and as the glucoside was produced in relatively large amounts by the resistant insect, the evidence that the benzoic acids are a genuine insect product from DDT seems to be fairly good although the mechanism of formation is unknown.[260] The nature of the oxidation products of DDT formed by the $F_c$ strain of houseflies originally investigated by Oppenoorth[397] which carries the DDT-*md* gene for microsomal oxidation on chromosome 5 (Vol. II, Chap. 2B.2a) has not yet been clarified but according to Leeling, at least seven compounds are formed, 60% of a 1 μg dose per fly being metabolized in 3 days.[533] There is some indication from experiments of Oppenoorth and Houx using both [36]Cl and [14]C-DDT, that *p*-chlorine is lost during the process.[534]

The carbon atom adjacent to the benzylic carbon in dicofol is electron deficient and in such situations decomposition to DBP occurs fairly readily under a variety of conditions. DBP, the corresponding alcohol DBH (Figure 8) and DDE are among the biotransformation products of dicofol. The conversion to DDE can be effected chemically under reductive dechlorination conditions if the tertiary alcohol is first derivatized by acetylation to reduce DBP formation (Vol. I, Chap. 2B.3). This has led recently to speculation that DDE might be formed from dicofol in vivo by an enzyme catalyzed loss of HOCl.[535] There are certainly some interesting areas for exploration here; apart from the link between the oxidizing and dehydrochlorinating enzymes mentioned by Agosin,[531] there is the well-known inhibition of DDT-dehydrochlorinase (DDT-ase) by piperonyl cyclonene, an inhibitor of microsomal oxidations which ought not to affect DDT-ase according to current views.

The direct reductive dechlorination of DDT to DDD has been much discussed and the circumstances of its formation do not always permit easy distinction between enzymatic processes and chemical processes that can occur in the presence of the pigments which are available in biological media. Thus, a recent report indicates the presence *in pigeon liver* preparations of both nonenzymatic and anaerobic enzymatic mechanisms for the formation of DDD and heat denatured preparations apparently dechlorinate effectively when NADPH and riboflavin are present.[536] That the bacterial production of DDD occurs without intermediate formation of DDE appears to be well established, but there is still controversy about the direct conversion in mammals (Figure 8), some investigators maintaining that the compound is only formed anaerobically by bacterial action after death. The problem is complicated by the conversion of DDT into DDD under some of the GC-conditions used for analysis,[150] but the weight of evidence seems to favor direct DDD formation in live mammals and there is evidence that both gut flora and the liver are involved in man.[530] The pathway for DDT metabolism in pigeons only appears to be similar to that in rats as far as DDMU; DDMS is converted (half-life 8 days) into DDMU and DDNU rapidly forms DDNS, an additional, so far unidentified metabolite being formed in each case. DDA formation has not been detected from any of the compounds fed. Both DDE and DDMU cause an increase in liver weight in pigeons and with DDMU some further unknown metabolite is thought to be the causal agent.[526] Following the sequence of reductions and dehydrochlorinations leading to DDNU, a hydration step is required to give DDOH. A crystalline liver alcohol dehydrogenase has been shown to convert DDOH into the corresponding aldehyde, which has been synthesized and shown to convert readily into DBP. The hydration step merits further exploration since the above evidence suggests its absence in the pigeon. There appears to be a close relationship (or identity) between the enzymes that hydrate double bonds and those that hydrate epoxide rings to give the corresponding *trans*-diols, and it may be significant in this context that pigeons and some other avians have rather poor ability to hydrate the epoxide rings of certain labile epoxides of the cyclodiene series that are fairly readily attacked by mammals (see section B.3a, Vol. 2).

A number of early investigations indicated the excretion of DDA in the urine of rabbits, rats, and man, and the acid has been isolated in crystalline form from the urine and feces of rats treated with DDT.[537] The acid was also excreted as a conjugate with amino acids and this last investigation provided chromatographic evidence for the presence in the feces of DBH, DDM, DBP, and DDE. The DDA formed in cabbage plants treated with DDT is present both free and as two unidentified conjugates which degrade to DBH on hydrolysis with methanolic hydrogen chloride; another conjugate gives DDE under these conditions and several more remain to be identified.[486] Thus, the formation of DDE, DDD, DBH, and DDMU, plus DDA and its various conjugates in plants superficially resembles the situation in mammals, although the detailed pathways will take some time to work out. The appearance of measurable amounts of *o,p'*-DDT in feeding trials with DDT has aroused interest because of the estrogenic activity of the *o,p'*-isomer, but there seems to be no real evidence for the chemical isomerization of DDT to this isomer in vivo, its concentration in biological samples being usually appreciably less than is expected from the amount present in the DDT diets used.[537]

Interest has recently revived in certain analogues of DDT that have biodegradable substituents in place of chlorine atoms on the aromatic rings or in the aliphatic portion of the molecule.[38,524] Some of these have been available for many years but have been largely eclipsed by the low cost of DDT and its broad spectrum of insecticidal activity.

Methoxychlor is a good example of such a compound and has been known for a long time to have little tendency to store in fat. It is detoxified by oxidative cleavage of the 4- and 4'-methoxy-groups to give the corresponding mono- and diphenolic compounds and both parent compound and products can be dehydrochlorinated to the corresponding DDE analogues, actual metabolic patterns depending on the species considered. Other examples are the analogues of methoxychlor having 4- and 4'-ethoxy-groups, methylthio-groups (CH₃S) or methyl groups, called ethoxychlor, methiochlor and methylchlor, respectively. Ethoxychlor resembles methoxychlor from a metabolic standpoint. It is dehydrochlorinated and the ethoxy-groups are also oxidatively cleaved; the diphenolic compound produced is itself dehydrochlorinated and a terminal product is 4,4'-dihydroxybenzophenone, resulting from loss of the β-carbon atom. The methylthio-groups of methiochlor are sequentially oxidized in vivo to give the corresponding mono- and di-sulfoxide (CH₃SO) analogues, the sulfoxide-sulfone

(CH$_3$SO, CH$_3$SO$_2$) and the bis-sulfone. Some dehydrochlorination also occurs to give the corresponding DDE analogue, the amount formed depending on the species involved. With methylchlor also, some dehydrochlorination occurs depending on the species, but the methyl groups offer opportunities for oxidation and are converted through the mono- and bis-hydroxymethyl derivatives into the corresponding 4,4'-dicarboxylic acid. These conversions are in accordance with expectations from current knowledge of metabolic processes in living organisms and the extent to which they occur in a particular species determines the level of unchanged toxicant accumulated by that species on exposure. The extent to which biodegradability of DDT analogues may be manipulated by molecular modification is further revealed in Metcalf's investigations of asymmetrical derivatives incorporating various combinations of the para-substituents just discussed.[80a] It is a happy circumstance that the insect is frequently less capable than other organisms of degrading these analogues in vivo, and so is much more sensitive to their toxic action. Thus, the acute oral LD50's of methoxychlor to rats and houseflies are 6,000 mg/kg and 9.0 mg/kg respectively, a more than 600-fold difference (selectivity ratio) in favor of the rat. Other biodegradable DDT-analogues with favorable selectivity ratios have been described by Holan (see section C.1a).

The ultimate breakdown of these metabolic products from higher organisms depends on the ability of microorganisms such as those in soil and sewage to open the rings of complex polycyclic structures and to break them down into simpler structures and eventually into carbon dioxide and water as part of their energy producing processes. According to Wedemeyer,[538] *Aerobacter aerogenes* under anaerobic conditions converts DDT into DDE and also via DDD into DDA and then DBP by a sequence similar to that found in higher organisms, provided that an exogenous energy source such as glucose is available. It also appears that if some of the chlorine atoms are removed from DDT analogues, aerobic organisms can effect the further conversions which result eventually in aromatic ring fission. Thus, a *Hydrogenomonas* sp. isolated from sewage can be grown aerobically on diphenylmethane and diphenylmethanol as sole energy sources, and in the presence of such sources, DDM, DBH,

benzophenone, *p*-chlorobenzophenone, and 1,1-diphenyl-2,2,2-trichloroethane can be cometabolized.[539] The latter gave a product believed to be 2-phenyl-1,1,1-trichloropropionic acid, analogous to the phenylacetic acid produced from diphenylmethane. Thus, a combination of anaerobic sequences permitting chlorine removal with subsequent aerobic ones for ring cleavage may permit the conversion of chlorinated analogues of DDT into simpler structures.[539a] However, the inability of the *Hydrogenomonas* sp. to grow on some of the dechlorinated intermediates, although it cometabolized them, suggests an inability to degrade them completely into carbon dioxide and water. For the reasons outlined, the chemical and biological mechanisms of reductive dechlorination are of considerable importance and merit exploration. Of particular interest is the observation that the chromous chloride reagent employed by Cochrane[64] for derivatization of organochlorine insecticides (Vol. I, Chap. 2B.3) differs from some other dechlorinating agents, such as zinc dust and acid in that it effects some of the conversions observed in living organisms, such as the production of DDMU and DDNU from DDT. Bis (*p*-chlorophenyl) aceto-nitrile (*p,p'*-DDCN) is a significant degradation product when DDT is incubated with anaerobic sewage sludge, but it is not clear whether this product results from microbiological or chemical conversion. DDCN has also been found in the sediment of a Swedish lake.[539b,539c]

Concern about the effects of DDT on small organisms of aquatic food chains has prompted investigations of its uptake and fate in both marine and freshwater algae. All of the six species of marine algae examined by Rice and Sikka converted DDT to DDE in amounts, depending on the species, of from 0.03 to 12% of the total DDT in the cells.[508a]

### b. Lindane (γ-HCH)

The metabolism of lindane and some of its isomers has been studied in relation to insect resistance and mammalian toxicology, and somewhat more recently in relation to the question of persistence of its residues in the environment; even now, much remains to be learned about the metabolic pathways involved. In an extensive comparative investigation with lindane-susceptible and resistant houseflies, Oppenoorth concluded that the rate of breakdown of the isomers decreases in the order $\gamma > \alpha >> \delta$ (other investigations show that $\beta$-HCH is most stable of all) and that in general there is a positive correlation between the ability of a strain to metabolize the nontoxic $\alpha$- and $\delta$-HCH isomers and its resistance to the toxic $\gamma$-isomer.[540] Following their intravenous injection into mice, Van Asperen found that the $\gamma$- and $\delta$-isomers were metabolized quite rapidly and at about the same rate, whereas $\alpha$-HCH was metabolized much more slowly, which agrees with the results of more recently conducted autoradiographic studies on distribution patterns in vivo.[358]

The questions of interest in regard to metabolism concern firstly the mechanisms of aromatization to give various chlorobenzenes and their derivatives and second, the ultimate fate of these aromatic products. The second question clearly relates to the fate of other chlorinated aromatic compounds such as polychlorobiphenyls and certain herbicides based on chlorinated phenol. There has been much interest in the formation of $\gamma$-1,3,4,5,6,-pentachlorocyclohexene ($\gamma$-PCCH; Figure 9) on the insect metabolic pathway for lindane and it is still not entirely clear whether this compound is a labile intermediate or secondary product; the weight of evidence seems to favor the latter alternative for houseflies, but important species differences are likely. The situation has become more complex recently following the detection of a further housefly metabolite, said to be another PCCH isomer, which Reed and Forgash called "Iso-PCCH."[541] $\gamma$-PCCH itself is metabolized quite rapidly by houseflies, 1,2,4-trichlorobenzene (Figure 9, structure 11; 1,2,4-TCB) and 1,2,4,5-tetrachlorobenzene (1,2,4,5-TECB) being the only internal lipophilic metabolites detectable following topical application; 1,2,4-TCB; 1,2,3-TCB; 1,2,4,5-TECB; 1,2,3,4-TECB and pentachlorobenzene (PCB) are products from "iso-PCCH," which is metabolized more slowly.[542] The tetrachlorobenzenes, 1,2,4-TCB, and PCB appear to be stable when injected. Small amounts of the same compounds are produced from lindane by resistant houseflies and the appearance of these chlorobenzenes retaining much of the chlorine seems to favor the occurrence of direct aromatization (desaturation) of PCCH (giving PCB) or following the formation from PCCH of the various isomers of tetrachlorocyclohexadiene that can arise from it by dehydrochlorination.

In blowflies, bis($N,N$-dimethylaminophenyl)methane blocks the metabolism of $\gamma$-PCCH without apparent effect on lindane degradation (see section B.3b), indicating that the former is not an intermediate in lindane metabolism in this insect. Houseflies and grass grubs (*Costelytra zealandica*) and their tissue homogenates plus glutathione (GSH) convert $\gamma$-HCH into substantial amounts of a compound which appears to be 2,4-dichlorophenylglutathione; it is also produced from $\gamma$-PCCH by housefly homogenates and is a major $\gamma$-HCH metabolite in cattle ticks and locusts. All of this suggests that in the insects there is direct enzyme mediated attack by GSH on lindane to give $S$-pentachlorocyclohexylglutathione (not so far shown to be formed) which may subsequently lose chlorine to give the various $S$-dichlorophenylglutathiones, the 2,4-dichloro-derivative usually predominating. Neither chlorocyclohexane nor the HCH isomers are substrates for an enzyme, thought to be a glutathione $S$-transferase, which metabolizes $\gamma$-PCCH in the rat and the $\gamma$-PCCH metabolizing enzyme in houseflies differs from an $S$-aryltransferase, also present, which conjugates GSH with 1-chloro-2,4-dinitrobenzene.[484]

The nature of the aromatization process in houseflies is intriguing. Soluble enzymes from houseflies are reported to liberate appreciable amounts of chlorine from $\gamma$-HCH without consumption of glutathione or conjugate formation. On the other hand, the glutathione conjugate formed by such housefly preparations from $\gamma$-PCCH was indistinguishable by paper chromatography from the conjugate formed by rat liver soluble enzymes with replacement of one chlorine atom. Thus, assuming that the paper chromatographic evidence of identity is valid, it seems that the same housefly preparation shows very different behavior towards $\gamma$-PCCH and $\gamma$-HCH; it might be supposed that the former, having two allylic chlorine atoms, would be rather more labile than $\gamma$-HCH. Clearly, this particular aspect of HCH metabolism merits further examination. The low activity of rabbit and rat liver $S$-transferase enzymes toward lindane suggests that in these mammals $\gamma$-PCCH is formed initially and may be converted via 1,2,4-trichlorobenzene and an epoxide intermediate (Figure 9, structure 15) into the 2,3,5- and 2,4,5-trichlorophenols which are excreted both free and as sulfates and glucuronic acid conjugates.[543] 2,4-Dichloro-

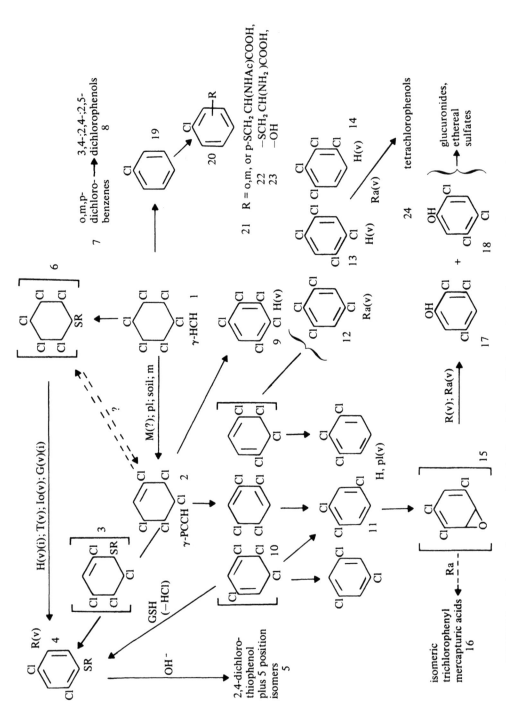

FIGURE 9.   Pathways (some speculative) of metabolism of lindane and related compounds. G = grass grub; H = housefly; lo = locust; M = mammal; m = microorganisms; pl = plants; Ra = rabbit; R = rat; T = cattle tick; v = in vivo; i = in vitro; in structure 4, –SR = glutathione for H, T, lo and G; mercapturic acid for the rat.

phenylmercapturic acid is also formed from γ-PCCH and probably also from γ-HCH in rats and might arise through the conjugation of GSH with either γ-PCCH or its dehydrochlorination product 1,3,5,6-tetrachlorocyclohexa-1,3-diene (Figure 9, structure 10). In contrast, rabbits treated with 1,2,4-trichlorobenzene, which appears to be on the metabolic pathway from lindane, excrete only traces of trichlorophenylmercapturic acids in addition to the trichlorophenols (TCP) and their conjugates. Recent work suggests that γ-PCCH (and another isomeric PCCH), trichlorobenzene, 1,2,3,4-tetrachlorobenzene, 1,2,3,5-tetrachlorobenzene and 1,2,4,5-tetrachlorobenzene are formed along with di-, tri-, and tetrachlorophenols (TECP).[544] Thus, the possibility exists of aromatization of partly dechlorinated intermediates, as appears to be the case in insects.

The production of tetrachlorophenols is partic-ularly interesting since it shows that oxidations (presumably via epoxides) can occur even with the tetrachloro-intermediates, and indications from other work that γ-PCCH (but not γ-HCH) is metabolized oxidatively by rabbit and rat liver microsomal preparations may connect with these findings. Thus, phenol formation from lindane or its isomers need not necessarily follow aromatization as Figure 9 implies. Oxygen might first be introduced by the epoxidation of alicyclic intermediates such as γ-PCCH. Alternatively, processes such as allylic oxidation (hydroxylation) with double bond shift, or the hydrolytic removal of allylic chlorine atoms, might lead to the formation of 2,3,4,5,6-pentachloro-2-cyclohexen-1-ol (PCCOL), 2,4,6-TCP, 2,3,4,5-TECP and 2,3,4,6-TECP as shown in Figure 9a. All of these compounds have recently been claimed as lindane metabolites in rats.[544a] The schemes in Figure 9a

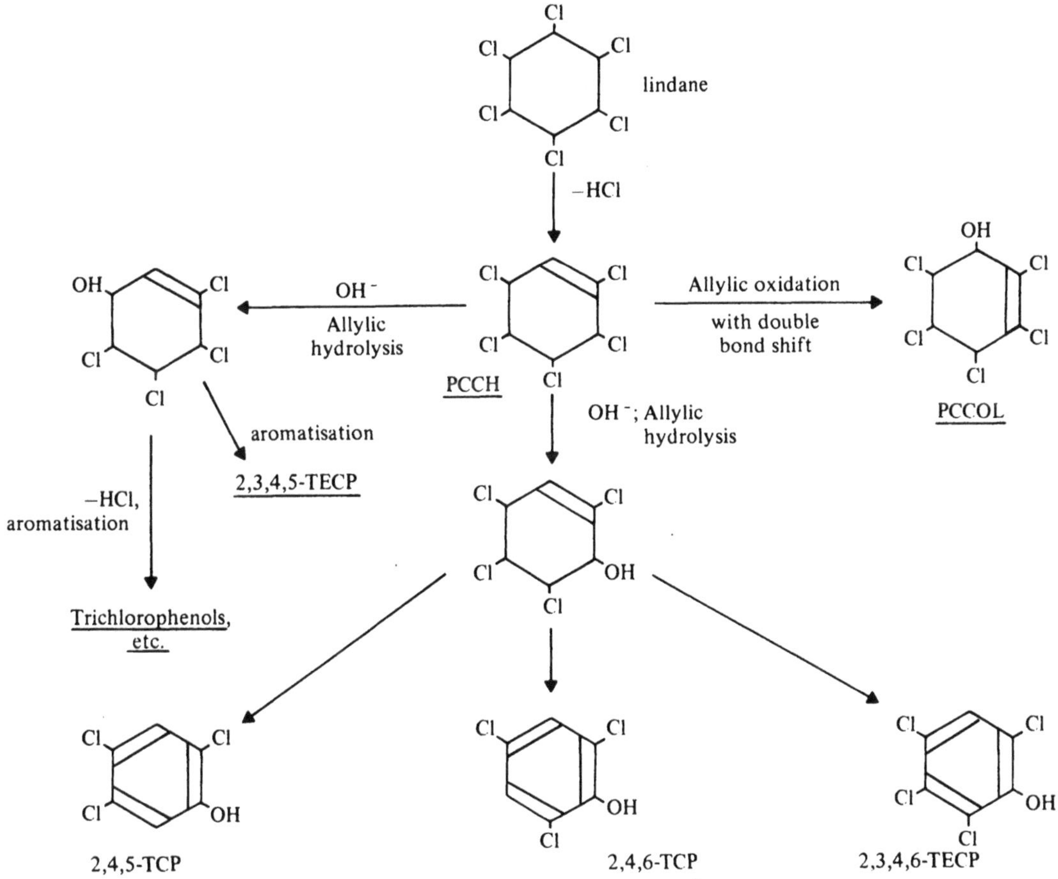

FIGURE 9a.  Possible mechanisms of lindane degradation in mammals. Such mechanisms provide alternative routes to some of the metabolites in Figure 9. However, PCCH has not yet been proved to be an intermediate in lindane metabolism.

favor γ-PCCH as an intermediate in lindane metabolism in mammals but there is so far no direct evidence for its formation.

There is evidence for the formation of monochloro- and dichlorobenzene as metabolites of HCH in insects, and the locust, for example, is known to form acid labile precursors of the various chlorophenyl-cysteine conjugates derived from chlorobenzene; mammals form the corresponding mercapturic acids (Figure 9, structure 21). Rabbits treated with dichlorobenzenes excrete conjugates of the various derived monophenols (Figure 9, structure 8) and the fate of a number of likely terminal lindane metabolites has long been known. Therefore, as far as mammals and insects are concerned, the elimination of the lower chlorinated aromatic metabolites apparently presents few problems and recent work on mammalian metabolism suggests that even the more highly chlorinated molecules can be oxidized although this must inevitably be more difficult.

Information about the nature of HCH metabolites in plants or microorganisms appears to be fairly limited at present, but γ-PCCH and 1,2,4-TCB have been identified as lindane metabolites in maize and beans grown in a nutrient medium containing it, and 1,2,3-TCB is also produced by young maize plants under these conditions.[244] γ-PCCH and 1,2,3,5-tetrachlorobenzene are among the products of metabolism by microorganisms,[485] and it is important to determine the extent to which they produce and can degrade pentachlorobenzene and the tetrachlorobenzenes, which seem to be the most likely environmental contaminants from lindane. The simple chlorophenols suffer hydroxylation to catechols followed by benzene ring cleavage. For example, 2,4-dichlorophenol, which is a likely degradation product of lindane and also a degradation product of the herbicide 2,4-dichlorophenoxyacetic acid, is converted first into 3,4-dichlorocatechol and then undergoes ring cleavage by soil bacteria with final liberation of the chlorine as chloride and the formation of succinic acid.[545]

What is known about the metabolism of the other HCH isomers indicates that they follow pathways similar to lindane. The main environmental problem arises from the use of technical HCH, which may contain 10% of the persistent β-HCH, the isomer which lacks the *trans*-relationships between adjacent hydrogen and chlorine that facilitate dehydrochlorination. Therefore, this isomer is degraded only slowly and is the main residue found in crops and in the tissues of living organisms. The problem has become particularly acute in Japan where, for economic reasons, technical HCH has been used at rates up to 28 lb/acre/year in order to achieve the desired concentration of the effective γ-isomer. Irradiation with light of various wavelengths down to 230 nm shows that small amounts of α-HCH are formed as the major transformation product from the other isomers, so that there is no apparent addition to the β-isomer residues by isomerization.

*c. Heptachlor, Chlordane, Isobenzan, and Endosulfan*

The metabolism of these insecticides is a complex matter because of the stereochemical possibilities afforded by the three-dimensional molecules (Vol. I, Chap. 3B.4). Figures 10 and 11 show the salient pathways of metabolism for heptachlor, chlordane, and the heterocyclic molecules isobenzan and endosulfan. The formation of a stable epoxide from heptachlor (Figure 10, structure 1) in vivo was indicated by Davidow in 1951 and the product (Figure 10, structure 2; mp 160°C, called HE 160) was soon isolated from rats and dogs and chemically characterized.[233] This epoxidation was soon found to be of general occurrence in the biosphere and seems to have attracted attention to the exclusion of other possible reactions of the molecule, although it is well known that allylic chlorine atoms such as that at $C_1$ of heptachlor are subject to replacement by other groups, such as hydroxyl. Accordingly, there appears to have been little interest in this aspect of heptachlor metabolism until the middle 1960's, when hydrolytic replacement of the allylic chlorine atom was shown to occur by chemical hydrolysis in water and also probably in the tissues of mosquito larvae in the water.[546] Experiments about the same time showed that the conversion to 1-hydroxychlordene (Figure 10, structure 8) also occurred in soils and investigations of the structure-activity relations of chlordene (structure 10) relatives in houseflies demonstrated the series of oxidative conversions[243] which have since proved generally relevant to the fate of heptachlor as a soil insecticide.

Chlordene (10) is 50 to 100 times less toxic than heptachlor to houseflies, and the reason for this appears to relate largely to differences in

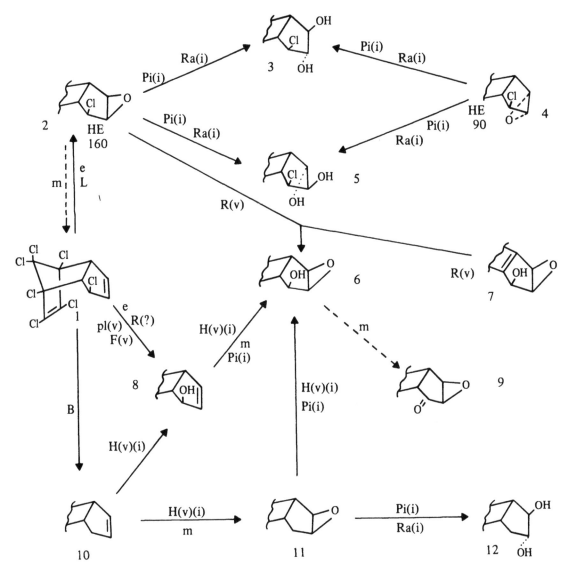

FIGURE 10. Environmental and metabolic conversions of chlordene and heptachlor and their relatives. B = soil bacteria; e = environmental (atmospheric oxidation, etc.); F = fish; H = housefly; m = microorganisms; Pi = pig; pl = plants; Ra = rabbit; R = rat; L = living organisms in general; i = in vitro; v = in vivo.

metabolism due to the presence of the allylic chlorine atom in heptachlor. In its absence, the molecule is oxidized to the hydroxy-epoxide (6) via the epoxide (11) or via 1-hydroxy-chlordene (8); all three compounds appear in the tissues and the hydroxy-compounds are not toxic to houseflies. An analogy with environmental transformations is evident since exposure of chlordene to air and light gives a complex mixture from which some of these compounds can be isolated.[243] The epoxide (11) is toxic but its true toxicity (intrinsic toxicity) is not shown in ordinary toxicity evalua-

tions on account of the enzymatic oxidative detoxication to hydroxy-epoxide (6). The methylenedioxy-benzene derived pyrethroid synergist sesoxane (sesamex; 2-(3,4-methylenedioxyphenoxy)-3,6,9-trioxaundecane) blocks these enzymatic oxidations (effected by microsomal mixed function oxidase enzymes) in vivo so that the epoxide accumulates in the tissues when both are applied to houseflies together and there is a consequent tenfold synergism (potentiation of toxicity) of the epoxide. Synergized chlordene epoxide is less toxic than HE 160 (but more toxic

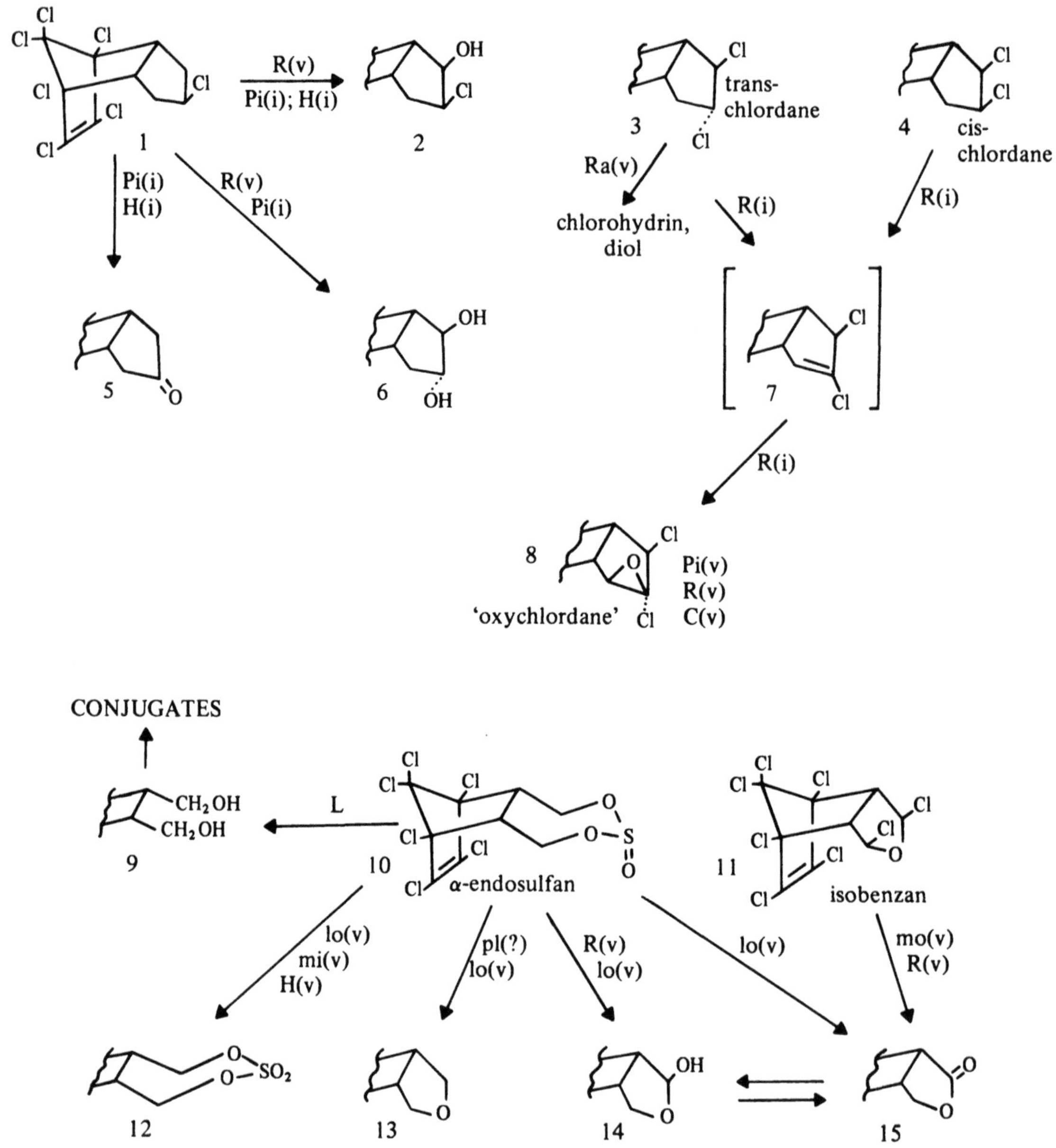

FIGURE 11. Metabolic conversions of β-dihydroheptachlor, cis- and trans-chlordane, the endosulfan isomers (only the α-isomer is shown), and isobenzan. H = housefly; lo = locust; mi = mice; mo = mosquito larvae; Pi = pig; pl = plants; C = cow; Ra = rabbit; R = rat; L = living organisms generally; i = in vitro; v = in vivo.

than the stereoisomer HE 90; Figure 10, structure 4) and this might be associated with some residual degree of oxidative detoxication or ring hydration to give the *trans*-diol (12), or a contribution of the $C_1$ chlorine atom in HE 160 to toxicity due to other factors (steric ones for example) unrelated to metabolism. It seems clear, however, that the major contribution of this chlorine atom is to prevent hydroxylation at $C_1$ in vivo.

Hydrolytic opening of the epoxide ring in (11) to give the *trans*-diol (12) occurs to a very limited extent if at all in the housefly and not being oxidative, this reaction is little affected by sesamex (see section B.3a). Such epoxide ring opening is effected by mammalian liver preparations in vitro, however, and it is likely that this reaction occurs in vivo although such experiments have not been performed with chlordene epoxide.[547] The insertion of chlorine at $C_1$ prevents the hydration of HE 160 by liver preparations from pig or rabbit and slows the similar reaction of HE 90 very considerably (in comparison with the corresponding molecule lacking $C_1$-chlorine). Accordingly, it is not surprising that HE 160 is rather stable in insects, vertebrates, and other living organisms. Following intravenous injection of [14]C-heptachlor, male rats excreted 22% of it within 72 hr and male rabbits 23% within 144 hr. The excreta of these animals contained HE 160 and larger quantities of a metabolite indicated to be the hydroxy-epoxide (Figure 10, structure 6). This may arise by hydrolytic replacement of chlorine in HE 160 or through 1-hydroxychlordene as a transient intermediate, and in these experiments appeared to be the terminal product of heptachlor or HE 160 metabolism.[548] In a different investigation, rats fed a total of 5 mg of HE 160 at a dietary level of 10 ppm excreted about 20% of the dose in the feces during 30 days and most of this appeared to be a different epoxide assigned the desaturated structure (7) on the basis of physical methods and catalytic hydrogenation to a compound chromatographically similar to epoxide (6).[549]

The formation of heptachlor epoxide as a residue following the application of heptachlor to plants was first observed many years ago, and later work showed that heptachlor is absorbed through the roots and can be epoxidized in the plant. Studies of the fate of heptachlor applied to the leaves of cabbage have revealed the formation of 1-hydroxychlordene, heptachlor epoxide, and another, unidentified conversion product having polarity intermediate between the two. Of a number of microorganism isolates from soil examined for their ability to transform heptachlor, 35 out of 47 fungi and 26 out of 45 bacteria and actinomycetes effected epoxidation.[284] 1-Hydroxychlordene produced by the chemical hydrolysis of heptachlor in aqueous culture media was further converted into the hydroxy-epoxide (Figure 10, structure 6) by 43 of the fungi and four of the bacteria and actinomycetes examined; HE 160 does not appear to be hydrolyzed directly to this compound, but a mixed culture of soil microorganisms in aqueous medium converted it into 1-hydroxychlordene (about 1% per week), possibly by initial deoxygenation to heptachlor followed by hydrolysis.[287] Soil bacterial activity also converts an appreciable amount of heptachlor into chlordene by reductive dechlorination and in laboratory experiments there is a lag period before the epoxidation of heptachlor and chlordene begins.

Recent surveys of soil residues in the United States indicate 1-hydroxy-chlordene to be a major residue in soils from five areas, with only small amounts of heptachlor epoxide present, and the hydroxy-epoxide (6), which is likely to be the terminal oxidation product of all these reactions (although further conversion to the corresponding ketone (9) is a possibility) has been found in Oregon soil that had been treated with technical heptachlor.[285] Since 1-hydroxychlordene production from heptachlor is presumably an example of allytic halide hydrolysis, its production instead of heptachlor epoxide may correlate with rainfall. Bonderman and Slach, who recently found 1-hydroxychlordene residues in soil, crop, and fish tissue samples, indicated a half-life of 3 weeks for this derivative in soil.[286] Its occurrence in fish is interesting since there was no evidence for its presence in human tissue, air, or water samples. Neither 1-hydroxychlordene nor the hydroxy-epoxide (6) is toxic to houseflies, nor the former to crickets; the mammalian toxicity of the latter has not been recorded but 1-hydroxychlordene (acute oral LD50 2400 to 4600 mg/kg) is clearly a detoxication product relative to heptachlor (LD50 90 to 135 mg/kg) and heptachlor epoxide (LD50 47 to 61 mg/kg) in rats. In total, these investigations reveal the existence of "opportunity factors" for the degradation of heptachlor before its epoxidation occurs, show that even the epoxide is not entirely inert, and indicate the need for sophisticated investigations under a wide range of environmental conditions before final judgments are passed on persistence.

If the double bond of heptachlor is absent as in the reduced position isomer β-dihydroheptachlor (β-DHC; Figure 11, structure 1), detoxicative hydroxylation mechanisms replace epoxidation, so that chlorohydrins and ultimately diols and probably ketones can be formed; the saturated ring system is quite vulnerable to such attack, as shown by the short half-life of β-DHC in rats. With additional chlorine substitution to give trans- and cis-chlordane (Figure 11, structures 3 and 4, respectively), enzymatic attack becomes more difficult, but by no means impossible. Thus, rabbits fed with $^{14}$C-*trans*-chlordane (14.3 mg daily) for 10 weeks had excreted 70% of the administered compound in urine (47%) and feces (23%) by the twelfth week. Of the two hydrophilic compounds appearing in the urine, one is thought to be the diol produced by replacement of the $C_1$ and $C_2$ chlorine atoms by hydroxyl groups and the other is assigned the structure of a 1-hydroxy-2-chlorodihydrochlordene, which was not further characterized.[508] Similar changes are likely to occur with "open chain" analogues of the chlordane isomers such as Alodan® or Bromodan® (Figure 16, structure 1; x = y = $CH_2Cl$ and x = H, y = $CH_2Br$, respectively), and in such cases, the ultimate conversion of halomethyl groups into carboxyl groups is possible. In accordance with this view, acidic metabolites have been found in the urine of rabbits, rats, and guinea pigs treated with Alodan, but no metabolite identities have been reported. An interesting development in relation to metabolism of the chlordane isomers is the recent observation that both are converted in vivo in mammals into the epoxide called "oxychlordane" (Figure 11, structure 8) which is stored in fat, the conversion and storage being rather higher when the *trans*-isomer is fed. With rats, the usual sex difference is apparent; females store considerably more than males, but in any case, the tendency to storage is not high. In the presence of NADPH and oxygen, rat liver homogenates produce this metabolite from either of the chlordane isomers and also from 1-*exo*,2-dichlorochlordene (Figure 11, structure 7), which appears to be an intermediate in the conversion. This intermediate presumably results from an enzymatic removal of hydrogen (desaturation) from (3) or (4) (easiest from structure 3), which has exposed *cis*-hydrogens and is converted into the metabolite by epoxidation. The metabolite isolated from pig fat is optically active when the *trans*-, but not the *cis*-, isomer is fed, indicating at least partial stereospecificity of enzymatic attack on *trans*-chlordane.[550-552]

The colorimetric procedures used for HE 160 (Vol. I, Chap. 3C.) give similar but less intense colors with "oxychlordane" (8), which may explain Davidow's original observation[553] that according to this method the individual components of technical chlordane were all stored in fat as the same metabolite. This early observation once led

the author to an inconclusive search for epoxide formation as a cause of toxicity of the chordane isomers toward houseflies. Since oxychlordane is reported to be rather more toxic than *trans*-chlordane,[552] the question might bear reexamination.

The nature of the water-soluble metabolites formed from chlordane in plants is yet to be determined, and the situation in this whole series of compounds is complicated by the photoisomerization possibilities discussed in the chemical sections (Vol. I, Chap.3B.4b). A number of these products are said to arise under practical conditions; little is known about their toxicity or further metabolism in most cases so that their toxicological significance is uncertain. With products such as photoheptachlor (Figure 13, structure 4), which have secondary chlorine atoms, oxidative dechlorination to the corresponding ketones has been indicated (compare photodieldrin metabolism, Figure 12), and has been suggested as a cause of the higher toxicity and rapid action of photoheptachlor and photodieldrin (as compared with heptachlor and dieldrin) to insects and aquatic organisms.[554]

Isobenzan and endosulfan are heterocyclic molecules that have the bis-hydroxymethyl derivative of hexachloronorbornene (Figure 11, structure 9) as a common precursor and some aspects of their metabolism relate to this common origin. However, endosulfan is the cyclic sulfite ester of the diol; this makes it biodegradable and completely different toxicologically from isobenzan, which resembles heptachlor epoxide in some respects. Since isobenzan is no longer marketed, its metabolism has only been of passing interest. Larvae of the mosquito *Aedes aegypti* kept in an aqueous suspension of $^{14}$C-isobenzan converted most of it into hydrophilic material which was found in both larvae and medium. One of the hydrophilic metabolites gave on hydrolysis the lactone (Figure 11, structure 15), which was similarly obtained from a rat fecal metabolite of $^{14}$C-isobenzan.[508] This implies the formation of the corresponding γ-hydroxy-acid following hydrolytic replacement of the chlorine atoms, but other mechanisms are possible and no information is available on this point.

The endosulfan isomers are oxidized in houseflies to the sulfate (12), which is as toxic as β-endosulfan and the α-isomer is said to be oxidized faster than the β-isomer. Thus, the situation is somewhat analogous to the aldrin-dieldrin conversion except that the sulfate is further metabolized and excreted. Hydrolytic and oxidative mechanisms are evident in adults of the locust (*Pachytilus migratorius migratorioides*) which excrete endosulfan sulfate, together with endosulfan ether (Figure 11, structure 13), α-hydroxy-endosulfan ether (14), and its oxidation product, the lactone (15) already referred to. Following its intraperitoneal administration to rats, endosulfan diol and an unidentified compound were excreted as conjugates. Other work has indicated the excretion of the α-hydroxy-ether (14) in rat urine. Mice treated with endosulfan stored the sulfate transiently in their fat and excreted endosulfan, the sulfate and the diol (9) in the feces. Dogs given the endosulfan isomers orally for 28 days excreted 13 to 25% of the unchanged compounds in the feces and a little in the urine. Endosulfan sulfate appeared in the fat but other tissues were residue-free and no metabolites other than the sulfate were found. The general picture seems to be that when endosulfan is ingested the sulfate appears briefly in the tissues, especially fat, and may appear in the milk of the animals producing it, but such residues disappear rapidly when exposure ceases; the half-life of excreted products in the urine and feces of sheep given a single treatment of 14 mg/kg of [14]C-endosulfan was about 2 days and the radioisotope level in milk fell to negligible proportions within 4 days. It is evident that endosulfan is a nonpersistent compound in mammals. Endosulfan sulfate and its ether (13) have been reported in extracts of foliage treated with the isomeric mixture and hydrolysis to the diol has been claimed but not substantiated. Laboratory experiments have shown that the sulfate is formed when endosulfan on plant surfaces is exposed to UV-light in the presence of water and so this oxidation is quite likely to occur in sunlight under moist conditions.[248,486]

The stability of endosulfan in aqueous environments is greatly affected by the presence of the cyclic sulfite ester group and hydrolysis to the diol seems to be the dominant chemical and biological reaction. Hydrolysis to this diol, which is nontoxic to rats (acute oral LD50 > 15,000 mg/kg) and guppies (*Lebistes reticulatus*), occurs in water and increases as the pH increases. Bacterial degradation is indicated by experiments simulating sewage treatment plants, which showed that the organisms present tolerated 100 to 250 ppm of endosulfan and converted 70% of it almost exclusively into the diol within 7 days of treatment; a 97% recovery of applied [14]C-endosulfan in such experiments shows that the formation of other metabolites is minor. The effect of endosulfan on aquatic organisms is of special interest because it is highly toxic to some species of fish and is extensively used for the protection of rice against rice stem borer in a situation which involves a complex relationship (ecosystem) between plant and aquatic life. Toxic concentrations in water to various species of freshwater fish are (24-hr LC100 ppm) Trout (*Salmo gairdnerii*) 0.01; Pike (*Esox lucius*) 0.005; Carp (*Cyprinus carpio*) 0.01, 48 hr; Goldfish (*Carassius auratus*) 0.01; Guppy (*Lebistes reticulatus*) 0.009; "Tawas" (*Puntius javanicus*) 0.0013; and Swordtail (*Xiphophorus helleri*) 0.005 to 0.01. Goldfish exposed for 5 days to a daily change of water containing 0.001 ppm of endosulfan acquired an average tissue residue of 0.35 ppm and when exposure ceased, 98% of the initial tissue endosulfan was excreted within 14 days as the free diol.[298] Evidence has also been presented that the diol is excreted as a glucuronide in the bile of some fishes. Other aquatic species are generally much less sensitive than fish. Thus, for various aquatic insect larvae the 24-hour LC100 lies between 0.015 and 1.5 ppm, the last figure referring to mosquito (*Aedes aegypti*) larvae; amoebas and ciliates tolerate 20 to 30 ppm in the medium, while various green algae such as *Chlorella* and *Scenedesmus* have LC100's of 40 to 80 ppm, and crustacea such as *Daphnea, Cyclops,* and *Asillus* spp. of 0.1 to 1.5 ppm. Algae rapidly absorb endosulfan and, like other aquatic species, appear to produce the diol rather than the sulfate. Various appraisals of the environmental toxicology of endosulfan have been made recently.[296,298,299,555]

### d. Aldrin and Dieldrin

The biotransformations of aldrin and dieldrin are summarized in Figure 12 and there is evidently some overlap with the chemical transformations effected by oxygen, light, or heat that were mentioned in the chemical sections of Volume I (see also Figure 13). Stability of the hexachloronorbornene nucleus has always been a distinguishing feature of the cyclodienes, but these molecules, like steroids, are reduced polycyclic systems and it would be surprising if they were not

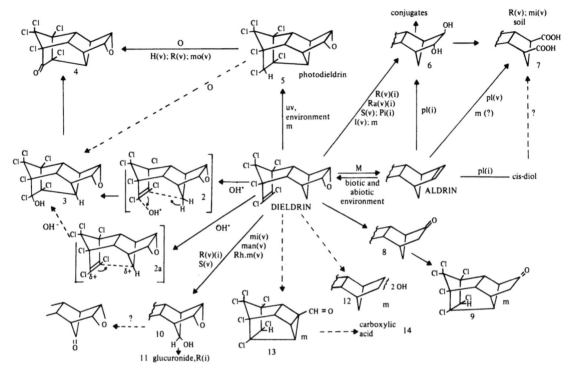

FIGURE 12. Metabolic conversions of aldrin, dieldrin and photodieldrin. H = housefly; I = other insects; m = microorganisms; mi = mice; mo = mosquito; Pi = pig; pl = plants; R = rat; Ra = rabbit; Rh.m = Rhesus monkey; S = sheep; i = in vitro; v = in vivo.

attacked in some way by the microsomal mixed function oxidases that are implicated in the primary biotransformation of so many other compounds.[556] In fact, the main metabolic routes are now seen to involve simple oxidation of the skeleton, rearrangement leading to or concurrent with oxidative or reductive dechlorination of the hexachloronorbornene nucleus, isomerization without further change, hydrolysis of epoxide rings, and cleavage of the unchlorinated ring system.

The well-known epoxidation reaction of aldrin is a typical mixed function oxidation that is effected, for example, by enzymes of the smooth endoplasmic reticulum of mammalian liver (see section B.3a) but this was not established until 1965,[557,558] some 10 years after the epoxidation was found to be of general occurrence in living organisms. For a long time, the epoxide HEOD (dieldrin) was popularly considered to be the terminal metabolite, although indication of its further breakdown in rats was obtained in 1953. In 1962, Heath[559] reported that mice injected with $^{36}$Cl-dieldrin subsequently excreted hydrophilic metabolites and the metabolism was con-

firmed by several other investigations around that time. The 1964 report of Heath and Vandekar contained accurate speculations regarding the general nature of the main metabolites, but these were not investigated in detail.[560] In 1965, the *trans*-diol (Figure 12, structure 6) was isolated in optically active form from the urine of dieldrin-treated rabbits;[561] 3 years later, a monohydroxy-derivative was isolated from rat feces[562] and the ketone (4) from the urine.[562,563] At first, the monohydroxy compound was thought to be a tertiary alcohol formed by hydroxylation of dieldrin at $C_{4a}$ or $C_5$, but a similar compound found in the urine of dieldrin-treated sheep is oxidized to a ketone by chromium trioxide;[564] reexamination of the rat metabolite confirmed this reaction and showed that the same ketone is formed in each case so that the original metabolite must be one or the other of the two possible methano-bridge ($C_9$ using the dimethanonaphthalene nomenclature) hydroxylation products, the 9-*syn*-hydroxy-structure shown (Figure 12, structure 10) being indicated by physical methods and by synthesis.[564a]

It now seems clear that 9-hydroxy-HEOD is a

major excretory product in rat, mouse, sheep, Rhesus monkey, and man.[565,565a] Moreover, the *trans*-diol (6) has now been found in the feces of both rats and mice, and the dicarboxylic acid (7) is known to occur in their urine. Rats given the diol, which is the most obvious precursor of this acid, excrete mostly diol but also an unidentified conjugate of the diacid (6% conversion of a 210 mg dose), whereas rabbits simply excrete the diol in either free or conjugated form.[566] The formation of this acid represents the first example of the cleavage of a norbornene ring system in this series of compounds, and the same acid is found among the hydrophilic metabolites associated with potatoes, maize, and sugar beet treated with aldrin and in the drainage water from aldrin treated soil.[486] It is interesting that *cis*-diol formation from dieldrin has recently been observed in rat liver preparations.[567]

9-Hydroxy-HEOD (Figure 12, structure 10) is readily conjugated with glucuronic acid in the presence of a rat liver glucuronyl-transferase in vitro and the product is difficult to hydrolyze with either β-glucuronidase or aqueous hydrochloric acid. The close proximity of $C_9$ and the dichloroethylene moiety in dieldrin suggests that 9-hydroxy-HEOD and the ketone (4) arise from alternative modes of attack of an electron deficient species such as $OH^+$ or OH on the molecule, which would be consistent with the involvement of mixed function oxidases and the inhibition of the conversion by sesamex in vitro. One possible mechanism involves the sequence dieldrin → (2), (3), and (4) in Figure 12. The alternative suggestion of McKinney[568] that a resonating carbonium ion (Figure 12, structure 2a) is a common precursor might be partly reconciled with microsomal oxidase involvement if the electron deficient species (say $OH^+$) instead removes hydride ion from $C_9$, the resulting carbonium ion then being neutralized by attachment of hydroxyl ($OH^-$) ion at $C_9$, or at $C_2$ with rearrangement. However, the ketonic oxygen comes from molecular oxygen in the first mechanism only; in the second it would come from water. Photodieldrin (Figure 12, structure 5) is produced by the rearrangement of dieldrin in sunlight, as on plant surfaces, for example,[569] and also by microorganisms isolated from soil, water, rat intestines, and rumen stomach contents of a cow.[570] It is oxidatively dechlorinated in vivo in rats, mosquitoes and houseflies to give the ketone (4) and this conversion in houseflies is inhibited by sesamex, supporting the idea that it is mediated by mixed function oxidases.[571,572] This conversion has not yet been demonstrated in vitro, although ketone formation from dieldrin occurs in rat liver microsomes plus the soluble fraction of the cells.[573]

Photodieldrin has a biological half-life of 2 to 3 days in rat adipose tissue, as compared with 10 to 13 days for dieldrin (Table 15) and the compounds are evidently handled rather differently since for either sex, urinary excretion seems to be much more important with photodieldrin and its metabolites than with dieldrin and its metabolites. Extremely high levels of $^{14}C$ appear in the kidneys of male rats receiving $^{14}C$-photodieldrin and the principal metabolite is ketone (4), whereas the females excrete considerably less material which contains at least four hydrophilic metabolites but no ketone. Castrated males adopt the female excretion pattern; the male pattern is restored in the castrates or induced in normal females by testosterone treatment and the effect seems to relate both to enhanced ability to excrete the ketone and to stimulated production of the ketone from photodieldrin.[574,575] Testosterone is well known to have similar effects on the metabolism of other foreign compounds in rats, the sex difference in drug metabolism rates being well established for this species.

Hydration of dieldrin to give the *trans*-diol (Figure 12, structure 6) in small amounts seems to be quite general and besides occurring in mammals has now been indicated to occur in mosquitoes, houseflies, cockroaches, and some microorganisms. The conversion appears to be effected by microsomal enzymes and some avians such as the pigeon have poor ability to effect this enzymatic conversion.[576] Furthermore, the lack of an effective epoxide hydrase in some insects, such as tsetse flies, appears to correlate with their susceptibility to certain dieldrin analogues that are readily hydrated and detoxified by others (see section C.1b).[484]

The characterization of cyclodiene insecticide metabolites in mammals has been a long and tedious process and it is not surprising that apart from the epoxidation reaction, little is known of the metabolic fate of these insecticides in other vertebrates such as birds or in wildlife generally, although balance studies indicate that metabolism is occurring. Thus, HEOD is eliminated from the adipose tissue of pigeons (fed 50 ppm of HEOD in the diet for 6 months) with a half-life of 47 days when feeding ceased,[576a] and most of the dieldrin accumulated by fish from a 0.03-ppm concentration in water was eliminated within 16 days after the exposure; the half-life of DDT was more than 32 days in these circumstances. Some mammalian pharmacokinetic data are listed in Table 15. Detailed toxicological investigations have only been possible with photodieldrin so far,[514] since insufficient quantities of the other metabolites or transformation products have been available. The pentachloro-ketone (Figure 12, structure 4) appears to be more toxic than dieldrin to houseflies and mosquitoes but no other evaluations have been possible since no convenient chemical synthesis is available for this compound at present. Toxicity evaluations on the *trans*-diol (6), diacid (7), and 9-hydroxy-derivative (10) show that these are much less toxic than dieldrin and may be regarded as detoxication products (see section C.1b).

The microbial conversion of aldrin into unidentified hydrophilic metabolites as well as dieldrin was established by Korte and his associates in 1962,[577] although the epoxidation in soil was already known to be at least partly microbial. Dieldrin was apparently not further converted and a later screen of more than 500 soil isolates afforded ten cultures of *Trichoderma, Pseudomonas,* and *Bacillus* spp. that degraded 1 to 6% of added dieldrin to hydrophilic compounds within 30 days. A pseudomonad from the Shell factory yards in Denver gave a mixture of compounds including aldrin, hydroxylated derivatives, ketones, and acidic material (Figure 12, structures 8, 9, 12, 13, and 14), of which aldrin, and the ketones (8) and (9) are positively identified.[578] With the exception of the aldehyde (13) and acid (14), which have lost one carbon atom, these metabolites retain the full cyclodiene skeleton and their toxicological properties are unknown. The formation of photodieldrin from dieldrin by microorganisms was mentioned previously and these organisms obviously have an ability to effect molecular rearrangements which is not often seen in the higher forms.

Algae, protozoa, coelentrates, worms, arthropods and molluscs are able to convert aldrin into dieldrin in vivo, indicating that these organisms may possess enzymes of the microsomal mixed function oxidase type. Khan has characterized such a system from the green gland of the crayfish and has shown that for this animal, a snail (*Lymnaea palustris*) and the clam, 80% of the total oxidase activity of the livers is present in the microsomal subcellular fraction.[578a]

In the last few years, attempts have been made to isolate fractions from plant tissues that will effect the epoxidation reactions of cyclodienes which occur in vivo. In preparations from pea roots examined by Lichtenstein and Corbett,[504] about half the low epoxidase activity, specific for aldrin and not requiring pyridine nucleotide cofactors, appeared in the 105,000-g supernatant, the remainder being distributed between nuclei, mitochondria, and microsomes (defined only by the centrifugation technique). In a more recent investigation, the 22,000-g supernatant (expected to contain the microsomal fraction) from a bean root homogenate converted added aldrin into dieldrin (12.4%) and a compound chromatographically similar to the *trans*-diol (1.6%) after 6 hr incubation at 37°C.[505] Similar results were obtained with pea seedling root homogenates, although these were less active. The preparations have the general characteristics of mammalian mixed function oxidases (require oxygen and are inhibited by methylenedioxyphenyl compounds such as sesamex), although NADPH is only stimulatory at high enzyme levels. An interesting observation is that dieldrin is not apparently metabolized and so does not appear to be the precursor of the small amount of "trans-diol" formed, supporting numerous indications that aldrin conversion in plants and soils can follow pathways other than epoxidation to dieldrin. For example, a *cis*-diol (presumably *exo*-), as well as the *trans*-diol, has been detected among aldrin metabolites formed by pea or bean root preparations and the concurrent presence of a monohydroxylation product suggests that a possible mechanism is double bond hydration followed by nonstereoselective hydroxylation.[567] Preparations from some plant roots evidently contain inhibitors of oxidative reactions.[579]

From the standpoint of aldrin residues formed under practical conditions, photoaldrin (Figure 13, structure 10) is the only product apart from photodieldrin that is likely to be formed in sunlight and its further epoxidation to photodieldrin is to be expected in any case. Photoaldrin is remarkably toxic to mosquito larvae and is as toxic as photodieldrin to houseflies (*see section* C.1b); it is oxidized to photodieldrin (Figure 12,

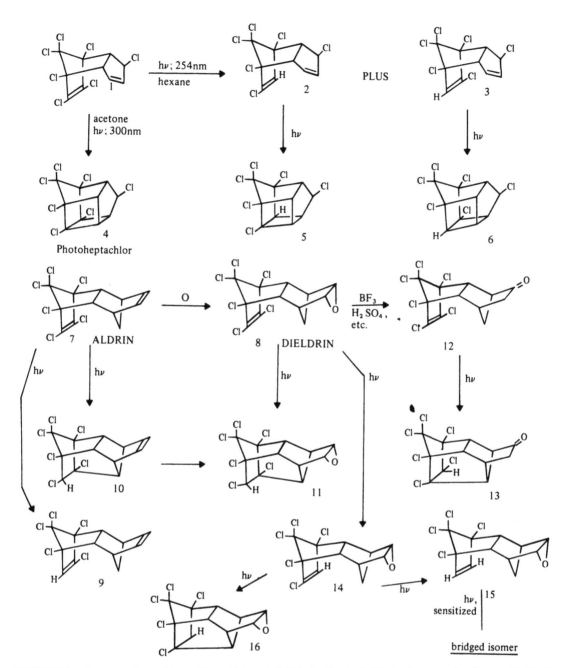

FIGURE 13. Some reactions of heptachlor, aldrin and dieldrin leading to dechlorination or molecular rearrangement.

structure 5) and the ketone (Figure 12, structure 4) by the latter insect but only the fully oxidized structure (4) appears in mosquito larvae. Successive removal of the vinylic chlorine atoms of aldrin or dieldrin by reductive photodechlorination (Figure 13) requires energetic short wave UV-radiation and has not been observed under field conditions so far. Toxicological data for the aldrin derivatives are lacking but they will undoubtedly be epoxidized to the corresponding dieldrin derivatives in vivo; monodechloro-dieldrin (Figure 13, structure 14) is more toxic to rats than dieldrin but less toxic to houseflies, and the bisdechloro-compound (15) is well known to be highly toxic to both mammals and insects (see section C.1b). These properties as well as the further conversion

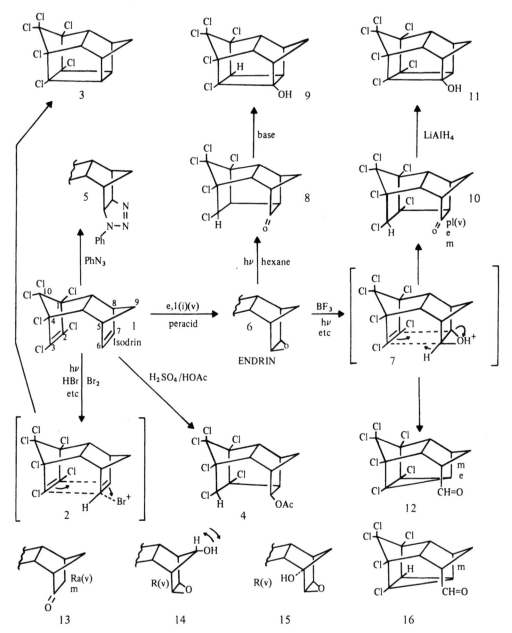

FIGURE 14. Some chemical, metabolic and environmental conversions of isodrin and endrin. e = environment; l = living organisms; m = microorganisms; pl = plants; Ra = rabbit; R = rat; i = in vitro; v = in vivo.

into corresponding bridged molecules by UV-irradiation with photosensitization are interesting theoretically but are not of practical significance since the initial dechlorinations do not occur in sunlight.

### e. Isodrin and Endrin

The epoxidative conversion of isodrin (Figure 14, structure 1) to endrin (6) was demonstrated in

the American cockroach in 1958 and was later shown to occur rapidly in houseflies.[580,581] Both investigations indicated the further conversion of endrin into more polar products. Although at least one report indicates the presence of toxic material in the tissues of rats fed isodrin (and this may have been endrin), it seems clear that endrin disappeared too rapidly to be detected. The epoxidation was later shown to occur in liver microsomal preparations from rats and rabbits as a result of mixed function oxidase action;[557,558] no further conversion of endrin was demonstrated but a later investigation by the author showed that when endrin was incubated with either rat liver or pig liver microsomes for several hours at 30°C (with provision of an NADPH source) a derivative behaving chromatographically and chemically as a monohydroxy-derivative was produced, and its formation was suppressed by an inhibitor of such reactions (sesamex).[556] Several subsequent investigations made it clear that the change in stereochemistry between the aldrin/dieldrin and isodrin/endrin series makes a profound difference to the rate of elimination of the last pair from mammalian tissues.[303] In contrast to the situation with dieldrin for example, there is no tendency for any progressive increase in the endrin level in the blood of dogs continually treated with it, nor is there any good correlation between tissue and blood endrin levels.

It is clear that endrin is nonpersistent in mammals, although generally more toxic than dieldrin (see section C.2b), but the reason for this difference awaits explanation. However, an inspection of the molecular structure suggests that it might be rather more vulnerable to enzymatic attack than dieldrin. Judged by the amount of the label in the excreta, the half-life of [14]C-endrin in rats following a single intravenous dose of 200 μg/kg is 2 to 3 days for males and 4 days for females.[517] As with dieldrin, methano-bridge hydroxylation (Figure 14, structure 14) occurs in the rat, but in contrast with dieldrin, the corresponding ketone is found in adipose tissue, liver, and brain.[582] The configuration of the 9-hydroxy-derivative, which is excreted in the feces along with a second monohydroxy-compound that is probably (15), has not yet been established unequivocally.[565] A preliminary report of metabolism studies in rabbits suggests that hydroxylation occurs without major skeletal rearrangement.[244] The urine, following administration of 0.5 mg/kg of [14]C-endrin for 3 to 4 days, is said to contain 40% of the excreted label as endrin hydroxylated at the junction of the norbornene ring systems, 40% as a conjugate of a hydroxy-derivative, 12% as a hydroxy-derivative of an unbridged isomeric ketone (Figure 14, structure 13, presumably), and another compound of undefined structure. In summary, the conversions so far indicated suggest that oxidation without skeletal rearrangement

preponderates in mammals although details remain to be worked out. There is no evidence for epoxide ring hydration so far, and trans-diol formation may be sterically difficult in any case.

Information regarding metabolism in birds, fish, or wildlife in general is scanty, although balance studies following oral ingestion by poultry or exposure to fish to concentrations in water indicate that metabolism occurs, while oysters are said not to store several cyclodienes including endrin.[482] Other reports indicate pronounced accumulation of endrin by aquatic organisms and it is well known to be very toxic to fish. a 96-hr LC50 of 0.0002 ppm being recorded for bluntnose minnows.[583]

The plant metabolism was discussed in connection with the biodynamics of endrin (Table 14) and beyond the certain formation of Δ-keto-endrin (Figure 14, structure 10) from endrin, and the epoxidation of isodrin, little information is currently available although other metabolites are certainly formed. Some reports indicate formation of traces of the aldehyde (12). Since Δ-keto-endrin is produced from endrin by UV-irradiation and by microorganisms, its precise origin is not easy to discern. In this connection, Terriere[505] has made the interesting observation that the 22,000-g supernatant derived from bean roots which epoxidizes aldrin also converts isodrin into substantial amounts of ketone (10) (according to chromatographic procedures) and the nonenzymatic rearrangement of endrin to this ketone which occurs under identical conditions is insufficient to account for the total amounts produced. The inference is that a significant level of epoxidation and subsequent rearrangement is occurring and that both are enzymatic. Korte[508] demonstrated the slow metabolism of [14]C-endrin incorporated at 0.13 ppm into nutrient medium containing growing Aspergillus flavus and Matsumura found that soil isolates which metabolized dieldrin also converted endrin.[485] The ketonic rearrangement product (Figure 14, structure 10) is a significant product in all cases. Other products, which have not been positively identified, may be the aldehyde (12) and reductive monodechlorination products of both aldehyde and ketone such as (16) and (8). Zabik has shown that the ketone (8) is a photochemical conversion product of endrin irradiated with sunlight in hydrocarbon solvents and it has been detected in endrin treated soil (up to 5% of a total residue of 20 ppm).[259] Photoisodrin

(3) does not appear to have been implicated as an isodrin conversion product in laboratory experiments on plant metabolism (isodrin is not used commercially), but it differs from the photoisomerization products of dieldrin in being less toxic than its precursor (see section C.1b) and there is evidence for its further metabolism in living organisms.[584] Indeed, a significant feature of all the photoconversion products of this series so far investigated toxicologically is their reduced toxicity, by normal routes of contact, compared with endrin. This property can be set against the apparent tendency to greater persistence of total endrin residues on plant surfaces.

The foregoing account of cyclodiene metabolism shows that most of the metabolites retain the largely intact molecular skeleton; the hexachloronorbornene nucleus is clearly rather stable, but when rearrangements occur, altering its structure, the products seem to be more readily degradable by mammals although they may be more toxic than the precursors to some species, as found in the dieldrin series. Hexachloronorbornene itself appears to be innocuous to insects, and its nearest insecticidal derivatives Alodan and the related compound Bromodan (see section C.1b) are remarkable for their safety to mammals (LD50/LD50 housefly = 1,000). That microorganisms can attack "stripped down" cyclodienes of this sort has only been demonstrated quite recently, and it may be that reductive dechlorinations effected by microorganisms under anaerobic conditions can be a prelude to aerobic oxidative degradation, as indicated for DDT. Some of these possibilities have been discussed elsewhere,[260] and are summarized below. It must also be remembered that the final residues of apparently degradable insecticides may not be particularly desirable; there may not be much to choose, for example, between a hexachloronorbornene residue from a cyclodiene and a chlorinated phenol or some other stable moiety arising as a hydrolysis product of an organophosphorus insecticide. Research on cyclodiene metabolism to date has been mainly concerned with the nature of the primary metabolites and these are frequently excreted in free form, but the indications are that the normal processes of acetate, glucuronide, and possibly amino acid conjugation are available for these products.

That reductive dechlorination of the dichloromethano-bridge of cyclodienes can occur in the presence of biological material is shown by the formation of the isomeric monodechlorocompounds when liver tissue is subjected to alkaline saponification before extraction to recover dieldrin ingested during feeding trials. This conversion may be catalyzed by liver pigments and may represent a combination of chemical methods previously discussed in Volume I (Chap. 3B.2b). From these results, it might be expected that under anaerobic conditions and in the presence of biological debris, microorganisms and trace metals, cyclodiene molecules would undergo reductive mono- or even bis-dechlorination of the bridge; ideal conditions might be found, for example, in sewage sludge, river mud, and so on.[260] They might also be found in organisms which are deficient in microsomal oxidases. In analogy with the conversion of photodieldrin into Klein's ketone (Figure 12, structure 4) in vivo, there is the possibility that an aerobic process may convert a partially dechlorinated methano-bridge into a keto-group (probably via an intermediate gem-chlorohydrin). Bridge carbonyl groups are fairly readily eliminated from such molecules, especially by heating. For example, the corresponding compound related to aldrin (that is, dichloromethano-bridge replaced by carbonyl) aromatized spontaneously with elimination of carbon monoxide to form 1,2,3,4-tetrachlorobenzene and cyclopentadiene (the elimination being accompanied in this case by a retro-Diels-Alder reaction). A similar reaction with dieldrin might afford 1,2,3,4-tetrachlorobenzene and cyclopentenone, although in this case the epoxynorbornane nucleus would probably be retained, it being attached, following the elimination of carbon monoxide, to a 1,2,3,4-tetrachlorocyclohexa-1,3-diene nucleus (aldrin numbering retained). Aromatization by loss of two hydrogens from this molecule would give the corresponding 1,2,3,4-tetrachlorobenzonorbornene epoxide.

As far as biological systems are concerned, these reactions have not been demonstrated and are speculative. Of interest here is a recent report that under anaerobic conditions, the bacterium *Clostridium butyricum* not only reduces the dichloromethano-bridge in hexachloronorbornene to $-CH_2-$, but further hydroxylates it to $-CH(OH)-$. However, tetrachlorocyclohexadiene is apparently formed only from the fully chlorinated molecule, implying the direct elimination of $>CCl_2$ rather than intermediate ketone formation.[586] Nevertheless, the demonstration that the unchlorinated methano-bridges of both dieldrin and endrin are hydroxylated in vivo, and that the compound from endrin is biologically oxidized to the corresponding bridge-ketone, which is found in fat, makes such possibilities rather less hypothetical. For example, it is possible that the endrin metabolite, if passed through a cooking process, would lose carbon monoxide to give a derivative of hexachlorobenzonorbornene, or, by retro-Diels-Alder reaction, hexachlorocyclopentadiene and phenol. If both bridges were to be eliminated from such molecules by biological degradation, derivatives of tetrachloronaphthalene would result.[260] The presence of an epoxide ring in the 6,7-position, rather than a double bond, makes such reactions energetically less likely, but it must be remembered that microorganisms appear also to be capable of generating olefins from epoxides, dieldrin being converted into aldrin, for example.

### 3. Biochemical Aspects
#### a. General Mechanisms of Organochlorine Insecticide Metabolism

The previous account of biotransformation products shows that they involve oxidation (hydroxylation), oxidative dechlorination, reductive dechlorination, dehydrochlorination, hydrolytic removal of chlorine, hydration of epoxide rings, and skeletal rearrangements. Information about these processes is essential if the action of organochlorine insecticides is to be understood, because transformation products rather than precursors may be the causal agents of toxicity. There is also the question of comparative metabolism in target and nontarget species; metabolic differences are a common cause of differing species sensitivity and information about them may help to answer some of the current questions regarding ecological effects. It is perhaps unfortunate that the principles of drug metabolism as established for mammals are generally valid in the insect world too, so that the number of readily exploitable differences is not great. Very frequently all that is involved is a quantitative difference in the level of some toxicant degrading enzyme that is common to both species.

In insects, plants, and bacteria, as in mammals, foreign molecules are initially modified in various ways (primary transformations) that make them more vulnerable to further conversion into water soluble compounds (conjugates) by enzyme catalyzed reactions with endogenous chemicals.[587] In this way, plants form glucosides and more complex glycosides, bacteria form glucosides, and insects form glucosides, sulfates or conjugates with amino acids and glutathione (which last may be degraded to cysteine conjugates). In mammals, conjugation is with glucuronic acid, amino acids, sulfuric acid, glutathione (to form mercapturic acids eventually) and some other moieties. Oxidations are primary transformations of major importance in all these forms and in mammals the submicroscopic tubule complex in liver cells known as the smooth endoplasmic reticulum (SER; called the microsomes when isolated by high speed centrifugal fractionation of disrupted cells) is an important site of these reactions.[484] In a very common type of transformation, the foreign molecule is conjugated with glucuronic acid after a hydroxyl group has been introduced by the enzymes of the SER, which are called mixed function oxidases from their ability to transfer one atom of molecular oxygen to the substrate in the presence of nicotinamide-adenine dinucleotide phosphate (NADPH) as a source of reducing equivalents;

$$NADPH + O_2 + \text{drug-H} + H^+ \xrightarrow{\text{enzyme}} NADP^+ + H_2O + \text{drug} - OH$$

The electron transport scheme involved in this process in mammalian liver has been extensively investigated and the transfer of oxygen to the foreign molecule is believed to occur as shown in Figure 15,[588] in which the haemoprotein cytochrome P450 is so named because the complex between its reduced form and carbon monoxide has an absorption maximum at 450 nm. These investigations date from about 1957 when it was noticed that steroid hydroxylation in bovine adrenocortical microsomes, which contain a similar system, was inhibited by carbon monoxide.

The benzylic hydroxylation of DDT to give dicofol was reported to occur in *Drosophila melanogaster* by Tsukamoto in 1959,[589] and in 1961 Agosin described an enzyme in the German cockroach which effected this hydroxylation in the presence of NADPH and oxygen.[531] After that, microsomal oxidation quickly came to be recognized as a highly important detoxication mechanism for insecticides in insects, especially as it explained the spectacular synergistic action of

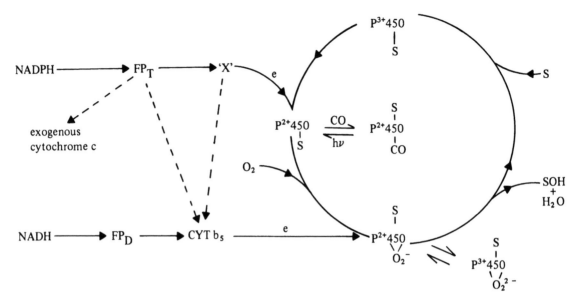

FIGURE 15. Current concept of the electron transport scheme for cytochrome P450 reduction and oxygen activation for hydroxylation reactions effected by mammalian liver microsomes. S = substrate xenobiotic (drug or pesticide). (Data adapted from Estabrook, Baron, and Hildebrandt.[588])

the methylenedioxyphenyl (1,3-benzodioxole) compounds that had been used for many years as potentiators of pyrethrum. The now classical 1960 paper of Sun and Johnson,[590] besides suggesting that pyrethroid synergism resulted from the inhibition of microsomal oxidation, showed that the 1,3-benzodioxole derivative sesamex inhibited aldrin epoxidation in vivo in houseflies. This discovery coincided with the author's interest in the great differences in toxicity between compounds such as chlordene and heptachlor and the low toxicity of some other cyclodiene molecules. As seen in section B.2c, oxidation in vivo later proved to be closely involved in some of these phenomena. In 1965, Ray showed that the microsomes from whole houseflies and cockroaches contained a pigment having several of the characteristics of the cytochrome P450 found in mammals.[591] The early work in this area was greatly hampered by the presence of endogenous inhibitors in the insect preparations, and more effective microsomal preparations were later made from housefly abdomens. Since carbon monoxide complexes with P450, it inhibits microsomal oxidations and it was possible to demonstrate directly its inhibitory effect on aldrin epoxidation in vivo in houseflies.[592] Such experiments are not possible in mammals because of the high affinity of carbon

monoxide for hemoglobin which results in respiratory failure.

The developments in research on insect microsomes have been concisely reviewed by Hodgson and Plapp,[593] and Hodgson has recently isolated and partly purified the NADPH-cytochrome $c$ reductase (equivalent to the NADPH-FP$_T$ system of Figure 15) from housefly microsomes and has shown its general resemblance, with differences in detail, to the corresponding mammalian liver enzyme isolated some years ago. Most of the difficulties of microsome investigations center around the particulate nature of the enzyme system and solubilization and reconstitution of the housefly system have not yet been achieved. The mode of action of mixed function oxidase (MFO) inhibitors such as methylenedioxybenzene compounds is of special interest, and there is now much evidence that these compounds act on the particulate enzyme system at the cytochrome P450 level. They appear to behave as alternative substrates for the toxicants whose metabolism they inhibit and generally have a high affinity for the enzyme, combined with a low rate of metabolism. However, the mechanism is still not clearly understood, and the binding to cytochrome P450 seems to differ from that of other substrates. It is fortunate that these compounds are much more stable in insects that in mammals; oxidative attack on the methylene group occurs rather readily in the latter.[594] Aldrin epoxidation, naphthalene *hydroxylation* and more recently, dihydroisodrin hydroxylation, have been used as an index of mixed function oxidase levels in insects. The cyclodiene conversions are particularly valuable because major single products are formed in each case. In insects, the major sites of microsomal oxidation appear to be the fat body, gut, and Malpighian tubules and particularly active aldrin epoxidase systems are found, for example, in the guts of lepidopterous larvae when endogenous enzyme inhibitors are removed.[595] In fact, the remarkably high and generalized insecticide tolerance of the larval stages of these phytophagous insects suggests that they should have high mixed function oxidase activity arising through constant exposure to a wide variety of natural substances in the diet. Accordingly, a recent survey of aldrin epoxidase activity in the guts of 35 species of caterpillars representing ten families of lepidoptera shows a distinct correlation between this activity and the range of plant species used for food, polyphagous larvae having about 14 times the activity of those feeding on only one plant species.[596] Microsomal enzyme preparations from insect tissues can provide valuable information of this sort regarding the availability of primary oxidative detoxication mechanisms but it must be remembered that, as in the case of mammalian systems, the observed enzyme activity may depend on the stage of larval development, age of adult insect, stage of some particular biological rhythm and so on.[597]

Plant tissues are rich in peroxidases and the abundance of phenolic derivatives found in higher plants suggests generally high oxidase activity. However, the nature of the enzymes involved in insecticide biotransformation is obscure, although the products frequently resemble those found in mammals. A microsomal type fraction from *Echinocystis macrocarpa* contains cytochrome P450 and is involved in the series of oxidations converting the gibberellin precursor (-)-kaurene into kaurenoic acid; whether such systems are specific for hormone synthesis or can also metabolize pesticides remains to be seen. Present information regarding the plant enzymes oxidatively metabolizing organochlorine insecticides is restricted to the work on cyclodiene epoxidases already mentioned in connection with biotransformation products (see section B.2d). The position

with regard to fungal and bacterial metabolism of these compounds is even more obscure, although cytochrome P450 has been implicated in alkaloid biosynthesis and fatty acid hydroxylation, and crystalline P450 has been isolated from the camphor hydroxylating organism *Pseudomonas putida*.[484] This area is of considerable importance because the final oxidation of complex ring systems to carbon dioxide and water usually depends on microbial action, and the extensive information available regarding the oxidative degradation of aromatic compounds such as the chlorophenoxyacetic acids[598] is obviously relevant to the ultimate degradation of DDT analogues and terminal residues of the HCH isomers.

Primary conversions involving dechlorinations of various types are obviously specific for organochlorines, but their mechanisms can sometimes be related to known conversions of other compounds. Thus, mammalian liver and kidney contain microsomal and soluble enzymes that can reduce nitro-groups anaerobically in the presence of NADPH or NADH,[587] and the involvement of cytochrome P450 has been indicated. The reductive dechlorination of DDT to give DDD may have a relationship to this process, although the extent of enzymatic versus nonenzymatic dechlorination is not always clear in this case. Dilute solutions of iron porphyrin complexes effect the conversion chemically, and it has been linked with the anaerobic peroxidation of unsaturated fats by iron-porphyrin enzymes in live wheat germ.[599] A recent study suggests that in anaerobic soils DDT can accept an electron from ferrous iron to give a free radical and a chloride ion; reduction to DDD then occurs in the presence of natural hydrogen donors in the soil.[600] It seems that several processes are available for this conversion, depending on the circumstances, but that the principles are similar. Rat liver microsomes are also claimed to produce a DDD-like compound from DDT with the normal mixed function oxidase requirement of NADPH and oxygen, and a similar system reductively removes chlorine and bromine from the volatile anaesthetics halothane ($CF_3CHBrCl$) and methoxyflurane ($CHCl_2CF_2OCH_3$), while carbon tetrachloride is converted into chloroform.[587]

Esterase activity is a common form of hydrolytic activity involved in the degradation of pesticides of appropriate structure, but is not applicable in the metabolism of the majority of organochlorine insecticides as defined here. Some of the pathways for cyclodiene metabolism outlined previously indicate apparent hydrolytic replacement of chlorine atoms by hydroxyl groups, but another mechanism, involving oxidative hydroxylation of a carbon atom carrying chlorine, followed by pyridine nucleotide mediated reduction of the resulting carbonyl compound, gives the same results and explains some of the cofactor requirements and products of dihydroheptachlor metabolism, for example, and the oxidative dechlorination of photodieldrin:

Similar reactions may explain the oxidative dechlorination of a series of chloroethanes and chloropropanes by mammalian liver microsomal enzymes requiring NADPH and oxygen.

A process that is probably more related to esterase activity is that of epoxide ring hydration, which appears to be effected by particulate enzymes located in mammalian and insect microsomes.[601,602] The alkaline hydrolysis of organophosphates or of carboxylic esters results chemically from nucleophilic attack by hydroxyl ion ($OH^-$) on a positive phosphorus or carbon atom, and the addition of water to double bonds or epoxide rings can be viewed similarly:

The last two reactions are better regarded as hydrations than hydrolyses and there is evidence that both conversions may be catalyzed by the same enzyme. Thus, fumarase, which catalyzes the reversible interconversion of fumarate and L-malate, also stereospecifically hydrates trans-2,3-epoxysuccinate; only the L-epoxide is hydrated, to give mesotartaric acid, and the olefin-hydrase and epoxide hydrase activities are inseparable.[715]

Boyland's 20-year-old suggestion[603] that the intermediate formation of epoxides in the oxidation of aromatic hydrocarbons would explain the ultimate appearance of phenols, dihydrodiols, and mercapturic acids as metabolites has been fully substantiated in recent years, and the epoxide hydrases, which add water to epoxide rings to give the corresponding trans-diols, are currently attracting much attention in relation to the metabolism of both aromatic and alicyclic compounds. Diol formation from cyclodiene epoxides is an obvious route for detoxication and elimination and several species are now known to hydrate dieldrin in vivo. The hydration of dieldrin and the heptachlor epoxides occurs slowly in liver microsomal preparations from pigs and rabbits, and rat liver microsomes appear to be even less active.

Hydration rate depends markedly on molecular structure and there are species variations in ability to effect the hydration (see section C. lb). These microsomal enzymes can be at least partly solubilized and are inhibited by a number of other epoxides as well as by organophosphorus compounds that are known to inhibit esterase activity.[235] Among the inhibitory epoxides, styrene oxide, 1,2-epoxy-1,2,3,4-tetrahydronaphthalene, cyclohexene oxide, 1,1,1-trichloro-2,3-epoxypropane, and a range of aryl-2,3-epoxypropyl ethers (glycidyl ethers), including 1-naphthyl glycidyl ether, are the most effective.[235,604] It is interesting that the juvenile hormone of the Cecropia moth, which has a terpenoid chain terminating in an epoxide ring, inhibits cyclodiene epoxide hydrase from the blowfly, *Calliphora erythocephala*, as do a number of hormone mimics having structures with terminal epoxide rings; they are probably alternative substrates for the enzyme. As a hydrase, the cyclodiene epoxide hydrating enzyme requires neither oxygen nor cofactors and usually is only affected by somewhat high concentrations of mixed function oxidase inhibitors; SKF525A is a moderately good inhibitor, possibly because it is also a carboxylic ester. Like fumarase, the enzyme from pig or rabbit liver is stereospecific since it hydrates only one of the two enantiomers of asymmetric epoxides such as chlordene epoxide (Figure 10, structure 11) and heptachlor-2,3-endo-epoxide (Figure 10, structure 4), the unchanged enantiomer being recoverable from the incubation mixtures.[547] Microsomal preparations from avians such as the pigeon have rather poor hydrating ability (although some hydration can be detected in vivo),[576] and it is tempting to speculate that they may also be unable to effect the hydration of DDNU, which is required for its conversion to DDA according to the sequence of Figure 8; this pathway is obscure beyond DDMU in the pigeon and DDA formation has not been detected.

There is only fragmentary information about the secondary conjugative transformations of organochlorine insecticide metabolites, except that DDA is known to be conjugated with amino acids and the chlorobenzenes and chlorophenols formed from HCH isomers are known to be excreted as glutathione conjugates, mercapturic acids, glucuronides and ethereal sulfates, as appropriate to the species involved. Acetate and glucuronic acid conjugation have been indicated for the dieldrin trans-diol (Figure 12, structure 6) and glucuronic acid conjugation for 9-hydroxydieldrin (Figure 12, structure 10), other conjugates not being characterized so far. Conjugation reactions normally occur in the soluble fraction of cells, although the enzymes effecting glucuronic acid conjugation (transglucuronylases) in mammals are found in the microsomal fraction of liver. The products are excreted by mammals but the glycoside conjugates produced in plants from pesticides usually remain in the tissues, except that hydrophilic metabolites from endrin applied to foliage appear to be translocated in plants and eventually enter the soil (see section B.1b). The mechanisms of these processes in insects merit further attention since their inhibition might result in "feedback" effects having far-reaching consequences for earlier parts of the metabolic sequence.

Glutathione (GSH) has a particularly important role in the formation of water soluble derivatives that can be eliminated by excretion and the various glutathione S-transferases catalyze its transfer to double bonds or epoxide rings (S-alkene-, S-aryl- and S-epoxide-transferases). These enzymes are believed to increase the ionization of GSH bound to them, so activating it for reaction with electrophilic reagents whose presence might cause cell damage. Glutathione conjugates are normally degraded to mercapturic acids (e.g., Figure 9, structure 21) in mammals, but may be excreted as the nonacetylated cysteine derivatives (22) or even as the original glutathione conjugates by insects such as the locust.

### b. Enzymatic Detoxication and Insect Resistance

The number of enzymatic mechanisms associated with the tolerance or acquired resistance of insects to chlorinated insecticides is quite limited. In the immature forms of reduviid bugs, *Triatoma infestans*, natural tolerance to DDT is apparently related to its detoxication by microsomal oxidases (MFO) and the natural tolerance of other insects to cyclodienes is also associated with oxidation. DDT-tolerance in the Mexican bean beetle (*Epilachna varivestis*) relates to its naturally high level of DDT-dehydrochlorinase and any natural tolerance to lindane may be attributed at least partly to the action of dehydrochlorinating enzymes. Where resistance mechanisms arising by selection with toxicant are due to metabolism the same mechanisms appear to be involved, and the genetic origins of DDT-resistance due to oxidation (controlled by gene DDT-*md*) and dehydrochlorination (controlled by gene *Deh*) in houseflies were discussed in Chapter 2 (section B.2a).

The increased level of oxidizing enzymes (enzyme induction) that results when these insects are exposed to various drugs and to toxicants such as the organochlorine insecticides themselves was also referred to earlier (Vol. II, Chap. 2B.3). There is a frequent association between dieldrin resistance and elevated MFO activity which in the past has posed the question whether dieldrin resistance (mechanism unknown) is a cause or a consequence of this activity, but the genetic "synthesis" of strains lacking the elevated MFO activity but retaining the dieldrin resistance shows that the two phenomena are genetically separable.[606] The epoxidative conversion of aldrin into dieldrin, widely used as an index of MFO activity, is evidently associated with a second chromosomal gene and this chromosome also carries resistance to carbamate insecticides (Figure 5). Ability to hydroxylate naphthalene is apparently correlated with aldrin epoxidase levels in 14 housefly strains with varying MFO activity, and the position of the epoxidase gene (32 units from the DDT-dehydrochlorinase gene *Deh*) locates it near to a gene associated with resistance to the organophosphorus insecticide diazinon, so that there seems to be a close genetic relationship between these oxidative mechanisms.[607] Sublethal doses of dieldrin applied to dieldrin-resistant houseflies induce MFO activity and produce increased levels of cytochrome P450; increased MFO activity is associated with increased P450 content in several housefly strains but differences in enzyme activity are generally much greater than differences in cytochrome P450, indicating that other factors modify the observed enzyme activity.[608]

In the previous section, reference was made to the importance of glutathione (GSH) and the GSH-transferases in the conjugation and elimination of foreign compounds from living organisms, and GSH is required for the dehydrochlorination of both DDT and the HCH isomers. With DDT there is only one metabolite, DDE, to consider and GSH is required but not consumed, whereas with lindane the metabolic picture (at least as far as the initial steps are concerned) is not entirely clear and GSH is certainly consumed by some of the metabolic steps. Some of these complexities were discussed in relation to the known lindane metabolites (see section B.2b). The possible formation of $\gamma$-PCCH (Figure 9) as an intermediate in lindane metabolism forms an attractive analogy with DDE formation from DDT, while for the latter, replacement of one trichloromethyl chlorine by GSH, followed by direct dissociation of the product into GSH and DDE, would explain the catalytic function of GSH.

Early work provides conflicting evidence as to whether the same or similar enzymes are involved in both reactions but more recent investigations indicate that DDT is readily converted into DDE by housefly enzyme systems that metabolize lindane. Furthermore, DDT and lindane inhibit each other's metabolism by these systems, while WARF-antiresistant (*N*-di-*n*-butyl-*p*-chlorobenzenesulfonamide) and bromophenol blue (3′,3″,5′,5″-tetrabromophenolsulphonphthalein) inhibit both metabolic processes. $\gamma$-PCCH, but not chlorocyclohexane or the HCH isomers, is a substrate for rat liver enzyme thought to be a GSH *S*-aryltransferase, whereas a housefly enzyme (*S*-aryltransferase) effecting the GSH conjugation

of 1-chloro-2,4-dinitrobenzene (CDNB) differs from that metabolizing $\gamma$-PCCH, though they have similar molecular sizes. In fact, the enzymes in housefly homogenates and rat liver that metabolize $\alpha$- and $\gamma$-HCH, $\gamma$-, and $\delta$-PCCH, DDT and CDNB have molecular weights of 36,000 to 38,000 by gel filtration.[609] Other work with houseflies also distinguishes GSH-aryltransferase, which behaves as a single protein, from DDT-ase, which appears to exist in more than one form,[610] and it has been suggested that active DDT-ase is actually trimeric or tetrameric (molecular weight 120,000), polymer formation being initiated by DDT and stabilized by GSH.[611] These observations may help to explain the differences in substrate affinity, catalytic activity, and susceptibility to inhibitors that are found in different strains of houseflies. They also provoke the speculation that aggregation to give active enzyme might occur in resistant but not in susceptible strains and that selection for resistance could involve selection for those individuals readily able to effect the aggregation, which differs somewhat from the viewpoint that selection is for individuals having a permanently elevated titer of DDT-ase. The result is a genetically controlled propensity for a form of enzyme induction which, since it involves aggregation of enzyme units, would result in elevated DDT-ase in either the living insect or its in vitro preparations when DDT is present. This situation would be indistinguishable from the classical one because DDT-ase has to be measured in the presence of DDT.

DDT-dehydrochlorinase was first isolated from DDT-resistant houseflies and partly purified by Sternburg and Kearns,[612] who showed that the activity of an extract from a single fly could dehydrochlorinate 5 to 7 $\mu$g of DDT per hr at 37°C. Chemically, the base catalyzed reaction involves removal of the benzylic proton in an $E_2$-type elimination (Vol. I, Chap. 2B.3), followed by elimination of a chlorine atom, reaction rates of various DDT-analogues being in the order 4,4′-diodo->4,4′-dibromo->4,4′-dichloro->4,4′-difluoro->>4,4′-dimethoxy, showing the influence of the 4,4′-substituents (Table 16). Early work on the enzymatic dehydrochlorination indicated no obvious relationship, or even an inverse relationship, between it and the chemical reaction since the 4,4′-diodo-compound is much less reactive enzymatically than DDT and is less reactive even than methoxychlor.[613] Further, the more readily enzymatically dehydro-

TABLE 16

Chemical Dehydrochlorination Rates of Some DDT Analogues and Resistance Ratios Measured for Houseflies

| $R_1$ | $R_2$ | $R_3$ | $R_4$ | $R_5$ | $R_6$ | Dehydrochlorination rate ($K \times 10^5$; 1/min/mol; 30°C) | $\Sigma^a$ ($R_1 + R_6$) | Resistance ratio[b] |
|---|---|---|---|---|---|---|---|---|
| F | H | Cl | Cl | Cl | F | $1.03 \times 10^3$ | +0.124 | 24 |
| Cl | H | Cl | Cl | Cl | Cl(DDT) | $6.1 \times 10^3$ | +0.454 | 276 |
| Br | H | Cl | Cl | Cl | Br | $8.9 \times 10^3$ | +0.464 | 515 |
| I | H | Cl | Cl | Cl | I | $1.04 \times 10^4$ | +0.552 | – |
| $NO_2$ | H | Cl | Cl | Cl | $NO_2$ | $6.9 \times 10^6$ | +1.556 | – |
| H | H | Cl | Cl | Cl | H | $1.22 \times 10^2$ | 0.0 | – |
| $CH_3$ | H | Cl | Cl | Cl | $CH_3$ | 36.2 | −0.34 | 1.65 |
| $C_2H_5$ | H | Cl | Cl | Cl | $C_2H_5$ | 27.0 | −0.302 | 4.0 |
| $SCH_3$ | H | Cl | Cl | Cl | $SCH_3$ | $6.1 \times 10^2$ | 0.0 | – |
| $OCH_3$ | H | Cl | Cl | Cl | $OCH_3$ | 29.5 | −0.536 | 5.0 |
| Cl | Cl | Cl | Cl | Cl | Cl(o-Cl-DDT) | $1.07 \times 10^3$ | – | – |
| H | Cl | Cl | Cl | Cl | Cl(o,p'-DDT) | $1.10 \times 10^2$ | – | – |
| Cl | H | Cl | Cl | H | Cl(DDD) | $2.98 \times 10^3$ | – | 460 |
| $C_2H_5$ | H | Cl | Cl | H | $C_2H_5$ | 16.7 | – | – |

Data adapted from Metcalf and Fukuto.[6 7]

[a]Summation of Hammett's sigma values for the para-substituents.
[b]Ratio of LD50 for the DDT-resistant Bellflower strain to LD50 for the susceptible NAIDM strain.

chlorinated compounds appeared to be more toxic, methoxychlor being quite a good toxicant. This situation immediately suggested a strict steric requirement for fit to the enzyme which strongly outweighs substituent influence in the chemical reaction. The correlation between DDT-ase activity and toxicity indicated a similar affinity of DDT analogues for their site of action and for the enzyme, which led Mullins to suggest that the enzyme is an integral part of an altered membrane structure arising as a result of the selection of resistant individuals by the toxicant.[614] However, it is now clear that dehydrochlorination is effected by a separate enzyme and Metcalf has shown that a plot of log (LD50 of 4,4'-DDT analogues to a resistant housefly strain/ LD50 to a susceptible strain) against the sum of Hammett's sigma values for the aromatic substituents gives a linear relationship parallel to that between log (chemical rate constant) and the summed sigma values.[67] The ratio of LD50's is the "resistance ratio (RR)" and is a measure of the enzymatic dehydrochlorination rate of the various analogues in vivo. Thus, lower resistance ratios (Table 16) are found for those 4,4'-analogues which have electron donating rather than withdrawing substituents and a consequently reduced chemical dehydrochlorination rate, while the similarity in slope of

the plots for chemical and for in vivo dehydrochlorination suggests that for some compounds the chemical and enzymatic reactions are, in fact, related. An in vivo measurement of this sort obviously avoids all the problems attending enzyme isolation and measurement.

The vital role of this enzyme in some cases of DDT-resistance is illustrated by the striking synergistic effect of blocking the action of DDT-ase with various inhibitors such as DMC and WARF-antiresistant. Various chemical changes in the DDT molecule also make it refractory towards DDT-ase, the best known of these being one or both of replacement of the benzylic hydrogen by deuterium or the insertion of one ortho-chlorine atom.[82,615] These changes immediately highlight species differences, since although deutero-DDT is still, disappointingly, dehydrochlorinated by houseflies and is no more toxic than DDT either to these or to the mosquito *Culex tarsalis*, it is 50 to 100 times more toxic to resistant *Aedes aegypti* and is not readily dehydrochlorinated by them. Ortho-chloro-DDT (Vol. II, Chap. 2B.2a) on the other hand is poorly dehydrochlorinated by resistant houseflies and larvae of *C. tarsalis* and is quite toxic to them, while resistant *Ae. aegypti* dehydrochlorinate it and are resistant; α-deutero-2-chloro-DDT is better in this case. DDT-ase in

*Culex fatigans* is more active toward DDT than DDD, but otherwise resembles the enzyme in *Ae. aegypti*, whereas the enzyme of *Culex tarsalis* resembles the one in R-flies.[616,617]

Although the synergistic effects of compounds such as DMC or WARF-antiresistant with DDT are usually very spectacular to begin with, when DDT-ase is the cause of resistance, it is well known that resistance to the mixtures soon develops; this may be due to further elevation of DDT-ase levels and probably also to enhanced ability to metabolize the synergist, rendering it ineffective. Thus, DMC may be expected to lose water to give the corresponding olefin and the methyl group may be susceptible to oxidative attack resulting eventually in the formation of DDA. The complexity of these situations is illustrated by the effects of selection of mosquito larvae (*Ae. aegypti*) with DDT-WARF mixtures, Prolan, or malathion. Normal larvae acquired typical resistance due to DDT-ase when selected with each of these toxicants but reduced penetration of Prolan as well as increased resistance to dieldrin and lindane also resulted from Prolan selection, while malathion produced resistance to itself associated with chromosomes 2 and 3, as well as selecting for the second chromosomal DDT-ase. Selection of resistant strains with DDT-WARF mixture produced higher resistance to DDT and to the mixture, with the appearance of an additional resistance mechanism for the mixture associated with chromosome 3.[618] These effects illustrate the flexibility of the genetic background of these insects which enables them to respond to a variety of challenges of increasing severity.

It has been difficult to assess the contribution to lindane resistance made by enzymatic detoxication and in all investigations in which resistant and susceptible insect strains are compared, problems may arise due to toxic effects in the susceptible ones. This difficulty has been avoided in houseflies by using the nontoxic α-HCH, which is degraded more slowly than lindane by pathways which are probably similar, to assess interstrain differences in penetration rate and metabolism. In this way, Oppenoorth showed a negative correlation between the varying levels of lindane resistance in eight different housefly strains and the rate of absorption of α-HCH.[540] There were similar tendencies with δ-HCH, from which he concluded that the absorption of the three isomers (α,γ,δ) probably depends on the same mechanisms and

that resistance to lindane also is due to reduced absorption. In the same investigation, a correlation was found between the capacity to degrade α-HCH and resistance to lindane in these strains and with similar reasoning, metabolism of lindane was concluded to be a resistance mechanism. Later, Busvine[619] found the metabolism of α-HCH to be significantly elevated in two other lindane resistant housefly strains, as compared with a normal strain, whereas lindane resistant and susceptible blowflies, (*Lucilia cuprina*) metabolized α-HCH at the same rate. This observation has a bearing on the cross-resistance patterns found between the cyclodiene compounds and lindane. Although Decker[620] obtained insect strains that were most resistant to the selecting toxicant when he selected with either lindane, chlordane, or dieldrin, Busvine has presented evidence to support a generalized resistance pattern for several insects which is independent of the selecting agent and in which resistance is lowest to lindane, intermediate to isodrin and endrin and highest towards aldrin, dieldrin, and *cis*-chlordane.[218] This suggests a common resistance mechanism, but houseflies are anomalous in having a higher than expected resistance to lindane and Busvine attributes the difference to their having, in addition to the general "cyclodiene type" resistance of unknown mechanism, an enhanced metabolic capacity for the toxicant that is absent in some other resistant insects.[619] Thus, the weight of evidence appears to support the involvement of enhanced metabolism in housefly resistance to lindane, although doubt has been cast on the validity of using α-HCH as a yardstick on account of metabolic differences between the isomers, and Bridges found a susceptible strain of housefly to be intermediate between two derived dieldrin resistant strains in its ability to metabolize α-HCH.[621]

Substantial amounts of the HCH isomers are degraded by houseflies in a few hours, which makes the cause of the high toxicity of lindane a particularly fascinating problem. A significant handicap in these investigations has been the lack of good inhibitors for the detoxication in vivo. GSH-*S*-aryltransferase in insects can be inhibited by certain dicarboxylic acids and phthaleins and such compounds might inhibit DDT-ase by virtue of an ability to complex with a GSH binding site on this enzyme, rather than with a DDT binding site as is believed to be the case with the classical

DDT-analogue synergism.[622] Thus, bromophenol blue inhibits DDT-ase as well as the S-aryltransferase that conjugates CDNB with GSH and the enzymes detoxifying lindane and PCCH isomers. DDT-ase and S-aryltransferase are also inhibited by bis(3,5-dibromo-4-hydroxyphenyl)methane, but only the former by compounds similar to bis(N-dimethylaminophenyl)methane. This compound is also effective in vivo and therefore synergizes DDT when DDT-ase is present; it inhibits the metabolism of PCCH isomers in blowflies, but not in grass grubs (Costelytra zealandica) in vivo, and does not alter the metabolism of the HCH isomers. These relationships are obviously complex and a number of the inhibitors are only of value in vitro, since their water solubility results in rapid elimination from live insects.

Indications of the hydrative opening of the epoxide ring of dieldrin have now been obtained in mosquitoes, houseflies, and cockroaches, and there are certainly enzymes in their tissues that are capable of effecting the hydration of related compounds.[484] The formation of hydroxyl derivatives, either by this means, or by oxidative hydroxylation, has been assumed to result in detoxication and there is certainly evidence that these compounds are less toxic by the usual routes of administration (see section C.1b). Furthermore, the inhibition of oxidative hydroxylation frequently results in cyclodiene synergism against normal houseflies, although not against resistant ones.[697] For dieldrin, the simplicity of the mode of inheritance of resistance (Vol. II, Chap.2B.1) is intriguing and there has always existed the possibility that some highly localized detoxication is involved near to the site of action. Therefore, the recent observation that trans-6,7-dihydroxydihydroaldrin and a related compound lacking the unchlorinated methano-bridge destabilize the cockroach synapse much more rapidly than dieldrin does, requires apparent inversion of previous thinking.[478] If diol formation is a toxication reaction, then blockade of the hydration process should result in antagonism of dieldrin action. The inhibition of aldrin epoxidation results in antagonism but this is believed to be due to the slower action of aldrin in the absence of epoxidation;[584] 6,7-dihydroaldrin is mono-hydroxylated in vivo and the suppression of this conversion results in synergism rather than antagonism.[623] At the time of writing it is not known whether the neurotoxicity of the diols extends to other hydroxy-derivatives and a number of interesting questions remain unanswered (see Section D.2b for further discussion).

c. Interactions with Microsomal Enzymes

One of the most important aspects of the biochemistry of organochlorine insecticides concerns their interaction with the enzymes that metabolize endogenous compounds or other lipophilic drugs and toxicants in living organisms. Compounds metabolized by liver microsomal enzymes include endogenous steroids, fatty acids, various sympathomimetic amines, thyroxin, N-acetyltyramine and N-acetylserotonin, indoles such as tryptamine, and methylated purines. Since organochlorine compounds are also metabolized by microsomal enzymes, they might be expected to compete with some of these endogenous substrates for the active sites of the enzymes as well as increasing the enzyme levels (induction). Enzyme induction by organochlorines was first noticed in the early 1960's in rats that had been kept in a rearing room treated with chlordane and it is now a well recognized effect of these compounds. Enzyme induction in insects was discussed in Chapter 2, section B.3.

The changes induced in mammalian liver during enzyme induction depend on the nature of the compound administered but the general effect is a proliferation of the smooth endoplasmic reticulum (SER) which is the site of metabolizing activity, and a consequent increase in the protein fraction of isolated microsomes, as well as an increase in the level of microsomal cytochrome P450. This effect is seen with phenobarbita, DDT and cyclodiene analogues, which cause nonspecific increases in microsomal enzyme activity, whereas polycyclic aromatic hydrocarbons such as benzopyrene and 3-methylcholanthrene increase some enzymes and not others, without proliferation of the SER, pronounced increase in microsomal protein, or the pronounced increase in liver size (hypertrophy) that may occur with the other compounds. The detailed origins of this effect, which usually results in enhanced metabolism of both inducing agent and other foreign compounds (xenobiotics), are beyond the scope of this discussion, but more superficially the elevated enzyme levels may result from increased protein synthesis or reduced protein degradation or a combination of the two.[624]

The changes in xenobiotic metabolizing activity are normally reversible, as are the effects on the SER, when the inducing agent is withdrawn and the duration of the effect thereafter depends on the type of compound administered and the species concerned. Since the liver is constantly exposed through the diet to a wide variety of foreign compounds unwanted by the body and generally capable of effecting enzyme induction, the change is undoubtedly adaptive but has been the subject of much discussion because of its possible effects on drug therapy or even on hormone regulation and because for certain compounds (such as carcinogens) liver changes may eventually be followed by others of pathological significance. During the experimental administration of a compound to mammals, a close watch is kept on the liver since enzyme induction is often the first sign of bodily response to low levels of exposure and the level of administration below which no change can be detected is the "no effect level" for enzyme induction. Clearly the "no effect level" of a treatment depends on what is measured and the sensitivity of the measurement and a change in some particular tissue may have no effect whatever on the health of the subject. The situation is somewhat akin to the pesticide residue one where "no residue" in a crop depends entirely on the sensitivity of the analytical technique used.

For workers occupationally exposed to cyclodiene insecticides in the manufacturing plants, dieldrin intakes may be from 85 to 175 times (590 to 1224 $\mu$g/man/day) those of the general population in the U.K. and the U.S.A.;[303] endrin intake cannot be quantified because the compound is rapidly metabolized and tissue levels are not measurable. In these individuals, levels of various serum enzymes are regularly measured and also the blood levels of DDE and the ratio of 6$\beta$-hydroxycortisol to 17-hydroxycorticosteroids excreted in the urine. Inducing agents increase the metabolism of DDE and hence reduce the blood level of it below that found in the general population and, by increasing cortisol hydroxylation but not the excretion of total 17-hydroxycorticosteroids, increase this steroid ratio in the urine. The interesting result of this exercise is that endrin plant workers have significantly lower blood DDE levels and significantly higher urinary steroid ratios than the aldrin-dieldrin plant workers, in whom these levels are normal, even at the highest intake mentioned above. Thus, there is no evidence of enzyme induction, measured by these two methods, at the highest levels of aldrin-dieldrin exposure, whereas endrin, or one of

the intermediates in its manufacture, is an inducing agent at undetectable blood levels. The effect, as measured only by blood DDE, in this case, apparently tends to persist in former endrin plant workers who are no longer exposed to this chemical or its intermediates. However, endrin is never detected in food, drinking water, or air, and DDE levels in human adipose tissue from areas where endrin is widely used do not differ from the levels found elsewhere.

The highest daily intake referred to above (1224 $\mu$g/man/day) corresponds to about 0.018 mg/kg/day so that exposures considerably below this level are the ones having real relevance to human health, a fact which should be remembered when reading about the results of animal experiments involving organochlorine treatments applied at levels of, say, 10 to 100 mg/kg in acute toxicity experiments. Rat liver appears to be very susceptible to the effects of enzyme inducers and a threshold level of 1 ppm of cyclodiene epoxides in the diet has been established for microsomal enzyme induction, with a highly significant increase in, for example, microsomal aldrin epoxidation at a dietary cyclodiene level of 5 ppm.[625] In the U.K., the daily human dietary intake of all groups of organochlorines totals about 0.04 ppm, which is 25 times less than the above threshold for cyclodiene epoxide induction in rats. Most of this intake is DDT-group residues and the dieldrin intake is actually about 170 times lower than the threshold for induction in rats. For rats given DDT or toxaphene in the diet, the lowest levels that caused significant induction in one or more microsomal enzymes were 1 ppm and 5 ppm, respectively,[626] so that the human DDT intake is well below this level. The actual threshold levels for induction in humans are not known.

It follows from this discussion that any effects of organochlorine insecticides on drug therapy in humans are most likely to be seen in occupationally exposed people, and they will appear mainly in the form of reduced half-lives of biological activity unless the biotransformation of the drug produces a metabolite of equal or greater activity. Thus, people occupationally exposed to a mixture of organochlorines (mainly lindane, chlordane and DDT) showed enhanced metabolism of antipyrine, and for factory workers having serum and fat levels of DDT-group compounds 20 to 30 times greater than those of a control population, the serum half-life of phenylbutazone was 19%

lower and the urinary excretion of 6β-hydroxy-cortisol 57% higher than in the controls. There was *considerable* variation between individuals and the investigators concluded that only small changes in these parameters would be likely to result from exposure levels found in the general population.[627] The converse effect was apparent in patients taking phenobarbital and/or diphenyl-hydantoin, who had lower tissue levels of DDE and dieldrin than the general population.[718]

Since the first observation that rats held in animal houses sprayed with chlordane exhibited reduced sleeping times when treated with various central nervous depressants, this effect has been clearly established for DDT, DDE, DDD, Perthane, methoxychlor, cyclodienes, and the HCH isomers, and is known to be associated with enhanced microsomal metabolism of the narcotic. The effects observed are dependent on the drug used and the species concerned and reflect not only species differences in metabolizing enzymes but also differing abilities of organochlorine compounds to stimulate the different enzyme systems involved. Thus, chlordane is usually a better inducer than DDT and stimulates (dose 50 mg/kg, injected) the metabolism of zoxazolamine both in vivo and in vitro in mice, whereas DDT (dose 50 mg/kg, injected) shortened zoxazolamine paralysis time and increased its hydroxylation in vitro less effectively, and had no effect on hexobarbital metabolism in either situation.[628] In other mammals, DDT also stimulates hexobarbital metabolism. Effects of dieldrin and DDT have been observed on microsomal enzymes of avian livers and an interesting species difference is seen in a study of the effect of DDT-analogues at 50 to 100 ppm in the diet on pentobarbital metabolism in rats and quail. Most lipophilic drugs appear able to inhibit each other's microsomal metabolism in vitro, and a similar effect often manifests itself in vivo as a period of increased sleeping time in mammals pretreated with the "inducer." As this effect is overtaken by enzyme induction, sleeping time returns to normal before a period of reduced sleeping time ensues. The first phase of this "biphasic" response is usually short with organo-chlorines but an intermediate situation is apparent in both male and female quail, since 100 ppm of DDT, DDE, or their *o,p'*-isomers in the diet produces an increased sleeping time with pento-barbital within the first few days of feeding, followed by a gradual return toward normal

without any stimulation phase. The toxicity of pentobarbital is correspondingly increased. In rats, all the isomers reduced sleeping time and the *o, p'*-isomers were as active in producing their effect as DDT or DDE in both species, despite their much lower accumulation in tissues; accumulation of the *p,p'*-isomers was about tenfold greater in quail than in rats and there is evidently a marked difference in the behavior and effects of these compounds in quail.[512] 2,4'-DDT or DDE pre-treatment also increases the sleeping time of domestic chickens ( 2 to 4 weeks old) treated with phenobarbital, whereas lindane, chlordane, or toxaphene reduce it.[629]

Since organophosphorus insecticides are so widely used, the effects of inducing agents, including organochlorines, on their metabolism has attracted much attention. For phosphorothioates such as parathion (containing the P=S moiety), metabolism is by activation (toxication) to the phosphate (P=O) and by detoxicative oxidation or dealkylation, while the phosphates are rather readily hydrolyzed. The effect on phosphoro-thioate toxicity of pretreatment with an organo-chlorine would therefore appear to depend on the final balance of its effects on these various enzymes involved in biotransformation in the liver and the other contacting tissues, a net increase in tissue phosphate level resulting in an increase in toxicity. However, the situation is complex and at least three recent investigations with parathion and paraoxon point to the lack of correlation between changes in enzyme levels (as measured in vitro after organochlorine administration) and observed toxic effects of these compounds.[630–632] Two of these favor a removal of free paraoxon by "binding" to increased plasma aliesterase induced by the cyclodienes as a mechanism for the protection often afforded by pretreatment with these compounds. The protective single doses of aldrin or dieldrin are about half their respective oral LD50's and it is interesting that while the treatment also protects against toxaphene and DDT, the toxicity of chlordane is increased, as it is by pretreatment with diets containing 400 ppm of DDT. Dieldrin pretreatment reduces dieldrin toxi-city in rats, but not in mice.[633]

Microsomal preparations from male rats fed 25 ppm of DDT in the diet by Gillett[634] showed large increases in epoxidase activity towards aldrin and heptachlor with little evidence for further conversion of the epoxides. Low levels of dieldrin metabolism have been demonstrated in vitro since then, and the rate is substantially increased by 200 ppm of dieldrin in the diet.[633] This last, quite recent observation is important since it provides one of the missing links between the in vitro situation and Street's discovery that rats fed on a diet containing 1 or 10 ppm of dieldrin plus 5 ppm of DDT stored significantly less dieldrin than when DDT was absent.[635] The storage of aldrin, endrin, chlordane, or heptachlor was later shown to be similarly reduced and the effect with dieldrin was accompanied by increased excretion of hydrophilic metabolites in urine and feces. This effect is also produced by the metabolites of DDT, namely, DDE, DDD, DDMU, and DDMS (Figure 8), by chlordane and by barbiturates and other drugs. The range has recently been extended by an examinaion of structure-activity relationships among DDT-analogues which appear to be generally more effective than other compounds. For rats, 4,4'-substitution with electron withdrawing substituents is required for optimal activity, methoxychlor, for example, being virtually ineffective; for a limited number of compounds the effect is correlated with the Hammett substituent constant for electron withdrawing ability. The effect is not related to acute toxicity, since highly active compounds such as ovex (p-chlorophenyl p-chlorobenzenesulfonate), α-chloro-DDT, dicofol and 2,4'-DDT are much less toxic than DDT, nor is it related to fat solubility.[636]

It is reasonable to suppose that the phenomenon is related to microsomal enzyme induction, and a similar correlation is found when structures are related to their effect on NADPH oxidation, cytochrome P450 level and aniline hydroxylation in mice,[637] although comparisons with the rat situation are complicated by species differences. However, there is the problem that inhibitors of protein synthesis known to prevent the effects of inducers did not alter their action on dieldrin storage in the early experiments and there may be some contribution to the effect by an enhanced release of dieldrin from storage sites to the microsomal enzymes.[635] The possibility of interspecific differences is emphasized by other experiments with dogs in which DDT fed with aldrin had no effect on storage of the dieldrin produced. The metabolites produced from dieldrin in these experiments have not been correlated with the known metabolites (Figure 12), but there seems to be a similar qualitative pattern (by chromatography)

when chlordane, heptabarbital, or DDT is the inducer, with quantitative changes in the metabolites that suggest differing influences on the pathways involved. Apart from effects on the primary metabolic processes, it is to be expected that conjugation steps will be influenced by these treatments.

In principle, these results show that the elimination of some organochlorine residues from tissues can be accelerated by treatment with other drugs that induce microsomal enzymes, and potent, short-lived inducers might find application for this purpose in cases in which such "cleansing" became necessary.[718] However, the use of such remedies might prove less desirable than the presence of organochlorine residues, so that such possibilities need to be viewed with caution. Rats chronically pretreated with small doses of dieldrin are protected against subsequent subacute treatment with larger doses, as might be expected from the enhanced metabolism. Pretreatment of rats with either lindane or DDT increases the metabolism of a subsequent single dose of lindane and there are both quantitative and qualitative differences (DDT being more effective for enhancement) in the metabolism induced by these treatments[638] so that this sort of interaction between organochlorines in regard to metabolism is well established for the rat. Little information is available regarding such interactions in other organisms, although dieldrin storage in rainbow trout is said to be reduced by feeding with DDT or methoxychlor, while feeding dieldrin with these compounds increased the storage of their residues in fat.[639]

The foregoing account of interactions mainly concerns those between organochlorines and the enzymes metabolizing other xenobiotics, but the dividing line, if such exists, between these enzymes and those metabolizing endogenous compounds such as steroids, is obscure.[640] Interest in the effects of DDT analogues on steroid metabolism began with the observation that DDD produces a severe adrenocortical atrophy in dogs that is not usually seen in other mammals or man. In humans, DDD affects adrenal metabolism of cortisol and suppresses adrenocortical secretion. The enhanced excretion of 6β-hydroxycortisol by workers occupationally exposed to DDT has already been mentioned and the enhanced hydroxylation of cortisol following 2,4'-DDD treatment is said to be of therapeutic value in the treatment of Cushing's syndrome. Since the early observations of these effects, numerous "DDT-like" compounds have been shown to affect adrenocortical enzyme systems and metyrapone, the well-known inhibitor of steroid 11β-hydroxylation and therapeutic agent, is a development from these discoveries.

In rats fed at 200 ppm of dieldrin in the diet, the adrenals increase in weight and the sectioned organs show an altered steroid metabolic pattern in vitro which has been attributed to the partial inhibition of a steroid 11β-hydroxylase. Chlordane and DDT inhibit the hydroxylation of testosterone in rat liver preparations but twice daily injections of 25 mg/kg i.p. of these compounds for 10 days stimulates its hydroxylation, as well as that of progesterone, estradiol-17β and deoxycorticosterone by microsomal preparations. Testosterone is hydroxylated in three different positions and these reactions in vitro respond differently to inhibition by organochlorine insecticides; similar selectivity is observed with regard to the stimulation following chronic pretreatment, 16β-hydroxylation being most enhanced in vitro.[641]

Most of the effects observed are seen at relatively high levels of exposure, but the accumulation of organochlorine compounds that is observed to occur in the tissues of some birds and wildlife has given rise to concern regarding the effects of these compounds on endocrine function and hence on reproduction. The uteri of rats, and oviducts of rats, white Leghorn chickens and Japanese quail treated by subcutaneous injection with 140, 136, and 190 mg/kg, respectively, of DDT or 2,4'-DDT during 3 days showed an estrogenic response to 2,4'-DDT but not to DDT, and quantities of 2,4'-DDT ten times lower were said to be active.[642] As seen in another comparative study with rats and quail,[512] 2,4'-DDT is a rather powerful inducer of microsomal enzymes in spite of its low tissue storage in these species and an obvious possibility is that the effect may be due to a metabolite of this molecule, which has a free para-position available for hydroxylation. This finding is of interest because technical DDT contains 15 to 20% of 2,4'-DDT. A recent survey of structure versus estrogenic activity shows that those diphenylethane, diphenylmethane and triphenylmethane derivatives having a para-position free or substituted by hydroxyl or methoxyl are active on rat uterus, and the activity disappears when these positions are substituted by halogen or alkyl groups. This activity seems to be well

explained by the similar distances between the para-hydroxy groups in such analogues and the oxygen functions in natural estrogens. Although 2,4'-DDT exhibits estrogenic activity, it differs from the natural estrogen in that it reduces the sleeping time of castrated mature male rats, whereas the female hormone has no effect. It should also be mentioned that polychlorobiphenyls and polychlorotriphenyls are estrogenic, although much less so than 2,4'-DDT.[643]

Reduced eggshell thickness and increased egg breakage have been cited as a cause of the decline in numbers of certain predatory birds and could result from an alteration in the metabolism of the steroids regulating deposition of eggshell calcium. Delayed ovulation and alterations in thyroid activity are said to occur in Bengalese finches and alterations in thyroid activity in homing pigeons treated with 4,4'-DDT. A low calcium diet containing 100 ppm of either 4,4'-DDT or 2,4'-DDT given to Japanese quail for 45 days caused a significant decrease in eggshell weight in both cases; the shells were also significantly thinner and the calcium content significantly lower than in controls.[644] Thin eggshells were also produced by Mallard fed on 10 or 40 ppm of dietary 4,4'-DDT or 4,4'-DDE for up to two breeding seasons.[645] The effects were less severe with 4,4'-DDT than with 4,4'-DDE, eggshells being 13% thinner with the latter compound and the incidence of breakages greater. In other experiments, breeding pairs of American sparrow hawks given diets containing either 0.84 ppm of dieldrin plus 4.7 ppm of 4,4'-DDT, or 0.28 ppm of dieldrin plus 1.4 ppm of 4,4'-DDT (based on wet weight of diet) for several breeding seasons produced eggs with shells significantly thinner than the controls.[646] On the other hand, hens fed 20 ppm each of DDT and DDE in the diet showed no effects on eggshell calcium levels,[629] while groups fed 10 or 20 ppm of dieldrin showed no decline in number of eggs laid or in their fertility, hatchability, or shell thickness, although chick survival was affected at the higher level of treatment; the eggs achieved steady concentrations of dieldrin with time of 12 ppm and 27 ppm, respectively.[647] Other experiments have examined the reproductive success of homing pigeons and Japanese quail treated chronically with dieldrin. Normal reproduction was experienced by the pigeons, treated with 2 mg/kg/week of the toxicant, with an 8 to 10% reduction (not always significant) in eggshell thickness (egg content 12 ppm). Slight effects were observed on the reproductive success of quail at a 10-ppm dietary level of dieldrin, which produced a steady concentration of 20.8 ppm in the eggs after the seventh week of feeding.[647]

Laboratory tests are subject to the objection that they cannot allow for the possible interactions between pesticide exposure and other adverse environmental factors affecting wild populations, and there are also evident species differences in susceptibility to the sublethal effects of organochlorines that make comparisons difficult. The observation[527] that in feeding experiments DDE is more toxic than DDT to pigeons may extend to other species not so far examined and possible effects due to the estrogenic 2,4'-DDT, which may not be readily metabolized by all species, have also to be considered. High residues of DDT group compounds in wildlife are therefore to be taken seriously and since this seems to be the case at the moment a better idea of their general significance will doubtless emerge in due course.

For dieldrin, the results seem generally encouraging since although there is the possibility of extreme differences in species sensitivity, the levels in eggs of experimental birds are much higher than those found in the eggs of most wild birds and residues of 10 ppm or less in the former appear to have little effect on survival.[647] The possibility also exists that different types of organochlorine residues (including the polychlorobiphenyls), when present together in the tissues, might synergize or antagonize each other's biological effect (if any), but there seems to be little information on this point so far. This comment refers, of course, to interactions at the sites of action, those at the level of the drug metabolizing enzymes being well established as we have seen.

## C. TOXICITY

Since the signs of acute poisoning in animals may give the first clues to mode of action, they are discussed later in that context. The present section considers only the quantitative toxicology of organochlorine poisoning in insects and nontarget organisms. For a given route of administration, toxicities can vary widely, being dependent on such factors as the vehicle of administration, age, sex and nutritional status of the animal, biological rhythms, etc. Therefore, interlaboratory comparisons are difficult and large collections of

comparable data from individual laboratories are of most value. This comment applies particularly to the data for mammals and other nontarget organisms; for insects there is often quite good agreement between different laboratories, probably because larger groups can be used. Insect toxicology is directed towards the development of effective insecticides and so is naturally approached from the standpoint of structure-activity relationships. In the case of mammals and other nontarget organisms, the main concern is with the recognition and prevention of adverse effects so that there is much concern with sublethal as well as with acute lethal toxic manifestations.

### 1. Toxicity to Insects and Structure-Activity Relations
*a. DDT Group*

Laboratory evaluations of acute toxicity to insects usually have two main objectives: (i) to provide information about the relationship between structure and activity *at the site of action* (intrinsic activity) so that other active compounds may be designed, and (ii) to provide some idea of the toxicity to be expected under practical conditions. Because of formulation and other problems, the extrapolation required in (ii) is complex and there is frequently a disappointing gap between high toxicity in the laboratory and efficiency in the field. Nevertheless, some mathematical correlations appear to be possible and have been discussed for DDT, cyclodienes, and some other compounds by Sun.[648] With regard to the first objective, laboratory evaluations appear to be of limited value unless there is a constant ratio within a structural series between the observed toxicity and intrinsic toxicity. This consideration is particularly important for organochlorine compounds since for these there is no simple measure of intrinsic toxicity equivalent to the measurement of organophosphate inhibition of cholinesterase preparations in vitro; this inhibition gives a direct measure of interaction between molecule and enzyme and therefore of intrinsic toxicity (since such inhibition appears to be the primary reason for toxicity in most cases) in the absence of the factors such as inert storage, metabolism, etc., considered previously.

The reasons for natural tolerance to DDT analogues were not examined until sometime after DDT-resistance appeared. Therefore the effect of metabolic detoxication on toxicity was not allowed for in early structure-activity investigations, although anomalous results were frequently attributed to "retarded penetration" and other undefined factors preventing access of the toxicant to the site of action. This situation changed dramatically with the discovery that resistant houseflies and several naturally DDT-tolerant insects rapidly detoxified DDT by an enzymatic dehydrochlorination which could be blocked by various inhibitors (see section B.3b). Such compounds, called synergists, stabilize the toxicant in vivo and hence cause restoration of toxicity in resistant insects or its appearance in some naturally tolerant ones when metabolic detoxication is involved;[594] a synergist, simply defined, is a compound which although of itself not toxic at the applied dose, enhances the toxicity of another compound with which it is applied. The role of detoxication in natural tolerance is well illustrated by the grasshopper (*Melanoplus femur-rubrum*) which dehydrochlorinates DDT in its gut and cuticle and has a topical (cuticle application) LD50 of >9000 $\mu g/g$, an oral LD50 of >2500 $\mu g/g$ and an LD50 by injection (by-passing both detoxication sites) of 2 $\mu g/g$.[649]

The discovery of dehydrochlorination and its inhibition by compounds such as WARF-antiresistant and the acaricide DMC led to the use of their mixtures with DDT as toxicants for resistant strains and to their use as diagnostics for resistance or natural tolerance due to DDT-dehydrochlorinase (DDT-ase).[41] Although DDT-ase (discovered about 1949) is an important resistance mechanism, it may be that oxidative detoxication is even more important for natural tolerance, as has proved to be the case with abandoned DDT analogues such as methylchlor (Table 17, compound 5) and ethoxychlor (compound 3). Although the benzylic oxidation to dicofol in vivo was discovered in 1959 (see section B.3a) and a similar enzyme was shown to be present in houseflies at about the same time, the involvement of oxidation in DDT resistance was not fully demonstrated until 1965 when the pyrethrins synergist sesamex, an inhibitor of microsomal mixed function oxidases, was found by Oppenoorth to suppress resistance in one strain of houseflies.[397] In the case of DDT, the products of oxidation have not been identified, but compounds containing alkyl or alkoxy groups

instead of chlorine are much more vulnerable to oxidation, even in normal insects, and the mechanisms, such as oxidative ether cleavage and the oxidation of alkyl groups to carboxylic acids, are predictable. This was recognized in 1964 when several DDT analogues, including methoxychlor, DDD, and Perthane were found to be synergized by sesamex in normal houseflies.[650] The recent investigations of Metcalf in this area illustrate the importance of biodegradability and species differences in toxicity and some of the data are shown in Table 17.[52]

It can be stated regarding DDT analogues in general, but with reservations due to the inevitable differences between species, that any alteration from the DDT structure reduces *apparent* toxicity. The 4,4'-difluoro-analogue of DDT, which acts rapidly and appears sometimes more and sometimes rather less toxic than DDT depending on the species and test method, is perhaps the nearest approach to DDT in the whole series. Besides the grasshoppers mentioned earlier, there are other insects such as the milkweed bug and the Mexican bean beetle in which natural tolerance is related to detoxication. Therefore the available information on structure-activity relations may give some idea of the likely efficiency of analogues in the field but needs to be regarded quite critically from the standpoint of theoretical interpretation. A good approach to the literature on structure-activity relations is through the summaries given by Metcalf for insects in general in 1955,[41] and for houseflies in 1968,[67] but virtually no allowance has been made for metabolic oxidation until quite recently.

Interest in DDT isosteres such as methylchlor dates almost from the beginning of the DDT story. Methylchlor itself ranges from about 0.03 to 0.2-fold as toxic as DDT to most insects but is as good as DDT against some. Piperonyl butoxide, another inhibitor of microsomal oxidations, is expected to suppress oxidation of the methyl groups to carboxyl in vivo, and methylchlor is nearly as toxic as DDT to houseflies in its presence (Table 17). It is ten times or more less toxic (orally) than DDT to various mammals and underwent field trials many years ago, but could hardly compete against the then highly valued residual toxicity of DDT. Dianisylneopentane (1,1-bis(4-methoxyphenyl)-2,2-dimethylpropane; DANP) is of particular interest as the first toxic unchlorinated analogue of DDT; it is 0.1 to 0.25-fold

TABLE 17

Toxicity of Various DDT Analogues to Houseflies, Blowflies, and Mosquitoes and the Effect of a Mixed Function Oxidase Inhibitor (Piperonyl Butoxide)

| | Substituents | | Houseflies[a] Topical LD50 (μg/g) | | | | | | Blowflies[b] | | | LC50 ppm Mosquito larvae[c] |
| | | | S[d] | | | R[d] | | | | | | |
| Compound | 4 | 4' | alone | PB[e] | SR[f] | alone | PB[e] | SR[f] | alone | PB[e] | SR | |
|---|---|---|---|---|---|---|---|---|---|---|---|---|
| (1) DDT | Cl | Cl | 14.0 | 5.5 | 2.5 | 170 | 40 | 4.3 | 11.5 | 8.25 | 1.4 | 0.07 |
| (2) Methoxychlor | $CH_3O$ | $CH_3O$ | 45 | 3.5 | 12.8 | 48 | 4.6 | 10.4 | 10.0 | 4.6 | 2.2 | 0.067 |
| (3) Ethoxychlor | $C_2H_5O$ | $C_2H_5O$ | 7.0 | 1.75 | 4.0 | 11.0 | 4.7 | 2.3 | 6.9 | 7.4 | 1.0 | 0.04 |
| (4) Methiochlor | $CH_3S$ | $CH_3S$ | 225 | 17.0 | 13.2 | 2800 | 500 | 5.6 | 36.4 | 14.5 | 2.5 | 0.21 |
| (5) Methylchlor | $CH_3$ | $CH_3$ | 100 | 17.5 | 5.7 | 135 | 29 | 4.7 | 61.2 | 21.5 | 2.8 | 0.081 |
| (6) | $CH_3$ | Cl | 62.5 | 6.5 | 9.6 | 140 | 35 | 4.0 | 11 | 9.25 | 1.2 | 0.032 |
| (7) | $CH_3$ | $C_2H_5O$ | 9 | 1.7 | 5.3 | 27 | 3.8 | 7.1 | 5.25 | 2.1 | 2.5 | 0.13 |
| (8) | $CH_3$ | $C_4H_9O$ | 33 | 1.95 | 16.9 | 65 | 18.5 | 3.5 | 70 | 28.8 | 2.5 | 0.063 |
| (9) | $CH_3O$ | $CH{=}C.CH_2O$ | 34 | 3.3 | 10.3 | 105 | 7.5 | 14 | 9.5 | 8.5 | 1.1 | 0.050 |
| (10) | H | Cl | 160 | 4.0 | 40 | 175 | 55 | 3.2 | 125 | 122.5 | 1.0 | 0.12 |

(Data adapted from Metcalf et al.[52])

[a] *Musca domestica.*
[b] *Phormia regina.*
[c] *Culex fatigans.*
[d] S=susceptible, R=resistant.
[e] Pretreated with piperonyl butoxide.
[f] Synergistic ratio (SR) = LD50 of compound/LD50 with synergist.

as toxic as methoxychlor to houseflies and can be synergized by mixed function oxidase inhibitors. The synergized toxicities of ethoxychlor and methiochlor (Table 17) nicely illustrate the extremes of biodegradability found with the two types of para-substituent; ethoxy- is less readily, and methylthio- more readily, oxidized by houseflies and so ethoxychlor is the most potent of the alkoxy-derivatives, especially when oxidative degradation is blocked by the synergist. An approximate order of biodegradability is $C_2H_5O < C_3H_7O < CH_3O < iso\text{-}C_3H_7O < CH_3S$.[52] Ethoxychlor is slightly less acutely toxic than DDT to mice and is a good housefly toxicant according to laboratory tests.

The data for compounds 6 to 10 in Table 17 show that various asymmetrical DDT analogues are remarkably active when some allowance is made for biodegradability of the para-substituents. Both monochloro-DDT (Table 17, compound 10) and the corresponding monomethyl-compound are strongly synergized and the results indicate that the receptor proposed by Holan (see Figure 18 in section D.1) can accommodate these asymmetrical molecules (even when one group is as large as n-butoxy; Table 17, compound 8), which raises new questions about interactions of DDT at the molecular level.* The approximate biodegradability of miscellaneous para-substituents is in the order $Cl < H < CH_3 < CH_3O < CH_3S$ and a further interesting feature of such molecules is that replacement of para-chlorine by para-alkyl or alkoxy reduces chemical dehydrochlorination rates[67] and probably also the rate of dehydrochlorination in vivo, so that ethoxychlor, for example, is quite toxic towards houseflies resistant

through DDT-ase action. An exception is methiochlor, for which the ease of dehydrochlorination may be increased in vivo because the increased electron withdrawing influence of the sulfoxide and sulfone groups produced by oxidation favors dehydrochlorination.[38] This may explain the rather poor toxicity and synergism of this compound against the resistant strain.[52] The interrelation between dehydrochlorination and oxidation is interesting and whichever is minor may become significant when the other is prevented, because of increased availability of substrate to the minor pathway.

Results with blowflies (Table 17) show lower and more uniform synergistic ratios and mixed function oxidase activity is known to be low in this insect so that toxicities measured without synergist give a better idea of intrinsic activity than those obtained with houseflies. The high activity of DDD as a mosquito larvicide is well known and many of the compounds examined by Metcalf,[52,67] especially the asymmetrical analogues, are equally active and certainly more active than DDT in laboratory tests. Insecticide combinations with synergists of the 1,3-benzodioxole type (for example piperonyl butoxide; PB) do not normally show equally enhanced toxicity towards mammals because, fortunately, the synergists are more readily degraded in the mammals so far investigated than in the insects. In consequence, the toxicities with synergist frequently show a shift in favor of the mammal as is found with Holan's series of DDT analogues (Table 18) which incorporates biodegradable aliphatic moieties as well as biodegradable aromatic substituents:[651,652]

(1)     $(Ar_1)(Ar_2)CHC(CH_3)_3$

neopentane type

(2)     $(Ar_1)(Ar_2)CHCH_2C(CH_3)_3$

neohexane type

(3)     $(Ar_1)(Ar_2)CH(CH_3)NO_2$

(4)     $(Ar_1)(Ar_2)CHCH(CH_2CH_3)NO_2$

(5)     $(Ar_1)(Ar_2)CHC(CH_3)_2NO_2$

(6)     $(Ar_1)(Ar_2)CC(CH_3)_2CH_2$ (with epoxide O)

*The enantiomeric forms of the monochloro- (Table 17, compound 10), monofluoro-, p-Br, p'-Cl-, and p-Br, p'-F-analogues of DDT have recently been separated. In the case of monofluoro-DDT, the l-isomer is somewhat more toxic than the d-isomer, but the enantiomeric forms of the other compounds are equitoxic.[650a]

TABLE 18

Toxicities of DDT and Several Biodegradable Analogues to Housefly and Mouse

| Compound | Housefly LD50 (μg/fly) | | Mouse LD50 (mg/kg) | Mouse LD50 / Fly LD50 (ppm) |
|---|---|---|---|---|
| | alone | synergized[a] | | |
| (1) DDT | 0.24 | 0.24 | 570 | 47 |
| (2) 1,1-bis(p-chlorophenyl)-2-nitrobutane (Bulan) | 1.69 | 1.78 | – | – |
| (3) 1,1-bis(p-ethoxyphenyl)-2-nitropropane | 0.48 | 0.065 | 1150 | 48 |
| (4) 1-(p-ethylthiophenyl)-1-(p-ethoxyphenyl)-2-nitropropane | 0.16 | 0.015 | 1040(360)[b] | 130(480)[c] |
| (5) 1,1-bis(p-ethoxyphenyl)-2-nitrobutane | 0.55 | 0.061 | 1160(980)[b] | 42(326)[c] |
| (6) 2,2-bis(p-chlorophenyl)-3,3-dimethyloxetane | 1.27 | 0.66 | 3500 | 55 |
| (7) 2,2-bis(p-ethoxyphenyl)-3,3-dimethyloxetane | 0.52 | 0.010 | 1200(380)[b] | 46(760)[c] |

(Data adapted from Holan.[651,652])

[a]5 μg sesamex per fly applied immediately following insecticide.
[b]Figure in parentheses is LD50 with the synergist sesamex at 1:1 or 2:1 ratio with the insecticide.
[c]Figure in parentheses is the ratio of synergized LD50's (mouse LD50's by intraperitoneal injection).

Types 1 and 2 resemble dianisyl neopentane (DANP), types 3, 4, and 5 resemble Prolan and Bulan in which alkoxy-groups replace aromatic chlorine, while type 6 incorporates a new aliphatic structure (3,3-dimethyloxetane) designed to simulate the optimal size of this portion of the molecules as deduced from structure-activity correlations. In the nitro-alkane and oxetane types, compounds containing one or more p-ethoxyphenyl groups instead of p-chlorophenyl have higher housefly toxicity, which appears to correlate with the DDT-ethoxychlor situation indicated in Table 17. A number of these compounds are more toxic than DDT to houseflies, even without synergist, and they have considerable biodegradability. These results show that when chlorine atoms are removed from the DDT molecule, it becomes vulnerable to enzymatic oxidation; the extent to which this situation differs between species and among the numerous DDT-analogues described in the literature is largely unknown and requires examination, if only as an aid to theories of action.

As a further departure from the basic DDT structure, Metcalf has shown that DDT-like toxic action resides in molecules of the type $Ar_1NHCH(CCl_3)Ar_2$ and $Ar_1OCH(CCl_3)Ar_2$, where $Ar_1$ and $Ar_2$ are benzene nuclei with the same or different parasubstituents. The α-trichloromethylbenzylaniline derivatives are generally more toxic than the α-trichloromethylbenzyl phenyl ethers and the toxicity of both types towards houseflies or blowflies is synergized by piperonyl butoxide, indicating degradation in vivo due to microsomal oxidase activity. The p,p'-diethoxy-compounds ($Ar_1 = Ar_2 = p$-ethoxyphenyl-) show high activity, as in the DDT series, while an unsymmetrically substituted α-trichloromethylbenzylaniline ($Ar_1 = p$-methylphenyl-, $Ar_2 = p$-ethoxyphenyl-) is more toxic than DDT to both houseflies and blowflies, and quite toxic to mosquito larvae. The α-trichloromethylbenzylanilines are much less toxic to mice than DDT and are biodegradable because the $-NC(CCl_3)-$ system is easily cleaved to give the corresponding substituted aniline and phenyl dichloromethyl ketone moieties, in addition to the usual possibilities of enzymic attack on the p,p'-substituents.[652a]

Information about intrinsic activity may be

obtained by electrophysiological techniques *measuring the* response of the intact insect nervous system to toxicants (which assumes no metabolism there); otherwise, the simplest method for organochlorines and their relatives is by the use of metabolic inhibitors and measurement of synergism. When using this technique, the possibility should be borne in mind that synergism might be due to causes other than inhibition of metabolic detoxication (examples are available from mammalian pharmacology) but its relationship to this phenomenon now seems well established.

With the exception of the reductive dechlorination product DDD, all the metabolites of DDT so far identified in insects are very much less acutely toxic than DDT and may be regarded as detoxication products from this standpoint. DDD may be up to tenfold less toxic than DDT, depending on the insect, but for certain lepidopterous larvae and mosquito larvae it is more toxic. It is also a constituent of technical DDT along with the estrogenic 2,4'-DDT, which is from 20 to 100 times less toxic than DDT to various insects. The 2,2'- and 2,3'-isomers of DDT are less toxic than 2,4'-DDT to houseflies, whereas the 3,4'-isomer is nearly as toxic as DDT.[41] However, DDD and 2,4'-DDT are synergized by certain mixed function oxidase inhibitors, indicating that their intrinsic toxicity is not normally reached; it is not known whether the other isomers, which have free para-positions that are possible sites of aromatic hydroxylation, can also be synergized.

The question of sublethal effects resulting in deformity, sterility, etc., has been mentioned in connection with resistance (in Chap. 2, section B.3) and it is often difficult to distinguish between truly sublethal effects and those due to selective mortality. For example, infertile eggs may result from either selective mortality in the ovaries following contact with insecticide present in the adult or from changes in adult reproductive physiology induced by sublethal amounts of the insecticide.[716]

## b. Cyclodiene Insecticides

Unlike the situation with DDT, few accounts of the insect toxicity of cyclodiene compounds have been given. Most of them relate to houseflies, but the comprehensive paper by Soloway gives relative toxicities of more than 100 compounds to six different species.[184] Before the discovery of the extensive range of photoisomers derived from cyclodienes, there seemed to be little scope for structural variations in this series, since the replacement of chlorine atoms by other groups such as hydrogen, methyl or methoxy- almost always reduces or abolishes toxicity.

The basic structures hexachloronorbornene (Figure 16, structure 1; X, Y=H) and hexachloronorbornadiene (structure 1; X, Y=double bond) lack measurable toxicity, which only appears when the latter molecule carries one or more halomethyl groups. Toxicity increases through the series (X=H; Y=$CH_2Cl,CH_2Br$ or $CH_2I$); the compounds Bromodan (X=H, Y=$CH_2Br$) and Alodan (X=Y=$CH_2Cl$) have housefly toxicities 0.05 to 0.1 that of dieldrin and very low mammalian toxicity.[475] When the X, Y addendum is a simple ring system such as cyclopentane (X-Y=-$(CH_2)_3$-) or cyclohexane (X-Y=-$(CH_2)_4$-) the housefly toxicity is 0.01 to 0.02 that of dieldrin, while the furan derivative (X-Y=$CH_2OCH_2$-) lacks toxicity. These molecules are vulnerable to microsomal oxidation in vivo; the low degree of synergism with sesamex suggests that the intrinsic toxicities of chlordene and dihydrochlordene (X-Y=-CH=$CHCH_2$- and -$(CH_2)_3$-, respectively) and the furan (X-Y=-$CH_2OCH_2$-) are low, but the analogous cyclohexene and cyclohexane derivatives are synergized tenfold, indicating higher intrinsic toxicity to houseflies for the larger rings.[192]

Further chlorination increases the toxicity of all these structures, but only chlorinated derivatives of chlordene and dihydrochlordene have been of practical interest. Additional chlorine atoms increase the spacial volume of the molecules and also serve to protect them from microsomal oxidation. The spectacular increase in toxicity (more than 50-fold) when chlordene (Figure 10, structure 10) is chemically converted into heptachlor (Figure 10, structure 1) appears to relate largely to protection of the latter molecule and its epoxide (Figure 10, structure 2) against microsomal oxidation as indicated elsewhere (see section

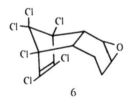

Hexachloronorbornene (X=Y=H)

Dihydroaldrin    Dihydroisodrin

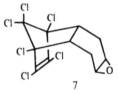

FIGURE 16. Hexachloronorbornene (1, X=Y=H) and some compounds derived from it by ring fusion through the *endo*-bonds (X and Y).

B.2c). The 1-fluoro-analogue of heptachlor is also highly toxic to several insects.[184] There is a similar toxicity difference between *trans*- and *cis*-chlordane (Table 19, structures 4 and 5, respectively) for most species examined which seems to parallel the difference between the two epoxides of heptachlor (Figure 10, structures 2 and 4); these four compounds do not seem to be significantly. metabolized in houseflies so that toxicity differences may be a genuine reflection of stereochemical influences at the site of action. The difference in toxicity between the nonachlor (Table 19, structure 9), which according to its melting point may actually be a *trans*-nonachlor (see footnote, Table 19) and the chlordane isomer (7) is noteworthy. The latter is virtually isosteric with isobenzan (Figure 11, structure 11), another compound that is highly toxic to both insects and mammals (Table 20).

Compounds 1 to 3 of Table 19 are the

TABLE 19

Toxicities to Housefly, Mosquito Larvae, and Mouse of Dihydroheptachlor and Chlordane Isomers and a Nonachlor Isomer

| Compound[a] | B | C | D | E | LD50, housefly (mg/kg) | Mosquito[b] larvae LC50 (ppm) | LD50, mouse (mg/kg) |
|---|---|---|---|---|---|---|---|
| 1 | exo-Cl | H | H | H | 13.0 | 0.3 | 1285 |
| 2 | H | exo-Cl | H | H | 8.0 | 0.15 | >9000 |
| 3 | H | H | endo-Cl | H | 90.0 | 2.4 | >6000 |
| 4 | exo-Cl | H | endo-Cl | H | 11.0 | 0.27 | 1100 |
| 5 | exo-Cl | exo-Cl | H | H | 4.0 | 0.12 | 500–600 |
| 6 | H | exo-Cl | endo-Cl | H | 2.0 | 0.075 | >600 |
| 7 | exo-Cl | H | H | exo-Cl | 2.0 | 0.045 | 31 |
| 8 | H | exo-Cl | H | H | 20.0 | 0.2 | – |
| 9[c] | exo-Cl | exo-Cl | H | exo-Cl | >100 | >3 | >400 |
| Dieldrin | – | – | – | – | 1.0 | – | 75–100 |

(Data adapted from Buchel et al.[205,206])

[a]In compound 8(A=Cl); others A=H.
[b]*Aedes aegypti*.
[c]This compound may be *exo, endo, exo*.

dihydroheptachlor (DHC) isomers (Vol. I, Chap. 3B.2a) investigated by Buchel.[205,206] They are oxidatively detoxified in vivo and are each synergized tenfold by sesamex, indicating intrinsic housefly toxicities correspondingly higher than are observed in the normal way.[234] The mammalian toxicities are low and β-DHC (Table 19, structure 2) is degraded quite rapidly by rats and by pig liver microsomes, but has not so far found practical application. Heptachlor epoxide (HE 160) formed biologically from heptachlor is the higher melting isomer; it is about sixfold more toxic to houseflies than the lower melting isomer (Figure 10, structure 4) and appears to be about as toxic to mammals as dieldrin. The unchlorinated analogue, chlordene epoxide (Figure 10, structure 11) is subject to microsomal oxidation in houseflies and is much less toxic;[243] it is quite toxic to the milkweed bug, however, raising the possibility of a species difference in metabolism.[184] Oxychlordane (Figure 11, structure 8) is now known to be one of the products formed from the chlordane isomers in several mammals and appears to be rather more toxic than *trans*-chlordane in both

acute and chronic toxicity tests, no insect toxicities having been reported so far. This remarkable transformation shows that in some circumstances fully saturated compounds can be converted into epoxides in vivo. In general, compounds of the chlordane and dihydroheptachlor series appear to be somewhat less effective as mosquito larvicides than the best DDT analogues although there are one or two exceptions.

Structurally, isobenzan (Figure 11, structure 11) is a compact version of heptachlor epoxide and this is reflected in the high insect toxicity. The high mammalian toxicity (Table 20) is unexplained, but certain related cage molecules have remarkably high toxicities and the indications are that such structures are approaching the ideal in terms of fit to a target site.[655] It is interesting but not really surprising that endosulfan, although derived from the same precursor as isobenzan, behaves quite differently. The insect toxicity of endosulfan is lower and the mammalian toxicity considerably lower, undoubtedly because the molecule can be attacked by detoxifying enzymes in a way that isobenzan cannot. An interesting

TABLE 20

A Selection of Toxicity Data for Some Cyclodiene Insecticides and Their Conversion Products

| Compound | Topical 24 hr LD50 to housefly ($\mu$g/g) | LC50 (ppm) to mosquito larvae (*Ae. aegypti*) | Rodent acute oral LD50 (mg/kg)[a] |
|---|---|---|---|
| Aldrin | 1.5 | 0.003 | 38–60 |
| Photoaldrin (Figure 13, structure 10) | 0.35 | 0.0005 | |
| Dieldrin (Figure 13) | 1.0 | 0.006 | 46.8 |
| | | | 77.3(m) |
| Photodieldrin (Figure 13, structure 11) | 0.3 | 0.003 | 9.6 |
| | | | 6.8(m) |
| Monodechlorodieldrin | 1.47 | | 15.8 |
| (Figure 13, structure 14) | | | |
| Bis-dechlorodieldrin | 0.20 | | 0.92 |
| (Figure 13, structure 15) | | | 1.4(m) |
| 9-Hydroxydieldrin | | | >400(m) |
| (Figure 12, structure 10) | | | |
| Klein ketone (Figure 12, structure 4) | <HEOD | | |
| *trans*-Dihydroaldrindiol | | | 1250(m) |
| (Figure 12, structure 6) | | | >850(m) |
| Isodrin (Figure 14) | 3.0 | 0.019 | 12–17 |
| Photoisodrin (Figure 14, structure 3) | 15.0 | 0.058 | >2000 |
| Endrin (Figure 14) | 2.0 | 0.017 | 7.5–43 |
| | | | 29(m) |
| Δ-Keto-endrin | >12.0 | >0.096 | 10–36[b] |
| (Figure 14, structure 10) | | | 120–280[c] |
| Endrin aldehyde | >12.0 | >0.096 | >2000 |
| (Figure 14, structure 12) | | | 500(m) |
| Heptachlor (Figure 13, structure 1) | 1.0 | 0.005 | 100–162 |
| Photoheptachlor | 0.28 | 0.002 | |
| (Figure 13, structure 4) | | | |
| Heptachlor epoxide | 1.0 | | 60 |
| (Figure 10, structure 2) | | | 18(m) |
| Unbridged photo-heptachlor epoxide | >2500 | | 36(m) |
| Bridged photo-heptachlor epoxide | 2.0 | | 6(m) |
| 1-Hydroxychlordene | low | low | 2400–4600 |
| (Figure 10, structure 8) | | | |
| Chlordene (Figure 10, structure 10) | 50 | | |
| Isobenzan (Figure 11, structure 11) | 1.0 | | 3–10 |
| | | | 6(m) |
| Chlordane (technical) | 7–8 | 0.21 | 457–590 |
| Chlordane (trans) | 11.0 | 0.27 | 1100 |
| Chlordane (cis) | 4.0 | 0.12 | 500–600 |
| Endosulfan (technical) | 6.0 | | 110 |
| Endosulfan (alpha) | 5.5 | | 76 |
| Endosulfan (beta) | 9.0 | | 240 |
| Endosulfan sulfate | 9.5 | | 76 |
| (Figure 11, structure 12) | | | |
| Endosulfan diol | >500 | | >15,000 |
| (Figure 11, structure 9) | | | |
| Endosulfan ether | >500 | | >15,000 |
| (Figure 11, structure 13) | | | |
| Endosulfan hydroxy-ether | >500 | | 1,750 |
| (Figure 11, structure 14) | | | |
| Endosulfan lactone | | | 306(m) |
| (Figure 11, structure 15) | | | |
| β-Dihydroheptachlor | 8.0 | 0.15 | >9,000(m) |
| (Figure 11, structure 1) | | | |

TABLE 20 (continued)

| Compound | Topical 24 hr LD50 to housefly ($\mu$g/g) | LC50 (ppm) to mosquito larvae (*Ae. aegypti*) | Rodent acute oral LD50 (mg/kg)[a] |
|---|---|---|---|
| Alodan (Figure 16, structure 1; X=Y=CH$_2$Cl) | 15.5 | | >15,000 |
| Bromodan (Figure 16, structure 1; X=H, Y=CH$_2$Br) | 11.5 | | 12,900 |

(Data adapted from Balwin,[565] Brooks and Harrison,[192] Buchel et al.,[204-206] Ingle,[269] Jager,[303] Khan et al.,[554,572] Korte,[508] Maier-Bode,[248] Miles et al.,[287] Soto and Deichmann,[480] Winteringham,[475] and Ivie et al.[244a])

[a]For rat unless otherwise indicated.
[b]female.
[c]male.
m=mouse.

point here is that the endosulfan isomers and isobenzan belong to the limited number of cyclodiene insecticides that retain some toxicity towards dieldrin resistant houseflies.[192,653]

Aldrin and isodrin (see Figures 13 and 14) are effectively the result of linking the 1 and 3 positions of the cyclopentane ring in dihydrochlordene (X-Y=-(CH$_2$)$_3$- in Figure 16, structure 1) with an ethylene bridge, thus making the whole structure more rigid. The biological conversions of these compounds have been considered in detail and the weight of evidence indicates that aldrin and isodrin are toxic per se, without requirement for epoxidation, but act more slowly than the corresponding epoxides, possibly because they reach the site of action more slowly.[654] The best evidence for this comes from a comparison of the toxicities of the molecules 6,7-dihydroaldrin, its oxygen-bridge analogue (structure 4) and 6,7-dihydroisodrin,[584] which are shown in Figure 16. Houseflies detoxify all three compounds by hydroxylation in vivo and the inhibition of this process results in synergism tenfold; with the first two- and 20-fold with the oxygen analogue.[192] The oxygen analogue, which has a "built in" oxygen bridge, then becomes as rapidly acting and nearly as toxic as dieldrin, whereas the dihydroderivatives, although potentiated, still exert their effect more slowly; oxygen bridges appear to have the role of either improving the rate of arrival at the site of action or of improving the interaction of the molecule with the site. The precise location of the oxygen function seems not to be too critical if the structures of all intrinsically toxic epoxides, including optical isomerides, are considered.[547]

In summary, the situation seems to be that when aldrin, isodrin (and heptachlor) enter houseflies and other insects, they are converted into their epoxides which are rather stable toxicants, so that the total internal level of poison is maintained (the effect is actually increased, taking into account the more rapid action of the epoxides). With dihydroaldrin and dihydroisodrin the internal level of toxicant is reduced due to oxidative detoxication, so that less remains of a slower acting toxicant, with obvious advantage for the organism. When metabolic detoxication is allowed for so that intrinsic toxicities can be considered, there are obviously other molecules in the chlordane and dihydroheptachlor series that possess no epoxide rings but have considerable activity, and the discovery of a number of photoisomers having high insect toxicity adds a new dimension to the picture.[655]

Although the removal of one vinylic chlorine from dieldrin reduces housefly toxicity, the removal of two increases it fourfold, indicating that these two chlorines are unnecessary for toxicity. All other changes reduce toxicity to houseflies, but Soloway's data[184] indicate some interesting species differences, and it seems remarkable that the aldrin analogue having chlorine atoms at only C$_1$ and C$_4$ not only has measurable toxicity to houseflies and blowflies, but is highly toxic to the German cockroach. Removal of the hexachloronorbornene chlorine atoms may be expected to expose the resulting molecules to attack by microsomal enzymes and information about intrinsic toxicities is totally lacking. The preparation of isosteres having methyl groups instead of chlorine atoms might well prove interesting, with due allowance for the associated metabolic possibilities, but such derivatives have scarcely been examined.

Considering the available information, Soloway concluded that a toxic cyclodiene molecule has two appropriately positioned electronegative centers (for example, the chlorinated system of the hexachloronorbornene nucleus together with an epoxide ring or the additional chlorines in the case of the chlordane isomers) which lie on or across the plane of symmetry defined by the dichloromethano-bridge. However, when intrinsic toxicities can be examined, it is evident that a second electronegative center is not essential for toxicity (for example, dihydroaldrin), nor need it lie across the plane of symmetry (for example, α-dihydroheptachlor). Nevertheless, the most *rapidly acting* molecules do have such a center and the aldrin analogue of Figure 16, having the rather electronegative azo-moiety instead of the carbon-carbon double bond, is rather more toxic than dieldrin to housefly, milkweed bug, and German cockroach.[184]

The photochemical conversion products of cyclodiene insecticides (Vol. I, Chap. 3B.4; see also Figures 13 and 14) have attracted considerable attention during the last few years because of the possibility that the "bridged" molecules may arise in the environment through the action of sunlight. Unfortunately, there is no comprehensive account of their toxicity and in most cases lack of sufficient material has so far prevented full toxicological evaluations. Table 20 lists some acute toxicity data reported for cyclodiene insecticides, their metabolites and photochemical conversion products.

In the isodrin-endrin series, the three rearrangement products, photoisodrin (photodrin), the endrin hexachloro-half-cage ketone (Δ-keto-endrin), and the endrin hexachloro-aldehyde (Figure 14, structures 3, 10, and 12, respectively), tend to be less toxic than their precursors. On the other hand, the conversion of aldrin into photoaldrin (Figure 13, structure 10), dieldrin into photodieldrin (structure 11) and heptachlor into photoheptachlor (structure 4) results in increased toxicity to houseflies and mosquito larvae (*Aedes aegypti*).[572] In these insects photoaldrin is converted into photodieldrin and finally into the Klein ketone (Figure 12, structure 4) which appears to be more toxic than the precursors. However, full toxicological evaluation of the ketone has not yet been possible and

it is important to avoid generalizations from these few results. Since the removal of vinylic chlorines requires high energy UV irradiation it is unlikely to occur in the lower atmosphere. Therefore, the formation of dechlorinated bridge molecules (such as structure 16 in Figure 13) which could arise from such dechlorinated precursors in normal sunlight, is unlikely unless quantities of these precursors are formed by the cycling of cyclodiene residues to the upper atmosphere and back to earth. However, current information suggests that microorganisms may be able to effect such conversions. With the exception of photoisodrin the rearrangement products are rather compact molecules having the two electronegative centers discussed by Soloway; more detailed discussion of the structure-activity relations of these compounds is available elsewhere.[655]

The *trans*-diol (Figure 12, structure 6) from dieldrin evidently destabilizes cockroach synapse more rapidly than dieldrin does,[656] but with this reservation, the general indications are that the various metabolic hydroxylation products formed from cyclodiene insecticides in vivo are detoxication products (Table 20). The isomeric oxygenated analogues (4), (6; HCE, and 7; HEOM) of Figure 15 illustrate interesting differences in biodegradability. Compound (4) and HCE suffer only oxidation in vivo (in houseflies), and there is no hydration of the epoxide ring of HCE to form the *trans*-diol. Chlordene epoxide behaves similarly. At the other extreme, HEOM is hydrated so rapidly that no oxidation products appear. Accordingly, compound (4) and HCE approach dieldrin in toxicity when the oxidations are inhibited by sesamex, but the hydration process is virtually unaffected by this inhibitor and HEOM is not toxic to this insect or to the blowfly, which also hydrates it rather rapidly.[484] Insects such as the tsetse fly and to a lesser degree the stable fly (*Stomoxys calcitrans*) appear to be deficient in certain of these enzyme systems. Compared with DDT, HCE is more toxic to tsetse, while HEOM is from about half as toxic to equitoxic, depending on the species. The low degree of synergism observed with sesamex is a further indication of the lack of oxidative enzymes in this insect. HCE is more toxic than DDT to *S. calcitrans* and about eightfold less toxic than dieldrin, but HEOM is considerably less toxic than DDT to this insect. These results illustrate the influence of interspecific metabolic differences on toxicity and show, incidentally, that the molecule HEOM is intrinsically toxic to at least one insect. HCE is both hydrated to the *trans*-diol and oxidized (hydroxylated) by mammals and is less toxic than dieldrin to rat and mouse (acute oral LD50s > 400 and 200 to 400 mg/kg, respectively).[484] HEOM has not been tested but is hydrated rapidly by several mammals in extreme contrast to dieldrin,[576] and for this reason is unlikely to have high oral toxicity. Thus, the preparation of biodegradable cyclodiene epoxides is possible in principle, but such compounds are likely to have a rather narrow spectrum of insecticidal activity.

### c. Hexachlorocyclohexane and Toxaphene

The remarkable feature of lindane is the uniqueness of its structural requirement for toxicity. Its stomach, fumigant or contact action ranges from about 48 to 10,000 times that of the $\alpha$-, $\delta$-, or $\eta$-isomers, while the $\beta$- and $\xi$-isomers have negligible activity. Among the relative toxicities listed by Metcalf,[41] the $\delta$-isomer is cited as having sevenfold lower toxicity than lindane to the confused flour beetle (*Tribolium confusum*) by residual contact and 48-fold lower oral toxicity to the southern armyworm (*Prodenia eridania*) and these citations seem to represent the closest approach by other isomers to the toxicity of lindane. According to an early report the presence of large excesses of the inactive isomers, either individually or in technical mixtures, does not affect the toxicity of the $\gamma$-isomer and this seems to be borne out by the usefulness in practice of technical HCH containing only about 14% of lindane. However, there is evidence of interaction between the isomers in mammals (see section D.2a).

There have been occasional reports of the toxicity of more highly halogenated HCH analogues such as a $\delta$-heptachlorocyclohexane and a fluorohexachlorocyclohexane and there is a recent report describing the presence of a new HCH isomer in the residue of the technical material after lindane isolation.[345] Little insecticidal activity was found among the numerous compounds prepared by Riemschneider; his weakly active $\gamma$-dibromotetrachlorocyclohexane is shown in Figure 17, together with the $\gamma$-iodopentachlorocyclohexanes recently prepared in Japan.[361] The most active mixed halogen compound discovered thus far is a $\gamma$-dibromotetrachlorocyclohexane

(Figure 17, last structure; X = Cl, atoms 1 and 4 are bromine), which is said to be more toxic than lindane to adult mosquitoes (*Culex pipiens pallens*), the iodo-compounds reported being very much less toxic (more than 1,000-fold).[358] It follows from this structural specificity for toxicity that dehydrochlorination destroys insecticidal activity, so that pentachlorocyclohexenes derived from HCH isomers are nontoxic, as are the various other derivatives (Figure 9) produced by insects in vivo.

It may be that isosteres of lindane that have high potential insecticidal potency can be made using sterically appropriate combinations of chlorine atoms with methyl-, methoxy-, or other groups (see also Vol.I, Chap.4C.), and the methoxypentachlorocyclohexane of Figure 17 (last structure) is said to approach lindane in activity toward mosquitoes (*C. pipiens pallens*).[657] Such groups may well confer additional biodegradability so that tests with synergists may be required to reveal intrinsic toxicity. There are interesting reports in the literature of the toxicity of mixtures of alkyl halogen derivatives of cyclohexane.[41] Thus, a mixture of 1-methyl- and 1-chloromethyl-1,2,3,4,5,6-hexachlorocyclohexane is said to be toxic to various ectoparasites. The hexamethyl ether of *meso*-inositol is not a true isostere of lindane, yet showed some toxicity towards houseflies, and hexaethylcyclohexane is weakly toxic to grain weevils. Since all these compounds contain substituents that are potentially vulnerable to microsomal oxidation a repetition of these tests with appropriate synergists might reveal that their intrinsic toxicities are higher than has been supposed. The hexamethyl ether of mucoinositol, which has the $\gamma$-configuration (aaaeee) is of some interest in this context.

The action of lindane is usually rapid and its high toxicity compared with other organochlorines is seen in Table 21. Since metabolic detoxication occurs, especially in houseflies, the high toxicity is remarkable and suggests an excellent intrinsic toxicity. Busvine[218] pointed out the resemblance between the chlorinated system of lindane and that of the cyclodienes and the cross-resistance between the two structural types (see section B.3b) strengthens the suggestion of a relationship. Interesting conclusions follow from the supposition that lindane offers the most appropriate steric arrangement of chlorine atoms for toxicity in these two series.[655] The molecules are consid-

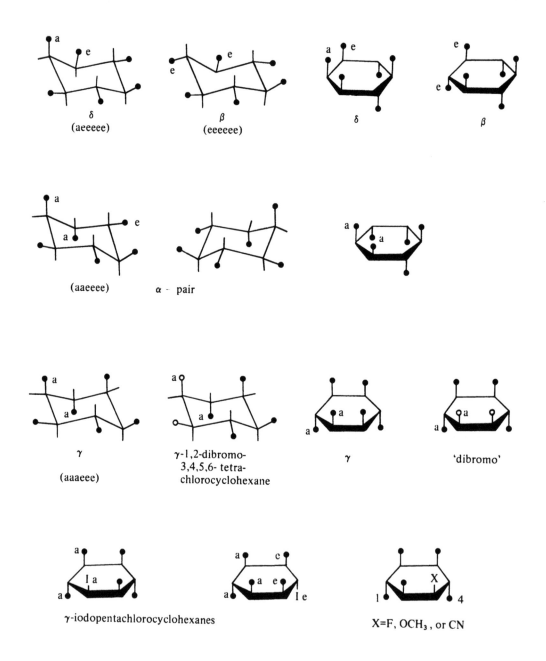

FIGURE 17. Conformations of the major isomers of hexachlorocyclohexane, and of a weakly insecticidal dibromo-analogue of γ-HCH. Also shown are the same structures derived from a "planar" cyclohexane ring. In the planar structures the plane of symmetry (absent in α-HCH) is in the plane of the paper. Two corresponding atoms are designated in the planar and conformational structures for orientation. The iodo-compound on the left may be in the ring conversion form in which the bulky iodine atom is equatorial.

TABLE 21

Insecticidal Potencies Relative to DDT (=1.0) of Several Organochlorine Compounds Toward Pests of Medical Importance

| Insect[a] | Methoxychlor | DDD | Chlordane | Aldrin | Dieldrin | γ-HCH | Toxaphene |
|---|---|---|---|---|---|---|---|
| Housefly (Musca domestica) | 0.35–0.92 | 0.25–0.5 | 1.2–3.0 | 7–11 | 10–17 | 3.3–17.0 | 0.15–0.44 |
| Mosquito (Aedes aegypti) | | | 0.34,0.66 | 2.6 | 4.0 | 2.0,2.3 | 0.05,0.14 |
| Bedbug (Cimex lectularius) | 0.97 | 0.44 | 0.60 | 1.9 | 3.7 | 13 | 0.40 |
| Argasid tick (Ornithodorus moubata) | | | 2.5 | | 25 | 25 | 2.4 |
| Body louse (Pediculus humanus) | 0.33 | 0.33 | 3.0 | | | 20 | |
| Aedes spp. | 0.09 | 1.0,3.3 | 0.40 | | 2.0 | 0.5 | 0.25 |
| Anopheles quadrimaculatus | 0.10 | 1.0 | 0.75,1.0 | 0.53 | 2.6 | 0.16,0.35 | 0.78 |
| Anopheles albimanus | 0.08 | 1.5 | | | | | |
| Culex fatigans | 1.05 | 1.8 | | | | | |

(Data adapted from Busvine;[282] Metcalf and Fukuto.[67])

[a]First five insects nymphs and adults, last four mosquito larvae.

ered in the manner of Soloway,[184] in which cyclodienes are projected in the hexachloronorbornene plane of symmetry containing the dichloromethano-bridge and lindane is projected in the plane containing its central axial and central equatorial chlorine atoms (Figure 18). If the central axial chlorine atom of lindane (in projection) overlaps the approximate position occupied by the epoxide ring of dieldrin (second electronegative center) in a hypothetical receptor, then lindane can be orientated so that its central equatorial chlorine falls approximately in the position of the dichloromethano-bridge chlorine anti- to the hexachloronorbornene double bond. In this position the syn-chlorine atom of the dieldrin bridge (b, in Figure 18) and the two vinylic chlorine atoms (v) are seen to be superfluous and for other orientations of lindane in which the central axial chlorine remains close to the epoxide ring the two vinylic chlorines project beyond the lindane profile; when these chlorine atoms are removed from dieldrin the molecule becomes more toxic (Table 20). If the chlorinated double bond is removed, for example by the formation of photodieldrin, the molecule becomes more compact and the formerly vinylic chlorines, which are now attached to a norbornane rather than a norbornene nucleus, lie in positions more closely corresponding to those occupied by the

outer axial chlorine atoms of lindane. Thus, if the configuration of lindane does indeed represent the ideal for this type of toxicant then some explanation may be found for the higher toxicity of certain cyclodiene photoisomers; photoisodrin appears exceptional but in fact is detoxified in vivo and can be synergized.[572,584] Further studies on the metabolism and toxicity of photoisomers should shed some light on this situation.[655]

Little can be added to the information about polychloroterpene insecticides (toxaphene group) given in Chapter 5 of Volume I and in the sections of Volume I on insecticide usage (e.g. Vol. 1, Chap. 2A.3b) since the individual constituents have not been isolated. Similar remarks apply to adamantane, which has not been used commercially in any case. Toxaphene is compared with DDT against insects of medical interest in Table 21.

### 2. Toxicity to Vertebrates and Other Nontarget Organisms
### a. DDT Group

Despite the widespread controversy concerning the environmental hazards attending the use of DDT in pest control,[717] it is indisputable that in relation to the quantities used and the benefits it has conferred in terms of increased food production and freedom from disease the safety record of

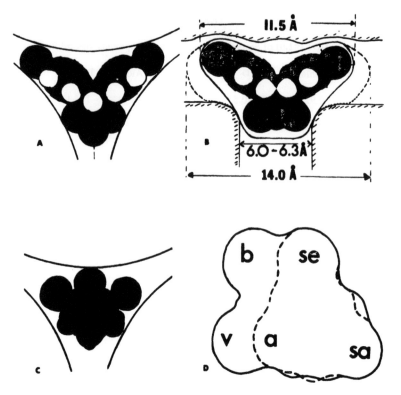

FIGURE 18. Hypothetical receptors for DDT and lindane, and structural relationships between lindane and dieldrin. Fit of DDT into: A, the membrane interspace postulated by Mullins,[614] and B, the protein-lipid interface of the receptor proposed by Holan.[671] Of the HCH isomers, only lindane fits into the Mullins receptor in planar orientation, as in C. In D, lindane (dotted curve) is projected in its plane of symmetry defined by the central axial chlorine *(sa)* and central equatorial chlorine *(se)* and placed upon dieldrin projected in the plane of symmetry of the hexachloronorbornene nucleus. Thus, *sa* overlaps the epoxide ring projection; *b* is the projection of the dieldrin methano-bridge chlorine closest to the double bond, *v* the combined projection of the two vinylic chlorines and *a* the combined projection of the remaining two axial chlorines of lindane.

DDT for man is quite outstanding (see Vol.I, Chap.2A.3d). By 1950, after some 10 years of intensive wartime development and subsequent postwar use, 14 deaths due to DDT poisoning had been confirmed, and all of them were the result of massive accidental ingestion or suicide.[24] Workers in DDT factories have been exposed to intakes of 0.25 mg/kg/day for up to 19 years without more effect than the expected increased storage and excretion of DDT and its metabolites and some elevation of liver microsomal enzyme levels; this daily intake is about 600 times greater than that of the general population of the United States.

The quantity of DDT required to kill 50% of a test group of mammals by a single oral dose (acute oral toxicity) varies widely according to species,

medium of administration, and so on, but the minimum recorded figure appears to be that given for dogs of 60 to 75 mg/kg.[658] For rats, the figure ranges from 113 to 400 mg/kg and an acute toxicity on dermal application of 2,500 mg/kg is recorded for females, while for rabbits and mice the dermal toxicities are rather higher (250 to 500 mg/kg). For goats, sheep, and chickens, oral toxicity is said to lie between 1,000 and 2,000 mg/kg. From a consideration of an early fatal poisoning episode, together with some parallel experiments on baboons, a lethal oral dose of 150 mg/kg (when administered in kerosene) was arrived at for man.[24] A single oral dose of 10 mg/kg is sufficient to cause *illness* in some but not all men.[658a] The compound is much more toxic by intravenous injection, the LD50's being about ten times lower by this route of administration. *DDD* has a much lower oral toxicity (3,400 mg/kg for rats) and the dermal toxicity for rabbits is also much lower (4,000 mg/kg), while methoxychlor is even safer, with an oral toxicity of 6,000 mg/kg for rats and 800 mg/kg for mice. Perthane, which is biodegradable and resembles methoxychlor in showing much less tendency than DDT to store in fat, has an oral LD50 of 8 to 9 g/kg for rats and mice. Acute oral toxicities for birds depend on the carrier used for administration; for DDT given in capsules the LD50's range from 841 to more than 4,000 mg/kg for young coturnix quail, lesser sandhill cranes, young pheasants, young mallards, and pigeons, with the sensitivity decreasing in that order.[583]

Wild mammals appear to be rather resistant to DDT, since forest spraying in Ontario at 6 lb per acre affected none of them; in fact no mortalities were observed until the treatment level was raised to 100 lb per acre. Birds are more susceptible and while 1 to 2 lb per acre had little effect, 5 lb per acre caused severe decimation of bird populations. Fish are highly susceptible to organochlorine insecticides in general and DDT (LC50 typically 0.005 ppm) is more toxic than methoxychlor, lindane and some of the cyclodienes such as chlordane but less toxic than endrin (LC50 0.0005 to 0.002 ppm).[2,24,583] Unfortunately, quite modest treatments at say 0.5 to 1.0 lb per acre can produce these concentrations in water. The acute effects of DDT on various forms of wildlife were well recognized in the 1940's and were extensively discussed by Brown in 1951.[24]

Assuming that the destruction of populations through acute effects can be avoided by carefully controlled application of the insecticides, the major concern is then with the effects of more extended exposures to lower levels of toxicants. For example, it is most important to determine the effects on living organisms of chronic exposure to the various organochlorine-containing solutions, powders, and emulsions which are the diluted forms of the toxicants used in practice. A great many investigations of this sort were done in Switzerland and later in England and America during World War II, and it soon became evident that large quantities of dusts or aqueous suspensions containing DDT could be applied to mammalian skin without hazard. Organic solvents facilitate entry of the toxicant through the skin and some cause skin irritation or are toxic in their own right, but even so, rather large quantities of the dilute solutions (0.1 to 0.3%) used in practice are required to produce manifestations of toxicity. As an example, rabbits receiving daily skin applications of a 1% kerosene solution of DDT (kerosene itself has toxic properties) corresponding to an application of 50 mg/kg of the organochlorine (say, 5 ml of solution applied to a 1-kg rabbit) survived no more than 16 applications; guinea pigs tolerated no more than 18 applications at this rate, but only one rat died after 20 applications.[12] Cattle and other farm animals can tolerate several gram quantities of DDT applied in emulsions, any hazard with animals arising from subsequent licking of the skin. A concentration of about 20 mg/l of air is required to produce signs of toxicity in most laboratory animals, although mice are more susceptible, and this concentration is some 4000-fold greater than is needed for insect control.[24]

There are evident species differences in oral toxicity; 20 daily doses of 50 mg/kg of DDT given to rabbits by stomach tube (in liquid paraffin-tragacanth suspension) killed four out of ten of them, while ten rats survived 30 such doses and monkeys are unaffected by 16 months at this daily dose rate. As mentioned in connection with enzyme interactions (see section B.3c), the no-effect level for liver microsomal enzyme induction appears to be about 1 ppm of DDT in the diet; beyond this level there is an increase in the smooth endoplasmic reticulum of hepatic cells (at 5 to 10 ppm) and a tendency to cell enlargement (hypertrophy) at 50 ppm becoming pronounced at 400 ppm. Cell enlargement is accompanied by peripheral distribution of basophilic granules (margination) and cytoplasmic changes. Prolonged feeding at higher levels (100 ppm upwards) produces liver necrosis similar to that caused by chloroform or carbon tetrachloride and it may be accompanied by considerable increases in liver weight and fatty degeneration. However, these changes are reversible in their early stages and at that point appear to be adaptations that simply reflect the increased work done by the liver.[633] Most of the rabbits in the subacute oral tests referred to earlier showed some degree of liver necrosis, with slight tubular degeneration of the kidney, but no effects on other organs. Hypertrophy of the kidney may occur in rats, and changes in other organs have occasionally been observed in laboratory animals;

cows, horses, and sheep have been observed to develop hemorrhages in the heart and gastrointestinal tract when given 200 mg/kg daily in the diet — a rather high level of treatment.[12,24]

DDD is more readily tolerated by mammals than DDT; multiple dermal applications of 100 mg/kg can be tolerated and a dietary level of about 2500 ppm is required to produce toxic effects. It affects steroid metabolism in the adrenals and causes an increase in their size in chickens, but the pathological change of adrenocortical atrophy seems to be limited to dogs. Liver changes resemble those seen with DDT. Methoxychlor is one of the safest insecticides known, as indicated by the low oral rat and dermal rabbit toxicities (each 6000 mg/kg or more) and the chronic dietary dose at which effects appear is said to be 5000 ppm. Rabbits given 200 mg/kg daily in the diet show anorexia and diarrhea and sometimes develop fatty hearts and livers, but dogs fed 300 mg/kg daily for 1 year showed no effects. Unlike DDT, methoxychlor shows little tendency to store in adipose tissue. The treatment of elm trees to control the beetle vectors of Dutch elm disease requires unusually large DDT treatments which are believed to have resulted in bird deaths (especially robins) through contamination of their food (earthworms) with insecticide.[583] For this reason there has been a tendency in recent years to use methoxychlor for such purposes.

The production of the long-lasting metabolite DDE from DDT in vivo presents a problem, since although it appears to be nontoxic (acutely) to most insects examined, the situation may well be different and variable among higher forms. In subacute feeding tests, DDE is said to be about half as toxic as DDT to coturnix quail but nearly as toxic to young pheasants; however, the last comparison is with technical DDT, which in this instance appears to be almost twofold less toxic (LC50 935 ppm in the diet) than 4,4'-DDT.[583] Feeding tests indicate that DDE is more toxic than DDT to the pigeon (*Columbia livia*) but less toxic to the blackbird (*Turdus merula*)[527] so that generalizations from limited information are to be avoided. The precise form of oral administration is critical, but for solid DDT the chronic dietary LC50 appears to range upward from about 200 ppm for mammals and birds so far investigated. The incidence of thin eggshells and increased egg breakage was mentioned earlier (see section B.3c) and there is now a large literature implicating DDT

and DDE as causal agents in the decline of certain avian species, especially the predators. There is certainly evidence from experimental treatments for eggshell thinning but care is needed when extending the results of limited experiments to the field situation, in which so many factors operate that are not properly understood at the present time. It may be that these and other organochlorines have contributed (possibly by joint action, since tissue and egg residues consist of mixed organochlorines, including polychlorobiphenyls) to the decline in predator populations observed in some areas, but there is no evidence for a general decline in bird populations attributable to this cause, and allowance needs to be made for man's nonchemical depredations.[659] These problems have been thoroughly and critically discussed by Robinson in relation to events in Britain and two of his reviews are especially recommended for further reading.[647,660] From all the discussion there emerges the need for continuous recording of natural fluctuations in bird populations in relation to pesticide usage and especially a need for tissue residue monitoring by, preferably, a single organization in each country, so that adequate and uniform supporting analytical data can be provided. The analytical problems are now known to be complicated by the almost universal presence in the environment of the widely used polychlorobiphenyls. These are not pesticides, but they give analytically similar responses and produce sublethal effects such as enzyme induction in vivo; their chronic effects are subject to debate at the present time.

Fish appear to be reasonably uniform in their response to particular organochlorine insecticides, so that laboratory studies of species response to different compounds are perhaps more meaningful from a practical standpoint than is often the case with other organisms. It is unfortunate that the compounds having the most favorable mammalian toxicity are still rather toxic to fish, although methoxychlor and lindane are frequently 100-fold safer than endrin in terms of LC50. As with birds and other wildlife, evidence of poisoning in the field relies on comparison of tissue levels in dead samples with the values found in experimentally poisoned fish. Thus 0.001 ppm of DDT in water killed 50% of exposed brown trout (*Salmo trutta*) within 100 days; dead fish had concentrations of DDT in muscle of 1.1 to 11.1 ppm (plus 0.04 to 6.3 ppm of DDE) and no survivor had more than 2 ppm of DDT in muscle, most having less than 1.1 ppm. Muscle of trout from English streams with only background DDT levels contains 0.004 to 0.13 ppm of DDT plus 0.006 to 2.7 ppm of DDE so that the likelihood that these compounds contributed to death can be assessed. Most trout muscle in English environments contains less than 0.2 ppm of dieldrin with a range of 0.004 to 0.42 ppm in fish from uncontaminated areas; muscle of experimentally poisoned fish contained 0.8 to 17.2 ppm and of fish poisoned by effluent from sheep dips, 0.7 to 7.0 ppm.[54] The threshold concentration seems to be about the same for DDT and dieldrin in experiments with this species. For others, deterioration in reproductive success is attributed to the sensitivity of fry to organochlorine concentrations that are sublethal to adults and eggs, the young being quickly killed when they hatch. However, there is evidence of increased organochlorine tolerance in various fishes from areas of high crop exposure.

### b. Cyclodiene Insecticides

When taken orally by humans, chlordane appears to be less toxic than DDT. The ingestion of 32 mg (2.2 g/70 kg) by an adult and 10 mg/kg by a child has produced convulsions and 6 g proved fatal to a woman. Symptoms of acute toxicity have appeared in adults following the ingestion of 70 mg and 1 g, respectively, of aldrin and 25 mg has caused death in a child. The lethal oral dose of dieldrin for man is estimated to be 4 to 5 g and endrin is expected to be more toxic than this; estimates from fatal cases of endrin poisoning range from 5 to 50 mg/kg (0.35 to 3.5 g per 70 kg adult).[658a]

Mammalian toxicity data for some cyclodiene insecticides and their conversion products were given in Table 20. Intravenous LD50's are expectedly higher than the acute oral values listed and for rats dermal LD50's of 690 to 840, 195 to 250, 98, 60 to 90, 23 to 35, 15 to 18 mg/kg have been given for chlordane, heptachlor, aldrin, dieldrin, isodrin, and endrin, respectively.[661] Toxicity trends among the photoisomers and other photoproducts need to be viewed cautiously, since, for example, photodieldrin is several times more toxic than dieldrin to mouse or rat but is not more toxic to dog, chicken, or pheasant; it is said to be more toxic to bluegills and minnows but less toxic to *Rasbora* spp.[514,662] The various hydroxylated metabolites of cyclodienes are evidently detoxica-

tion products as far as oral administration is concerned and this is the most significant route of administration from an environmental point of view.

The early toxicology of chlordane had an influence on its development (Vol. I, Chap.3C.1a) and has been thoroughly discussed by Ingle.[269] Present day technical chlordane is the least toxic of the commercially used insecticidal derivatives of hexachlorocyclopentadiene. A 2-year rat feeding trial described by Ingle in 1955 showed a "no-effect level" on growth and mortality of above 150 ppm in the diet and slight alteration in hepatic cells at 50 ppm; in a second investigation dietary levels up to 60 ppm for 100 days had no effect on subsequent reproductive capacity. Another report of 1957 suggests hepatic cell changes at 2.5 ppm, but from the critical discussion by Ingle it seems unlikely that any significant changes actually occurred below 25 ppm, which produced hypertrophy in male rats only. He criticizes the later description of these changes as "liver damage," and, indeed, as knowledge of enzyme induction phenomena has increased in recent years, there has been much discussion concerning the point at which damage actually begins in an organ constantly subject to adaptational challenge by foreign compounds.

There is no evidence to suggest that chlordane is carcinogenic and available information indicates it to be no more toxic than DDT on chronic ingestion and less toxic on an acute oral basis. At high dosage levels, the liver damage caused resembles that produced by DDT and other organochlorines. Since the material is metabolized, it accumulates in fat depots to a level determined by the balance between the rate of ingestion and of metabolic elimination. Toxicity data listed in Table 19 show that with one exception, the chlorinated derivatives of dihydrochlordene (including *cis-* and *trans*-chlordane isomers) have uniformly favorable toxicity to mice. Heptachlor is converted into heptachlor epoxide in vivo, and the no effect level for minimal alterations in rat liver cells is about 5 ppm in the diet for both; 10 ppm of DDT or of heptachlor given in the diet for 8 months caused similar degrees of proliferation of the smooth endoplasmic reticulum with the expected reversal when treatment with the inducing agent was discontinued.[269]

Aldrin is converted into dieldrin in vivo, so the two compounds can be considered together from a toxicological standpoint; a useful summary of pertinent information to 1967 is given by Hodge et al.[706] Female rats metabolize dieldrin less readily than males, tend to store higher levels in the fat, and are somewhat more susceptible to dieldrin poisoning. Neither males nor females showed changes in hematological parameters or liver histology at dietary dieldrin levels of 0.1 or 1.0 ppm; the females showed increased liver weight at 1.0 and 10 ppm and there was some evidence of lever cell changes at the 10 ppm level. For dogs, an intake of 0.2 mg/kg/day (8 ppm in the diet) of aldrin or dieldrin produced no clinical or histopathological abnormalities, but 20 ppm of either caused loss of weight and convulsions, and changes appeared in liver, kidney, and bone marrow with dieldrin, but not with aldrin. Beagles given 0.05 mg/kg/day (equivalent to 1 ppm in the diet) of dieldrin for 2 years showed some increase in the weight of their otherwise normal livers. No adverse effects were observed in rhesus monkeys receiving dietary levels of dieldrin up to 2.5 ppm (100 μg/kg/day) for 36 months, although levels above 1.0 ppm produced slight elevation in microsomal enzyme activity; all liver parameters measured were completely normal after 6 years at dietary levels up to 0.5 ppm (corresponding to about 200 times the daily dieldrin intake of the general population).[303]

On present information, there is no evidence that aldrin or dieldrin produces malignant tumors in rats, dogs or monkeys at any tolerated dose during long term feeding studies.[303,663] However, the life span of mice fed 10 ppm of dietary dieldrin is shortened and there is a significant increase in the number of apparently benign liver tumors over those arising spontaneously in the controls; in this respect mice are unusually susceptible to the action of enzyme inducing agents, and similar changes are seen with phenobarbitone and DDT.[633] A dietary no-effect level on reproduction of 1 ppm has been established for rats. Livestock appear to be rather susceptible to dieldrin poisoning, acute oral LD50's lying within the range 10 mg/kg to about 75 mg/kg for cows, horses, pigs and sheep, and sprays containing more than 0.25% are toxic to some animals. Subacute feeding with 2 mg/kg daily has killed cows, although data from several investigations indicate no effect of dieldrin at dosages from 0.003 to 2.66 mg/kg/day given for more than 100 days.

Endrin is the most acutely toxic of the cyclo-diene insecticides in practical use and yet, except for endosulfan, is least persistent in mammals (see section B.2e). Dietary levels of 50 or 100 ppm were lethal to rats within a few weeks and the mortality rate of females was increased at 25 ppm. At 5 ppm in the diet, increased liver weight was found in males and increased kidney weight in females. A no-effect level of 1 ppm has been established for rats, and also for beagle dogs. Because this compound is so rapidly metabolized and excreted, the tissue levels attained when excretion balances intake in chronic treatment are much lower than is the case with corresponding dieldrin treatments, and endrin cannot usually be detected in the blood. None of the animal studies so far have given any evidence that endrin is either tumorigenic or carcinogenic.[303] Mice exposed to endrin in the field have acquired a 12-fold resistance to it.

The α- and β-endosulfan isomers have different acute oral toxicities to rats (Table 20) and females are more susceptible to the technical mixture than males. The sulfate formed from both isomers is as toxic as α-endosulfan. Except for endosulfan lactone, which appears to have intermediate toxicity to mice, all other known conversion products have greatly reduced oral toxicity. An acute dermal LD50 of 359 mg/kg has been determined for endosulfan applied to rabbits in oil solution. In chronic feeding tests for 2 years with rats and for 1 year with dogs, a "no toxic effect" level of 30 ppm in the diet (corresponding to 0.75 mg/kg/day for dogs) has been established; the no-effect level on reproduction in rats is 50 ppm and no trial so far has given any evidence of an increase in the natural rate of tumor formation. The low persistence of endosulfan in mammalian tissues has been mentioned before; it disappears quickly from pigs, cattle, and sheep treated with amounts greatly in excess of the stipulated residues on their food and appears only transiently in the milk.[248]

Isobenzan (Telodrin®) has a high acute oral toxicity and is no longer used as a practical insecticide.[303] A dietary no-effect level of 5 ppm has been established for rats and a significant increase in mortalities is seen at the 10-ppm level. Dogs and mice are more susceptible, no-effect levels of 0.5 and 1 ppm, respectively, having been found for these animals in various feeding trials. Curiously, neither lethal nor sublethal doses of this toxicant appear to cause structural changes in any organs examined and there is no information regarding liver enzyme induction.

Among the cyclodiene insecticides, chlordane, endosulfan, and heptachlor appear to be least toxic to American partridge, ring-necked pheasant and mallard on a subacute or chronic basis, while aldrin and dieldrin also are only moderately toxic to the last species.[583] Subacute toxicities for mallard, pheasant, and coturnix quail indicate that technical chlordane has about the same or a slightly higher toxicity than DDT fed in a similar regime and that heptachlor is rather more toxic than chlordane. The comparable experiments with endosulfan indicate a more uniformly low toxicity, ranging from 900 to 2,250 ppm in the diet for these three birds, and also including bobwhite quail. Endrin appears to be generally about 100-fold more toxic on the same basis. There are evident species differences in the effects of these compounds on reproduction and the results of feeding trials with dieldrin were mentioned in relation to toxicant-enzyme interactions (section B.3c).

Fish and other aquatic organisms provide particularly good examples of the equilibrium situation discussed at the beginning of Section B, since they are continuously exposed to a large amount of water and must eventually equilibrate with any lipophilic pollutants it contains. Thus, indefinite exposure to a constant concentration of organochlorine insecticide in water results in tissue uptake until this process is balanced by any elimination process the fish possesses. Since a low concentration can cause mortality if exposure is long enough to produce the necessary internal level, LC50 is inversely related to exposure time. Chlordane and heptachlor are among the least toxic of the organochlorine compounds, having LC50's 10 to 100 times higher than that of endrin,[2] depending on the species, and there is evidence of chlordane resistance in highly exposed areas.[583] Aldrin and dieldrin are somewhat more toxic, although there is noticeable species variation, carp being most resistant (48-hr LC50 for dieldrin 0.067 ppm) and sunfish most susceptible (LC50, 0.009 ppm).[662] Various species of fish from cotton growing areas exposed to several insecticides are found to be resistant to aldrin (16- to 75-fold) and dieldrin (22- to 36-fold).[583] Plateau levels of dieldrin are attained in liver, gills and muscle of rainbow trout exposed to increasing concentrations, indicating either an approach to tissue saturation, with consequently reduced uptake or that intake is balanced by an elimination process.[664] The toxicant is taken up rapidly by trout muscle on constant exposure and is lost equally rapidly on transfer to clean water. Endrin and endosulfan are highly toxic to fish, but endosulfan is converted into the diol precursor by some fish (see section B.2c) and appears to be generally somewhat less toxic than endrin. In areas in which endrin and toxaphene have been extensively used on cotton, resistance levels of 10 to 200-fold for the former and 10 to 40-fold for the latter have been reported in various species of fish.[583]

The mammalian toxicity of the hexachlorocyclopentadiene derivatives Pentac®, mirex, and chlordecone (Kepone®) increases in that order, Pentac having low toxicity to rats (> 3,000 mg/kg), mirex about that of chlordane (300 to 600 mg/kg) and chlordecone about that of heptachlor (114 to 140 mg/kg). Mirex appears to be less toxic than chlordane to several birds and has been ranked with lindane and chlordecone in having relatively lower fish toxicity than some of the other organochlorine insecticides.

### c. Hexachlorocyclohexane and Toxaphene

From an outbreak of poisoning following the ingestion of a meal containing γ-HCH lindane, the lethal oral dose of this compound is estimated to be 0.3 g/kg (20 g for a 70-kg human adult), which compares favorably with the cyclodiene insecticides.[658a] Technical HCH (HCH) has low acute oral toxicity ranging from 1,000 to 15,000 mg/kg for a number of common laboratory animals, cats, sheep, chickens, and pigeons, while that of lindane appears to be fairly uniform, lying in the region of 100 to 130 mg/kg for a number of species.[24] Other acute oral toxicities to the rat (mg/kg) are α-HCH (500), δ-HCH (1,000), and β-HCH (> 6,000).[41] The acute dermal toxicity of lindane for rats is about 1,000 mg/kg for both sexes; cattle survive single dips in 5% technical HCH suspensions or repeated dips in 1.5% suspensions, and the latter are also well tolerated by goats, sheep, pigs, and horses. HCH has a much lower chronic toxicity than DDT and rats receiving 500 mg/kg daily for 57 days maintained normal health and growth. Mice survived daily doses of 40 mg/kg for four weeks and cattle tolerated sufficient HCH (125 mg/kg/week) to protect them from tsetse flies. It did not retard the growth of rats at below 800 ppm in the diet.

Of the individual isomers, lindane is rapidly metabolized, while $\beta$-HCH is most persistent, stores in adipose tissue and has the highest chronic toxicity; this isomer retards rat growth rate at 100 ppm in the diet whereas such effect is not seen with lindane below 400 ppm. The no effect dietary levels for tissue changes in rats during a 2-year feeding trial were less than 10 ppm for $\beta$-HCH, 10 ppm for $\alpha$-HCH, 10 ppm for technical HCH, and 50 ppm for lindane. Quite high levels of the $\delta$-isomer produce no adverse changes in the liver, whereas the other isomers and technical HCH at high levels produce enlargement, necrosis and other signs of organochlorine poisoning. Fat deposits of $\alpha$-, $\gamma$- and $\delta$-isomers disappear in a few weeks when feeding ceases, whereas the $\beta$-isomer persists for several months.

The acute oral and dermal rat toxicities of toxaphene are similar to those of lindane but it is more toxic to dogs (oral LD50 15 to 30 mg/kg).[41,583] Strobane is less than half as toxic as toxaphene to rats on an acute oral basis. Toxaphene has been widely used as a veterinary insecticide and adult cattle, horses, sheep, goats, and pigs tolerate repeated applications of a 2% spray or a single application of a 4% spray without toxic effect. In chronic feeding tests with rats the dietary no effect level for liver changes is 25 ppm; increasing levels above 50 ppm produce effects ranging from increases in the smooth endoplasmic reticulum to typical chlorinated hydrocarbon liver pathology.[665] Dog livers are said to show evidence of "slight degeneration" after 2 years on a dietary level of 40 ppm, while monkeys suffered no signs of intoxication or tissue damage on 10 ppm (about 0.70 mg/kg/day) for 2 years. Chronic ingestion results in fat storage proportional to the amount ingested and elimination occurs rapidly when exposure ceases. It must be remembered when considering toxaphene that the material is a mixture of unidentified compounds.

In subacute feeding tests on birds, the LC50 of lindane ranges from 400 to 1,100 ppm for coturnix quail, pheasant, and bobwhite quail while the LC50 for mallard is more than 5,000 ppm; the range for the four species is 470 to 900 ppm for toxaphene and strobane.[583] The acute oral LD50 for toxaphene administered by capsule lies between ten and 100 mg/kg for one species of grouse, for pheasants, for two species of ducks and

two of quail. Toxaphene is highly toxic to fish, with an LC50 for small species of about 0.003 ppm and for larger varieties of about 0.01 ppm; strobane is similar, but 300-fold resistance has been noted in mosquito fish regularly exposed to it in cotton-growing areas. Lindane is generally about tenfold less toxic than toxaphene and has a half-life in some fishes of less than 2 days.

All of the organochlorine compounds under discussion are toxic to a variety of aquatic and other nontarget organisms but there are wide species variations in sensitivity and each needs to be regarded as an individual problem from the standpoint of toxic effects. Undoubtedly, there is a basic requirement for an overall picture of the dynamics of populations within a given ecosystem that will enable the effects of all kinds of adverse factors to be recognized and predicted. Only when this is available will it be possible to discern the true long-term effects of man's environmental manipulations using pesticides.

## D. MODE AND MECHANISM OF ACTION

Studies of mode of action attempt to answer the closely related questions: "What do the toxicants do?" and "How do they do it?" For the organochlorine insecticides a great deal has been written in answer to the first question but much less that really provides any answers to the second. It is to be expected that the overt physiological effects of those compounds will be accompanied by biochemical changes, but the difficulty is to distinguish between changes caused by direct action of the insecticide and those that are secondary consequences of some other action (or actions) on the nervous system. Death must result from multiple effects and these will differ between species; failure of the respiratory muscles, for example, is a frequent cause of death in mammals, but not in insects since these do not depend on muscular activity for their oxygen supply.

### 1. DDT

The signs of poisoning in both insects and mammals indicate an action of DDT and analogues on the nervous system. Treated insects rapidly become hypersensitive to external stimuli and develop tremors of the body and appendages (DDT jitters). After a period of violent motion they fall on their backs and the continuous leg movement eventually becomes more spasmodic, being succeeded finally by paralysis.[218] The insect *heart may* continue to beat for some time after all other motion ceases.[666] Mammals become frightened, cold, and hyperexcitable, after which fine tremors develop in the face and rapidly become general.[12,41,666] The heart muscle is sensitized to external stimulation and ventricular fibrillation from this cause is a frequent cause of death. At a later stage tonic (rigidity) and clonic (frenzied movements) convulsions may occur and may be followed by death from cardiac or respiratory failure. Death may also be the end result of gradual depression and coma, following a series of convulsions. If the dose is sublethal the nervous and muscular effects eventually subside and recovery is complete within 18 to 48 hr, depending on the route of administration. Many of these effects are quite similar in insects and mammals given comparable treatments. In man, acute poisoning produces numbness and weakness of the extremities accompanied by severe gastrointestinal disturbances, but the effects appear to be quickly reversible.[666] Histological changes, some of which were referred to in relation to chronic toxicity, are frequently reversible and their study has been of little value in understanding DDT action in either insects or mammals.

Over the years, much evidence has accumulated that in arthropods DDT acts on peripheral nerve as distinct from central nerve, and the classical demonstrations that legs amputated from poisoned insects continue to tremble and convulse and that typical symptoms can be induced by applying DDT to amputated healthy legs, support this supposition. Motor nerves and muscle fibers are only affected by high concentrations (1,000 ppm) whereas low concentrations (0.01 ppm) affect sensory nerves. The symptoms in mammals result from disturbances in the central nervous system, where the motor area of the cerebrum and the cerebellum appear to be involved, and there is said to be a direct correlation between the severity of signs of both acute and chronic clinical toxicity in the adult rat and concentration in the brain.[667] The now well-known physiological signs of DDT poisoning of insect nerve were first demonstrated in 1945[668] and are such that a single nerve impulse arriving at an area of a nerve fiber treated with DDT gives rise to a multitude of impulses; the action is generally held to be on the nerve axon rather than on neuromuscular junctions or synapses (which may be an oversimplification) and it can be demonstrated on isolated nerve trunks.

Resting axons are polarized with the inside negative relative to the outside due to the presence of a higher potassium ion and a lower sodium ion concentration inside than outside. When an impulse arrives as the result of a stimulus applied to the nerve, the axonic membrane becomes permeable to sodium so that the resting potential (membrane potential) falls to zero and may become temporarily positive. The membrane then becomes exclusively permeable to potassium ions, which now pass out through the membrane, restoring the potential to its resting level; a mechanism referred to as the "sodium pump" is required to maintain the normally low internal concentration of sodium ions. The voltage change produced by the arrival of an impulse can be measured as a spike called the action potential on a recording of voltage relative to some outside position measured versus time with an electrode inserted into the axon. Added DDT appears to prolong the phase of sodium permeability, and to retard the onset of potassium permeability, the result being a delay in the return of the internal voltage to its resting negative value which results in a plateau on the tail of the spike. On this plateau may appear the trains of additional impulses (repetitive discharge) produced by the destabilizing effect of the toxicant. The details of these changes have been discussed by Narahashi in a recent review of the action of neurotoxic insecticides.[478]

The most remarkable feature of insect poisoning by DDT is the so-called "negative temperature coefficient of intoxication"; insects showing marked signs of poisoning at 15°C are restored to normality at 30°C, and this change can be repeatedly reproduced by alternately lowering and raising the temperature.[669,670] Since pertinent kinetic parameters such as penetration and detoxication have positive temperature coefficients, this is a most interesting situation. There is much evidence that insects contain and can tolerate larger amounts of DDT at high than at low temperatures and while it can be argued that the toxicant is continually arriving at the site of action and is simply detoxified more rapidly there at high than at low temperatures, it seems unlikely that this explanation accounts for more than a part of the effect. Narahashi[478] concludes that there is an increase in nerve sensitivity at low temperatures and the results seem to support the recently favored idea that DDT forms with nerve membrane a complex (possibly of the charge-transfer type) which is dissociated with increasing temperature, so that although more DDT may be present near the target at higher temperatures, it is less effective.[671]

Using American cockroaches, Eaton and Sternburg made the interesting observation that the abnormal afferent activity from DDT treated sensory nerves actually increases with temperature, while in the absence of sensory treatment with DDT no abnormal DDT-induced trains appeared in the central nervous system (CNS) itself until a large amount of the toxicant was actually injected into the sixth abdominal ganglion, in which event the abnormal activity was again positively related to rise in temperature.[672,673] According to these observations, abnormal sensory input persists (with some frequency changes) at all temperatures and seems to be unrelated to the signs of DDT poisoning, which seem rather to result from synaptic impairment in the CNS resulting at lower temperatures from the high abnormal sensory activity. Since only large amounts of DDT itself produce abnormal activity in the CNS, the implication is that a substance (or substances) other than DDT is liberated there as a consequence of extreme sensory activity and accumulates at low, but not at higher temperatures, to the detriment of synaptic transmission. In the course of poisoning, the sixth abdominal ganglion and the thoracic ganglia become increasingly unstable and complete block eventually develops in the latter. The condition is reversible with rise in temperature and the degree of impairment of the thoracic ganglia is thought to be closely related to the visible signs of poisoning. This situation has been further discussed by Narahashi,[478] and it seems difficult to distinguish between the alternatives of (i) a temperature change in the actual sensitivity threshold of the central synapses to abnormal incoming activity from sensory nerves or (ii) a temperature influenced reversible accumulation of unidentified neuroactive substance(s) such as that mentioned above. These results appear to conflict somewhat with the idea that sensory nerve disturbances are the primary cause of poisoning, although in this interpretation they are primary in the sense of being necessary to trigger the critical central changes.

There has been much discussion concerning the liberation of neurotoxic substances by DDT and other insecticides since the 1952 report that DDT has this effect in cockroaches.[674] Sustained activity for long periods also produces a factor in the blood which will cause paralysis or death in untreated cockroaches and is chromatographically similar to that produced by DDT in this insect and in crayfish.[2] The toxic factor appears to be different from acetylcholine, adrenalin, noradrenalin, γ-aminobutyric acid, 5-hydroxytryptamine, histamine, and γ-butyrobetaine, which eliminates most of the likely neurotransmitters. It is clear that the liberation of secondary substances as a consequence of DDT action may be the actual cause of death, which may therefore vary between different species. This situation has recently been highlighted by the observation that the malpighian tubules of fifth stage larvae of the triatomid bug (*Rhodnius prolixus*) are caused to secrete profusely at the paralytic stage of poisoning produced by the major insecticide types including DDT, cyclodienes, and lindane.[675] The effect, which is caused by the release of a diuretic factor, results in the accumulation of a large volume of fluid in the rectum and a corresponding depletion in the haemolymph volume; such alterations in water balance, although not necessarily arising from the same cause, have been observed before and may be the cause of death in some species or in others may contribute to a more general degeneration of normal function that results in death.

DDT poisoning in American cockroaches was early reported to result in the accumulation of acetylcholine in the nervous system at the prostrate stage but there appears to be no inhibition of cholinesterase in vitro in the nerve cord of this insect or in other preparations examined.[41] Acetylcholine synthesis occurs in homogenates of DDT poisoned houseflies made at the prostrate stage and there appears to be no real accumulation in vivo if this change in vitro is allowed for.[479] However, there is evidence that the increased level of acetylcholine in DDT-poisoned cockroaches is accompanied by protection of cholinesterase against inhibition by tetramethyl pyrophosphate, which has the useful property that its inhibitory effect is irreversible and that excess of it hydrolyzes rapidly in the intact insect, so causing no in vitro inhibition during cholinesterase assay operations.[676] The effect increases rapidly from the hyperexcitable stage of poisoning and is related to the poisoning phenomenon since it disappears if intoxication is relieved by raising the temperature. In contrast, lindane, which does not *inhibit cholinesterase* either, causes an increase in acetylcholine level without any associated protection against phosphate inhibition. It is tempting to link this phenomenon with the formation, consequent upon DDT action, of a degradable inhibitor of cholinesterase.

In the insects so far studied DDT causes a rapid increase in oxygen consumption which appears to peak during the period of maximal muscular activity and fall off when paralysis sets in;[24] the effect seems to be associated with this hyperactivity rather than with direct effects such as the uncoupling of oxidative phosphorylation, which would also create a requirement for additional oxygen intake. The depletion of carbohydrate reserves which occurs in American cockroaches during hyperactivity can be prevented if they are immobilized by cyclopropane anesthesia and so appears unrelated to the direct action of DDT;[41] removal of the anesthetic reveals the unusual signs of poisoning in spite of the preservation of these energy sources and cyclopropane anesthesia prevents neither the falls in thoracic ATP and α-glycerophosphate seen in DDT poisoned houseflies nor the enhanced desiccation associated with DDT poisoning.[677] DDT and its toxic relatives have been reported to have direct effects in vitro on various enzymes involved in oxidative metabolism but in some cases, as with oxidative phosphorylation, for example, nontoxic analogues give similar effects. Thus, the cytochrome oxidase, succinic oxidase and to some extent the succinic dehydrogenase systems of housefly thorax were found to be inhibited by $3.3 \times 10^{-4}M$ DDT, but comparable effects with DDE and other nontoxic structural analogues make it unlikely that such inhibitions are related to the basic action of DDT.[41]

Carbonic anhydrase, the enzyme which catalyzes the decomposition of carbonic acid into carbon dioxide and water in human blood is sensitive to inhibition by DDT to the extent that its inhibition in erythrocytes has been used to determine DDT residues, but there was little effect of DDT on this enzyme in the heads of American cockroaches, either in vivo or in vitro.[41] Moreover, results obtained in vitro need to be viewed with caution, since the apparent inhibition of bovine erythrocyte carbonic anhydrase by DDT, DDE and dieldrin is said to be an artifact resulting from co-precipitation of enzyme and insoluble toxicant from solution.[710]

DDT poisoning in houseflies or American cockroaches causes depletion in the endogenous amino acids tyrosine and proline and an accumulation of phenylalanine. In the poisoned cockroach, there is enhanced conversion of proline into carbon dioxide and into glutamine. O'Brien has pointed out a possible link between increased turnover of the latter and the enhanced production of uric acid and allantoin via formate in houseflies, suggesting that proline is used to replenish the depleted formate pool and that the process may reflect increased excretory activity by the organism.[2]

It must be admitted that although there has been much investigation of the effects of DDT at the biochemical level in insects and mammals, the results cast little light on its mode of action. The alterations in action potential referred to earlier indicate an effect on the $Na^+$, $K^+$ equilibrium of nerve membrane and nerve cords from DDT-poisoned cockroaches show increased ability to take up sodium or lose potassium compared with those from untreated insects.[678] These observations have led to an interest in the $Na^+$, $K^+$, $Mg^{2+}$, and $Mg^{2+}$, $Ca^{2+}$ dependent adenosine triphosphatases (ATPases) of the nervous system, which may be involved in the regulation of ion transport through the membrane and are partially inhibited by various organochlorine insecticides and related compounds. Matsumura reports that a portion of the ATP-ase complex from rat brain nerve endings which corresponds to $Na^+$, $K^+$, $Mg^{2+}$ ATP-ase is 1,000-fold more susceptible to DDT (I50 about $3 \times 10^{-7}M$) than to DDE, indicates a negative temperature coefficient for this inhibition, and provides evidence for a positive correlation between its intensity and the toxicity of various DDT-analogues to mosquito larvae in vivo.[679] However, another investigation gives the order of inhibitory effectiveness of DDT-analogues toward the $Na^+$, $K^+$, $Mg^{2+}$ ATP-ase from American cockroach nerve cord as dicofol > DDD > DDT = DDA > DDE.[680] Dicofol is a good inhibitor of both types of ATP-ase and yet is noninsecticidal (although a good acaricide) like DDA so that a number of questions remain to be answered regarding the possibility that these enzymes are major targets for DDT action. It was suggested many years ago that DDT exerts its effects on crayfish motor axon by disturbing the binding of calcium ions with nerve membrane phospholipids and the role of $Ca^{2+}$ in membrane function has been discussed in detail in relation to the action of membrane stabilizers (such as local anesthetics) and labilizers (such as DDT) by Weiss.[681] Species differences are apparent, since the DDT mimicking effects of $Ca^{2+}$ concentration changes on crayfish nerve are not evident with insect preparations.

Some of the early theories relating to the toxic action of DDT were discussed in relation to its history and development. Two of these, namely that the toxic combination of 4-chlorobenzene nuclei is carried to the site of action by the lipophilic -$CHCl_3$ moiety,[7] or that ability to liberate HCl near a sensitive site is critical (see section B.3b), stimulated much research.[682] However, the remarkably high insect toxicity of dianisyl neopentane,[65] which contains no chlorine atoms, and the recent development of a range of unchlorinated but toxic DDT-analogues (see section C.1a), shows that the steric characteristics of the molecule as a whole are important. Furthermore, the discovery of dehydrochlorination as a resistance mechanism is incompatible with the second idea. The importance of overall steric configuration is also implied in the extensive discussion of structure-activity relations given by Riemschneider, who concluded that good lipid solubility, a molecular weight of 270 to 450, moderate melting point ($< 180°C$), and above all an ability of the benzene rings to rotate freely, are required for toxicity.[683] Gunther et al. observed a linear relationship between the summed logarithms of the van der Waals bonding energies of the alkyl and aromatic substituents of various DDT-analogues and the negative logarithms of their LD50s to mosquitoes, and consequently favored a physical interaction of DDT with a cavity in some sensitive structure, suggested to be an enzyme.[684]

About the same time, Mullins, who had been considering the molecular basis of narcosis, noticed the gradation in bioactivity of HCH isomers from strong excitation (in lindane), through weak excitation ($\alpha$-HCH) and inactivity or weak depression ($\beta$-HCH) to strong depression ($\delta$-HCH) and showed that if a nerve membrane (for example, axonic membrane) is considered to consist of cylindrical lipoprotein macromolecules of diameter 40Å packed together (hexagonal array) so that they are 2Å apart, then the gaps between the cylinders form pores of such a size that only lindane (diameter 8.5Å in the plane of the cyclohexane ring) can enter in any orientation and then rotate so as to fit tightly into the "pore" in the planar orientation.[685] Interactions with the membrane in this "tight fit" orientation (Figure 18) were assumed to result in membrane distortion leading to ion leaks and hence to excitation. On the other hand the "end-on" entry of the other isomers would lead at most to narcotic effects such as were postulated to result when the interspaces become packed with small, loosely fitting molecules lacking any strong interactions with the pore that might result in membrane distortion. An interesting extension of this idea is that the same interspace accommodates DDT with its benzene rings in "end-on" configuration (which incidentally also allows maximal contact of both benzene rings and the 4 and 4'-chlorine atoms with the cylinders forming the pore) but not the nontoxic 4,4'-diodo-analogue (4,4'-substituents too large) or DDE (lack of fit due to loss of the tetrahedral bond angle on double bond formation).[614] When discussing the "fit" of molecules into such "receptors" in relation to toxicity, it is obvious that the arguments are only valid if accurate information about *intrinsic* toxicity is available, which has not usually been the case in the past. The recent results of Holan and of Metcalf provide such information (see section C.1a) and show that modifications are required in the size limitations indicated by the Mullins pore (Figure 18), but the basic hypothesis is unchallenged and the possibility that the primary action of chlorinated insecticides is by an interaction of this sort currently seems stronger than ever.

The $\pi$-electron systems of the two p-chlorobenzene rings of DDT are able to serve as electron donors to an accepting molecule such as tetracyanoethylene, as indicated by the formation of a yellow charge transfer complex with the latter in chloroform solution.[686] O'Brien and Matsumura speculated in 1964 that charge transfer complex formation between DDT and a component of nerve axon would provide a plausible mechanism for destabilization of the axon.[687] They provided some evidence for such interaction with components of cockroach nerve that was later brought into doubt by the observation of similar interactions with nontoxic DDT analogues and between DDT and non-nervous tissue. Furthermore, the solubility of DDT is a complex matter and it has been shown that changes of state occurring in aqueous-alcoholic solutions of DDT produce ultraviolet spectral changes resembling those that have been observed during DDT interaction with tissues.[686] Nevertheless, the formation of such complexes is compatible with the negative temperature coefficient of DDT toxicity to insects.[671] For example, a comparison of DDT with 1,1-di(*p*-chlorophenyl)-, 1,1-di(*p*-ethoxyphenyl)-, and 1,1-di(*p*-ethylphenyl) -2,2-dichlorocyclopropanes reveals that there is a very similar and uniform increase in LD50 with rising temperature for the first three compounds, whereas there is an enhanced rate of increase in the LD50 of the last compound above 21°C. This has been explained by the presence in each of the first three compounds of para-dipoles that enhance their ability to form complexes, whereas with the di(*p*-ethylphenyl)-analogue, complex formation relies solely on the weaker interaction of the ring π-electron systems and dissociation rate is expected to increase more rapidly with rising temperature.

Holan[671] examined the toxicity to houseflies and mosquito larvae of a number of variously substituted cyclopropane derivatives of the above type and thereby concluded that the aliphatic portion of the molecule (i.e. halogenated cyclopropane ring, trichloromethyl or analogous system) has a critical diameter for toxicity of 6.1 to 6.3Å and an area of 50 to 60Å$^2$, this diameter being remarkably similar to that derived for a hydrated sodium ion (6.2 to 6.6Å). The limiting distance between *p,p'*-negative atom dipoles (van der Waal's limits), was found to be 11.5Å with a somewhat higher limit of 14.0Å when these substituents contain alkyl-groups (Figure 18). Since ethoxy was the largest group falling within Holan's toxicity limits, a further extension of this last dimension is necessary following Metcalf's observation that certain compounds containing a *p-n*-butyloxy-substituent (Table 17) have high intrinsic toxicity.

On the basis of his observations. Holan proposed that DDT interacts with a bimolecular leaflet (unit membrane) consisting of a protein layer and a lipid layer separated by about half the height of the DDT molecule, in such a way that the aromatic rings form a complex with the protein layer while the aliphatic portion fits into a recess in the lipid layer having the dimensions of the hydrated sodium ion. If the DDT molecule is actually regarded as a wedge placed between the coils of a phospholipoprotein spring which is selectively permeable to nonsolvated potassium ions when compressed and to hydrated sodium ions when expanded, the spring is thereby held in the expanded, sodium permeable state with a resulting delay in the falling phase of the action potential. According to this concept, an increase in temperature results in dissociation of the aromatic rings from the protein layer so that the molecules are no longer held in position and the spring is free to revert to its compressed state. The resemblance between the molecular dimensions of DDT[713] and published unit membrane dimensions suggests the possibility of a disruptive interaction of DDT with other membranes, which could explain the scattered evidence that multiple effects of this toxicant contribute to ultimate mortality. This hypothesis is difficult to reconcile with Eaton and Sternburg's observation[673] of the positive temperature coefficient of sensory effects. Its significance in relation to nerve function remains to be determined, but it has certainly stimulated the synthesis of a number of toxic, biodegradable DDT-analogues lacking chlorine atoms (see section C.1a).

Elaborating a proposal originally made in 1954,[684] Metcalf and colleagues[685a] suggest that optimum effect of DDT analogues occurs when interaction between molecule and receptor site due to van der Waal's forces is maximal. This optimum interaction is a summation of the interactions of all substituents on the benzene nuclei and the α (benzylic) carbon atom, so that some loss of interaction due to a change in one substituent may be compensated by a balancing change in some other substituent, provided that the total interaction is unaltered. It follows that unsymmetrically substituted DDT analogues can be as effective as symmetrical ones having the same summed interaction with the receptor.

Accordingly, fit to the receptor site should be governed primarily by steric effects, with analogue interaction passing through a maximum and decreasing again as the summed size of the substituents increases. This suggests a parabolic relationship between toxicity and the summation of Taft's steric substituent parameter ($E_s$) for the groups X, Y, L, and Z, leading to the equation:

$$\log \text{LD50} = \alpha + \beta \Sigma E_s + \gamma \Sigma E_s^2$$

where $\alpha$ and $\beta$ are constants and $\Sigma E_s$ is the summation of steric parameters for the substituents. An examination of this relationship for 25 variously substituted $p,p'$-DDT analogues (L and Z constant), using toxicity data for houseflies, shows a reasonable correlation between toxicity and $\Sigma E_s$ for X and Y when piperonyl butoxide is used to minimize oxidative metabolism so that intrinsic toxicities can be measured.

For a further series of DDT analogues having variable $\alpha$-carbon substituents (X and Y constant), the inclusion of Taft's polar substituent constant $\sigma^*$ with $E_s$ is required for significant correlations with data from either houseflies or mosquito larvae. This indicates that both steric and polar effects contribute to the interaction of these groups with the receptor. The incorporation of a $\pi^2$ term (indicating the importance of the lipophilicity of L and Z) gave an improved correlation only in the case of mosquito larvae and in all these multiple regression analyses, the steric substituent constant $E_s$ was the single most important free energy parameter for correlating structure with toxicity. It has not yet been possible to examine the effect of simultaneous variations of all four types of substituent, but there is evidence that the effects of ring substituents are not independent of those on the $\alpha$-carbon atom.[685a]

The findings outlined here and in section C.1a show that the fascinating story of DDT analogues is by no means concluded. What seems strange is the gap of nearly 20 years between these developments and the observation in 1953 that the unchlorinated analogue dianisylneopentane has considerable insect toxicity. The reason presumably is that DDT has been so efficient that until the appearance of residue persistence problems, incentives for further exploration of this area have been lacking. It is tempting to wonder what the situation would now be if present day stimuli had been applicable in the 1950's.

## 2. Lindane and the Cyclodiene Insecticides
### a. Lindane

Lindane is undoubtedly the most intriguing of the organochlorine insecticides because of the structural specificity associated with its toxic action, which is generally more rapid than that of the cyclodienes, but otherwise similar. Links with DDT are implied by the fit of both molecules into the hypothetical Mullins nerve membrane interspace (see section D.1 and Figure 18) but evidence for a similar site of action is lacking. Overt signs of poisoning include tremors, ataxia, convulsions, and prostration. However, Baranyovits[688] compared DDT and lindane action on houseflies and found with lindane a lack of both the early irritability and later tremor of the tarsal extremities which are induced by DDT, while movement ceases in severed legs, suggesting that the effect is under ganglionic control.

The progress of poisoning is usually more rapid than with DDT and effects on the central nervous system appear to predominate over any on the peripheral one, although opinions regarding peripheral involvement differ. Treatment of cockroach central nerve cord with lindane significantly increases the number of spontaneous discharges, but the spurious spike (action potential) frequency and amplitude are lower than with DDT;[478] there is a negative temperature coefficient of toxicity to cockroaches but its magnitude is much smaller than that which is seen with DDT.[670] In contrast to lindane, the $\beta$- and $\delta$-isomers are depressants;[689] the latter causes flaccid paralysis in southern armyworm (*Prodenia eridania*) larvae and is said to be a strong paralytic poison in snails.[24] Increased respiration is observed during the hyperactive stage of lindane poisoning in several insects; the effect is more pronounced than with DDT and is probably again related to the violent muscular activity. Free acetylcholine levels increase substantially in the nerves of lindane-poisoned cockroaches, but the protection of cholinesterase against organophosphate inhibition that is found during DDT poisoning is absent in this case; lindane is not a cholinesterase inhibitor.[676]

In mammals, stimulation of the central nervous system results in a rise in blood pressure and a fall in the rate of heart beat (bradycardia). Increased respiratory rate and restlessness may be followed by coarse tremors of the whole body, salivation, grinding of the teeth, and convulsions. Slowing of respiration and paralysis follow and death may result from cardiac arrest or more usually from respiratory failure. In man, headache, dizziness, weakness, diarrhea, and epileptiform attacks have been described as symptoms of poisoning. Epileptiform seizures recorded by EEG measurements on poisoned animals are not modified by barbiturates, although these suppress the convulsions. Pentobarbital counteracts the rise in blood pressure and atropine prevents the bradycardia.[24,666]

Although only δ-HCH is a true depressant, the convulsive effect of lindane in rats is antagonized by the α-, β-, and δ-isomers and it has been suggested that the β- and γ-isomers act on the thalamus because they can protect against electrically induced convulsions.[666] Protection afforded by HCH isomers against strychnine convulsions has been advanced as evidence for their action in the spinal cord and a recent autoradiographic study using the [14]C-labeled isomers showed a particularly rapid accumulation of α-HCH in the white matter of the central nervous system (CNS) of mice, which persisted for more than 24 hr after its injection; lindane also was taken up rapidly by the brain, but the distribution was more uniform and its presence quite transient.[690] There is still no precise knowledge of the site of action in the central nervous system.[340a]

In autoradiographic studies conducted to follow the penetration of [14]C-lindane into American cockroaches, it appeared to accumulate in the peripheral region of the CNS, rather than inside.[490] Since insect cross-resistance spectra and structural similarities suggest a relationship between lindane action and that of the cyclodienes, a similar distribution of the two types in the CNS might be expected. In contrast, however, dieldrin has been located within the CNS of both houseflies and cockroaches, where it appears to localize on axonal membranes.[488,489,691]

So far, there is little evidence that lindane exerts its primary effect through interaction with a specific enzyme, although it is a better inhibitor of the $Na^+$, $K^+$, $Mg^{2+}$ ATP-ase (28% inhibition at $7 \times 10^{-5}$ $M$) from American cockroach nerve cord than DDT is and also inhibits the $Mg^{2+}$ ATP-ase from cockroach muscle;[680] it may be that such interactions form part of the total toxic effect of this compound. In 1945, Slade suggested that lindane had the same configuration as the B-vitamin, meso-, or myoinositol (one of the stereoisomers of hexahydroxycyclohexane) and might exert its toxic effect by antagonizing the action of this compound at some vital site.[321] In fact, myoinositol is the isostere of δ-HCH rather than of lindane and in view of the completely opposed physical properties of the hexahydroxy- and hexachlorocyclohexanes, it is difficult to imagine that they could actually compete with each other at a common site. However, myoinositol is a nutritional requirement for mammals and fungi so that

this suggestion stimulated much investigation. It was later shown that growth of the yeast *Saccharomyces cerevisiae* and some fungi is inhibited by lindane and that the effect is actually reversed by exogenous myoinositol.[2] It is noteworthy that a more recent investigation indicates δ-HCH to be more toxic than lindane to *S. cerevisiae* and the HCH isomers are thought to act on the cytoplasmic membrane, causing an inhibition of ATP-ase activity.[506] However, there is no effect of myoinositol on the acute toxicity of lindane to either insects or mammals, whereas the isosteric δ-HCH antagonizes the action of lindane in mammals.[2,666] Therefore, the weight of evidence is against the involvement of myoinositol antagonism in the acute insect or mammalian toxicity of lindane but there may conceivably be some significance for this interaction in relation to chronic toxicity or sublethal effects.

Recently, there has been renewed interest in lindane due to its high toxicity and relatively high biodegradability[340a] and it is certainly desirable that the properties of this and other organochlorine molecules should be reexamined in the light of improved experimental techniques. Lindane was the inspiration for the much discussed Mullins hypothesis (see section D.1) which Holan[651] recently modified to accommodate numerous new DDT analogues. It is not easy to envisage the reverse of Mullins' original approach, namely, the interaction of lindane with the Holan DDT receptor and yet there are certain similarities between its dimensions and those of lindane. Thus, lindane's molecular diameters measured at 60° intervals in the plane of the ring are each 8.5Å,[685] which is not far from the height (measured between the limits represented by the trichloromethyl group and the plane containing the 4,4'-substituents) of Holan's general model of DDT (Figure 18), but the molecule thickness (7.2Å) is rather larger than the diameter of the hydrated sodium ion (6.2 to 6.6Å).

An examination of the β- and δ-HCH isomers is of much interest in view of Weiss' suggestion[681] that the resting state of excitable membrane corresponds to the contracted, $K^+$ permeable state of a phospholipoprotein spring and that certain flat molecules (some steroids for example) can fit within the membrane in this form, with consequent stabilization leading to narcosis. β-HCH (said to be a weak depressant) has maximum surface (molecular dimensions each 9.5Å) and minimum thickness (5.4Å) but low thermodynamic activity and small or zero dipole moment. However, the δ-isomer has fairly similar molecular diameters (8.5, 9.5, 9.5Å), intermediate thickness (6.3Å; compare hydrated sodium ion), high thermodynamic activity and a significant dipole moment to assist interaction; it might therefore enter the membrane end-on and interact rather strongly to hold it in the contracted state. Although such arguments are speculative, they explain why a change of two axial chlorine atoms in the more bulky lindane (thickness 7.2Å) to the corresponding equatorial positions should produce the strong depressant activity found in the δ-isomer.[655]

### b. Cyclodiene Insecticides

The signs of poisoning produced in insects by cyclodiene insecticides appear to be generally similar for all members of the group and are more similar to those of lindane than of DDT although they usually develop more slowly. With houseflies or blowflies, a period of normal activity is often followed by a more quiescent period toward the end of which wing tremors become increasingly frequent; they increase in severity so that the insects have spasms of wing beating without "take off" which may culminate in sudden spells of violent uncoordinated flight ("flight convulsions"). When the convulsive phase subsides, the wings appear to become "fixed" in an abnormal position above the thorax and the insects may crawl with an ataxic gait for a time before falling on their backs. During this "knock-down" stage, the tremulous motor activity seen with DDT is lacking, but the legs are drawn up to the thorax with periods of rhythmic contraction and extension before movement finally ceases.

With compounds such as dieldrin which are highly stable in vivo, this process is irreversible and ends in death, but with related compounds that suffer metabolic detoxication, recovery can occur after up to 36 hr of total immobilization, although permanent wing damage is apparent.[192] This remarkable phenomenon indicates that the primary effects of cyclodiene poisoning (in houseflies) are reversible. In fact, the whole process of poisoning seems to resemble in some respects a prolonged version of the sequence of events leading to general anesthesia in mammals and Cherkin has emphasized the early excitatory effects of anesthetics.[692] Chlordane appears to act

as a depressant in the American cockroach and high dosages immobilize the insect very quickly. In this condition, external stimuli applied to the insect frequently result in bursts of violent tremors. Busvine[218] notes that the signs of acute poisoning by lindane, cyclodienes, and organophosphorus compounds are also produced during the excitatory stage following exposure to narcotic vapors such as benzene, chloroform, naphthalene, and p-dichlorobenzene, so that the use of signs of poisoning as a guide to specific mode of action is of fairly limited value.

In early experiments with aldrin- and dieldrin-poisoned American cockroaches, leg tremors were prevented by cutting the nerves supplying the legs, and application of the toxicants to legs amputated from untreated insects produced no tremors. It was therefore concluded that these compounds act on the central nervous system.[2] However, another investigation revealed that dieldrin, heptachlor, and chlordane cause spontaneous discharges from sensory neurons in the metathoracic leg, but only after a delay period longer than that observed with DDT. The central effects appeared to be confirmed by an increased frequency of spontaneous discharges and the prolonged synaptic after-discharge in cockroach central cord treated with dieldrin, but these effects were later attributed to impurities since they were not observed when purified dieldrin (that is, HEOD) was used.[478] A recent reexamination of this situation reveals that the time of observation is critical, there being no effect of dieldrin ($10^{-5}M$) applied directly to the metathoracic ganglion until about 60 min after treatment, while the last abdominal ganglion is relatively insensitive even then. However, photoaldrin and photodieldrin act more rapidly than dieldrin and 6,7-dihydroxydihydroaldrin acts most rapidly of all, producing prolonged synaptic after-discharge as early as 5 min after application.[693]

The effect on sensory neurons has been confirmed by injecting dieldrin, the photoisomers, and the trans-diol into the cockroach metathoracic leg. As before, there is long delay (45 min) with $10^{-5}M$ dieldrin, while photoaldrin and photodieldrin stimulate after 9 to 10 min and the trans-diol much more quickly (2 to 3 min), but the effects are much less intense than those produced by DDT. The trans-diol differs from the other compounds in that the spontaneous sensory discharge becomes completely blocked soon after hyperactivity commences. Accordingly, it seems that these compounds exert both central and sensory effects, although the latter appear to be less significant than is the case with DDT.[693] The ganglionic nature of the central effects is emphasized by Shankland, who found no evidence for any direct action of several cyclodiene insecticides (at up to $5 \times 10^{-4} M$) on central axonal membrane of *Periplaneta*, although there was an interesting joint action when individually subactive concentrations of DDT and dieldrin ($10^{-10} M$ and $2 \times 10^{-6} M$, respectively) were applied successively.[694] More recently, he has suggested that dieldrin causes excessive release of acetylcholine from its storage sites near the presynaptic membranes of cholinergic junctions.[694a]

When tested on the central nerve cord of the German cockroach (at $10^{-4} M$), dieldrin gave the most delayed effect (35 min); aldrin and isodrin stimulated after 30 min and endrin after 20 min. Heptachlor epoxide and DDT acted most rapidly (11 min), followed closely by lindane and isobenzan (12 to 13 min). However, there are some discrepancies to be explained since *trans*-chlordane and toxaphene show latent periods similar to those of lindane, isobenzan, and heptachlor, yet are several times less toxic than these, while DDT is about fourfold less toxic than heptachlor epoxide but acts equally rapidly in vitro. Of course, metabolic detoxication might account for some of these anomalies. The *trans*-diol is again said to act more rapidly than the other compounds (2 to 5 min at $10^{-5} M$) and on the basis of these interesting observations it has been suggested that *trans*-diol formation from dieldrin is a toxication reaction.[695]

Hydroxylated derivatives of cyclodienes are vulnerable to elimination by conjugation reactions and generally appear to be of lower toxicity than their precursors when topically applied or ingested (Table 20). In these circumstances, they may never reach a sensitive target and the situation might be quite different if such derivatives are actually liberated within the nervous system by metabolic conversion there of precursors such as dieldrin. An interesting observation is that although the dieldrin analogue HEOM (Figure 16, structure 7) is converted very rapidly into the corresponding *trans*-diol in insects such as houseflies and blowflies and is nontoxic to them (see section C.1b), the diol itself is neuroactive when applied to the cockroach ganglion.[696] In tsetse flies HEOM is much more stable and a toxic effect is evident,

indicating that the epoxide itself is toxic when present in this insect in finite amount. However, it might be argued that in this insect, some unchanged epoxide reaches the nervous system and there liberates the diol, whereas in the others insufficient epoxide is ever present for this to occur (except at very high doses) and the diol rapidly produced when HEOM is externally applied never reaches the target.

Other problems are evident. Thus, dieldrin is the obvious immediate precursor of the derived *trans*-diol, yet aldrin is somewhat more toxic to both German and American cockroaches.[41] The range of neuroactivity of hydroxylated derivatives in this series is unknown, but several compounds known to be oxidatively hydroxylated in vivo are synergized when these conversions are blocked, which is opposite to the effect expected if such transformations are toxicative.[697] The rapid action of photoaldrin and photodieldrin on cockroach ganglion could be due to an improved steric fit (compared with dieldrin) to the target (see section C.1c), to the formation of a neuroactive *trans*-diol or to the formation of Klein's ketone (Figure 12, structure 4); blocking of the last two processes should result in antagonism if they are toxicative. Dieldrin resistant houseflies appear to convert it into at least one product of oxidative hydroxylation as well as the *trans*-diol.[698] Mixed function oxidase inhibitors do not synergize dieldrin in these strains[697,698] but they are not particularly good inhibitors of *trans*-diol formation either,[235] so that this conversion may still occur in their presence.[484] In any case, if these conversions are toxications, the effect of their inhibition should be an antagonism of dieldrin's action that might be evident in susceptible strains but would not be noticeable in those already resistant.

Increased respiration is found in insects treated with cyclodienes,[24] but its onset follows different patterns, depending on the compound. German cockroaches poisoned with *cis*- or *trans*-chlordane experience a sudden fivefold increase over the normal level after a latent period of 4 to 8 hr, whereas heptachlor, which is a more efficient toxicant, causes a gradual rise from the time of treatment and the peak is less pronounced. Aldrin, dieldrin, and toxaphene are different again, in that they produce fairly steep rises after a latent period of 2 to 3 hr. Thus, the cyclodienes and toxaphene differ from DDT, methoxychlor, and lindane, which produce rapid increases in respiration shortly after administration, and this is a reflection of the differing speeds of onset of hyperactivity. There is evidence that the respiration of dieldrin poisoned houseflies falls off more rapidly than is explained by the exhaustion of endogenous energy reserves during hyperactivity and cyclopropane anesthesia does not restore the $\alpha$-glycerophosphate depletion observed in these insects.[677]

Like DDT, dieldrin produces an increase in acetylcholine level in American cockroach nerve cord and in either case the effect is independent of the intact peripheral nervous system or of blood factors since it occurs in nerve cords isolated from the insects at the early prostrate stage as well as in the nerve cords of intact treated insects; dieldrin is not a cholinesterase inhibitor.[699] It has been suggested that that there is action on the presynaptic vesicles storing acetylcholine at cholinergic junctions, resulting in an excessive release of acetylcholine.[694a] There is no evidence for the formation of $\gamma$-butyrobetaine or its esters in the nerve cords of dieldrin treated cockroaches, nor for its presence in those of normal ones, although this substance has been suggested as a factor in dieldrin poisoning in the rat. Altogether, there is little evidence that cyclodienes exert their effect by acting on a specific enzyme, although the possibility of interaction with ATP-ases continues to attract attention, as in the case of other organochlorines. Chlordane inhibits $Na^+$, $K^+$, $Mg^{2+}$ ATP-ase from American cockroach nerve cord and to a smaller extent than that from muscle: the $Mg^{2+}$ ATP-ase from both tissues is also inhibited, with the greater effect on that from muscle.[680]

Matsumura recently found a significantly lower ATP-ase activity in the cell membrane fractions of brain homogenates from normal German cockroaches than was present in several genetically pure cyclodiene resistant strains.[695] Dieldrin and heptachlor epoxide inhibited the activity but there was no apparent relationship between the degree of inhibition and of resistance to each compound. *It is interesting that phenylthiourea, mersalyl acid, and ouabain, which are known to be inhibitors of nerve ATP-ases, are more toxic to a resistant than to the corresponding normal strain and the isolated nerve cords of resistant insects are also more sensitive to these compounds.* The ATP-ase preparations are not completely inhibited by chlorinated insecticides but they are a complex mixture of different types of ATP-ase and only part of any

particular preparation may be critically involved in the processes of nerve conduction. This seems to be an area of investigation in which some advances may be expected, but the real significance of the observations in relation to mode of action has yet to emerge.

It can be seen from the previous discussion that the problem of dieldrin mode of action is intimately associated with that of dieldrin-resistance. Factors such as retarded penetration through the insect integument may contribute to resistance in some cases (Vol. II, Chap. 2B.2a), but the early observation[700] that the intact nervous system of resistant houseflies, for example, responded more slowly to contact with dieldrin than did that of a normal strain indicates that the major difference is at the level of the nervous tissue itself. However, penetration differences may still be involved there.

The contribution that might be made by penetration barriers at this level is highlighted by the intriguing investigations of Sun and colleagues,[487,701] who examined the different toxicities of dieldrin to houseflies when topically applied, injected, or infused (dose injected gradually during a predetermined time). The compound is more toxic by injection in acetone/dimethylsulfoxide (1:1) than by topical application in acetone, but surprisingly, slow infusion in acetone/dimethylsulfoxide (30 min) results in a 100-fold reduction in toxicity and the resulting dosage-mortality curves are flattened like those of the resistant strains. Since the only apparent difference between injection and infusion is the speed of toxicant entry into the insect, the observations suggest that infusion allows time for its storage in inert tissues, so sparing the CNS. In this context, it is interesting to note that when rats receive dieldrin by slow intravenous infusion it is deposited in the fat rather than in the CNS, while the higher the rate of injection, the higher the level that appears in the brain.[560]

If the insect nervous system itself has penetration barriers for the toxicant or a delayed response to its action, extra time is again available for storage of the poison at inert sites. The levels of dieldrin attained in housefly nervous tissue are indeed lower following infusion than when administration is by injection, but the difference is at most much smaller (twofold) than the observed difference in toxicity and in the experiments described, decreased with increasing dose, so that at about four times the LD50 by injection, which is sublethal by infusion, either technique produced the same level of dieldrin in the nervous system [701] It is difficult to understand how this level can be associated with toxicity in one case but not in the other. The implication is that the different techniques of administration produce different patterns of distribution within the nervous tissue, so that whole tissue levels do not accurately reflect the amounts of toxicant at the site of action. According to a widely held view (see section A.) topically applied toxicants penetrate the cuticle and enter the blood, and the question arises as to why topical application should be more effective rather than less effective than infusion. This problem recalls the suggestion by Gerolt (see section B.1a) that in topical application, other modes of entry are involved.

Sun[487] subsequently found little difference between the normal and dieldrin-resistant housefly nervous system in regard to uptake of dieldrin in vivo or in vitro, rate of loss of it (from nerve cords previously exposed in vivo) into saline in vitro, or distribution in the tissue. The results generally resemble those obtained by Ray in an earlier investigation on nerve cords of normal and resistant cockroaches.[702] There appeared to be a small and similar conversion of dieldrin into unidentified metabolites in both normal and resistant housefly nerve cords, and that metabolism can be detected in houseflies at extended times after treatment has been confirmed by Guthrie, who recently presented some evidence for the formation of compounds similar to trans-dihydroxydihydro-aldrin (Figure 12, structure 6) and 9-hydroxy-dieldrin (Figure 12, structure 10).[698] This last investigation also indicated reduced penetration into the rest of the housefly body from the point of topical application on the thoracic cuticle. The possibility that dieldrin might be in some way excluded from its site of action in resistant houseflies has been investigated using $^3$H-dieldrin combined with electron microscopic autoradiography to locate the toxicant in nervous tissue. In this way, labeled material is found to be deposited in the neural lamella of the thoracic ganglion and in the neuropile region along the axonal membranes, but not in the motor neuron region between neuropile and neural lamella.[488] The assertion that this material is largely dieldrin should perhaps be qualified by subsequent observations of metabolism in this strain, but at any rate, it seems clear that dieldrin initially reaches

the synaptic region of the resistant strain. Unfortunately, comparable experiments with normal strains have not yet been reported and problems arise because of the very small amounts of toxicant that can be tolerated by these insects.

Pursuing the idea that dieldrin might form complexes with some critical component of nerve tissue, Matsumura[489] examined the affinity of various fractions from German cockroach brain homogenate (crude nucleus, mitochondria, microsomes, and supernatant) for dieldrin and found a tendency for lower binding capacity in the resistant strains, although no significant interstrain differences could be found in the corresponding fractions prepared from whole bodies. Later, the nerve cords of normal cockroaches exposed to dieldrin in vivo were found to have taken up more of it than those of a resistant strain, although the latter can actually tolerate much higher levels in the nerve cord.[691,703] Fractions of the above type from the axonic regions of these nerve cords showed the greatest interstrain difference in uptake following exposure in vivo and these differences persisted when the resistant strain was purified genetically by repeated interbreeding with a normal strain, accompanied by selection of each generation with dieldrin. These experiments and later work utilizing autoradiographic techniques similar to those employed with houseflies have provided some biochemical and genetical evidence for a reduced uptake of dieldrin by nerve cell membranes of the resistant strain. However, these autoradiograms were made using $^{14}$C-labeled dieldrin and are difficult to interpret; further studies with $^3$H-dieldrin should provide valuable additional information.

The possibility has already been mentioned that cyclodienes could interact with the ATP-ases associated with ion transport through membranes and it is logical to extend this approach by investigating the effects of dieldrin on ion exchange in nervous tissue, as was done for DDT. Although the action of dieldrin appears to differ from that of DDT, Hayashi and Matsumura find a superficial resemblance between the effects of the two compounds on ion transport in cockroach nerve.[704] Dieldrin initially stimulates both uptake and efflux of sodium ions in nerve cords of American and German cockroaches, with little effect on potassium ion influx. Eventually there is an accumulation of sodium ions in the nerve cords, while potassium ion influx decreases, in similarity with the DDT poisoning situation. The rate of calcium exchange is also increased in the early stages of poisoning but decreases later on, similar although less marked effects being observed with resistant strains. Dieldrin is a mild inhibitor of calcium binding to phospholipids such as phosphatidyl L-serine in vitro and it has been suggested that this effect may occur in vivo with nerve phospholipids, so that calcium transport across the nerve membrane is eventually inhibited. This relates to the early view that DDT poisoning results from hypocalcemia within the membrane,[705] although this idea has not been substantiated. In any case, similar inhibition was found with the nerve phospholipids of both resistant and normal strains, so that the control of this process does not appear to be involved in resistance.

Numerous questions remain to be answered, and the role of metabolic transformation of cyclodienes is enigmatic, but current opinion favors a primary interaction of a physico-chemical nature (such as reversible complex formation), involving the parent compounds or possibly their biotransformation products. However, it is quite possible that a compound such as *trans*-dihydroxy-dihydroaldrin could be involved in covalent bonding as well as physical interactions. What little is known of the sort of receptor that might be involved in physical interactions is based on structure-activity investigations (see sections C.1b and c) and even now the available information on intrinsic toxicities (see section C.1) in this series is inadequate to permit a proper appraisal of the problem, which is complicated by the discovery of various active photochemical rearrangement products.[655]

Although some of the resistance studies indicate reduced access of dieldrin to resistant nervous tissue in general, there is no evidence so far for total exclusion of the toxicant from vital areas such as nerve synapses unless such exclusion is effected by highly localized detoxication. The solution of this problem will depend on very careful autoradiographic studies of the type already mentioned, coupled with some method of determining unequivocally whether the label so located is or is not the parent toxicant. If exclusion by one means or another from the site of action is not the answer, then the last resort is a change in the nature of the site so that it no longer responds to the toxicant.

Little can be added about the action of toxaphene on insects, except that the effects appear to be central and generally similar to those produced by the cyclodienes. Meaningful studies on this substance must await clarification of the structures of its active components.

The signs of acute poisoning produced by members of the cyclodiene group in mammals clearly indicate that they are neurotoxicants whose effects are generally similar although there may be differences in detail, especially between species.[706] Hypersensitivity to external stimuli is associated with generalized tremors, followed later by tonic-clonic convulsions culminating in death by respiratory failure. The symptoms in man include headache, dizziness, abdominal disturbances, nausea, vomiting, mental confusion, muscle twitching, and epileptiform convulsions which may occur suddenly and without prior warning.[303] Various barbiturates alleviate the convulsive symptoms; sodium amytal is said to be better than others for chlordane poisoning, while sodium thiopentone and phenobarbitone are recommended for other cyclodienes.

It will be recalled that increased acetylcholine levels have been found in the nervous systems of insects poisoned with cyclodienes, and aldrin poisoned cats were said to show peripheral parasympathetic disturbances such as salivation and slowing of the heart very like those produced by organophosphorus poisons.[2] However, these compounds are not anticholinesterases in vitro and there are obviously marked central effects separate from any peripheral ones, which are likely to be a consequence of central stimulation in any case. The detailed actions of aldrin and dieldrin appear to differ in some respects, although aldrin is converted into dieldrin in vivo. Endrin is said to resemble aldrin in producing in dogs certain parasympathomimetic effects,[2] but it is evident that the major changes arise from central stimulation.[303]

Toxaphene acts on the central nervous system to cause diffuse stimulation resulting in hyper-excitability, salivation, vomiting, tremors, clonic convulsions, and then tetanic contractions of the skeletal muscles culminating in death by respiratory failure if the dose is lethal. Pentobarbital and phenobarbital are used to alleviate convulsions and there is complete recovery from nonlethal doses.

In the search for biochemical changes that

might provide clues to the causes of cyclodiene poisoning, the familiar difficulty of distinguishing between causes and consequences of toxic action are evident. Several investigations have indicated an interference by dieldrin with mitochondria or mitochondrial enzymes. The first of them claimed that γ-trimethylaminobutyric acid (γ-butyrobetaine), and its $\alpha,\beta$-unsaturated and $\beta$-hydroxy-derivatives (crotonbetaine and carnitine, respectively) were liberated, as esters with co-enzyme A, from mitochondria in the brains of dieldrin poisoned rats. Several unrelated treatments including electroshock were said to produce this effect by damaging the mitochondria, thereby releasing the esters which then caused convulsions. However, it is equally likely that the convulsions were produced by other mechanisms and actually preceded the mitochondrial damage; the other difficulty is that only simple esters of these betaines have been positively shown to possess convulsive activity.[2]

From a detailed study of the changes in rat brain during dieldrin and isobenzan poisoning, Hathway concluded that the mitochondrial enzymes involved in glutamate metabolism are involved in the intoxication process.[707-709] Both compounds cause a significant increase in cerebral alanine levels *before* convulsions occur and isobenzan also causes an increase in glutamine. It was concluded that these toxicants cause the liberation of ammonia in the brain before and during convulsions, the ammonia being taken up by $\alpha$-ketoglutarate, glutamate or pyruvate to give glutamate, glutamine, or alanine, respectively. Later in the intoxication pattern when this mechanism becomes inadequate, free ammonia accumulates in the cerebral tissues, with convulsive effects. Picrotoxin, a convulsant unrelated to the cyclodienes, also causes elevated alanine and ammonia, but not glutamine levels and like dieldrin, it is thought to inhibit glutamine synthesis in the brain. The preconvulsive release of ammonia in brain is interesting, but the primary action of cyclodienes in brain and other nervous tissue remains obscure.

Whatever the primary event leading to poisoning, it is clear that the effects in the rat and in man are reversible.[560] There is no long-lasting lesion in the central nervous system; toxic effects are seen when dieldrin is present at appropriate sensitive sites and they disappear when the dieldrin does, which observation parallels to some extent

the results seen with degradable dieldrin analogues in houseflies.

There is no doubt that the problem of the primary mechanisms of action of the various types of chlorinated insecticides will continue to present a considerable scientific challenge. In particular, the seemingly unique structural requirement for toxicity found with lindane and the well defined and rigid stereochemistry of the cyclodiene molecules make them valuable tools for probing the structure and function of the nervous system. When their mechanism of action and that of DDT is fully understood, we shall also know much more about the action of anesthetics and natural neurotoxins.

## E. AN ADDENDUM ON MIREX

Mirex, the fully chlorinated cage molecule ($C_{10}Cl_{12}$) obtained by the self addition of two molecules of hexachlorocyclopentadiene (see Vol. I, Chap. 3.B.1) is an efficient insecticide when used in baits to control the imported fire ant (*Solenopsis richteri* Forel, *S. invicta* Burel), an agricultural and human pest in the southeastern United States. In Louisiana in 1970, 280,000 pounds of 0.3% mirex bait were distributed to help farmers control this ant pest.[709a] Furthermore, large areas of land in north Louisiana have been treated with this bait in a Federal/State cooperative program designed to retard the spread of the pest into Texas and Arkansas. Because of its evident chemical stability, the widespread use of this heavily chlorinated compound has caused much concern regarding its persistence in the biosphere and possible consequential effects on nontarget organisms. However, there have been no systematic studies of its behavior in living organisms until quite recently.

Compared with other pesticides, mirex has a low acute oral toxicity to vertebrates.[583] However, chronic effects such as parent mortality and reduced litter size have been reported in mice fed 5 ppm in the diet (for 120 days) and reduced litter size and cataracts in rats fed 25 ppm in the diet (for 45 days). The treated rats showed gains in liver weight and proliferation of the smooth endoplasmic reticulum. Mirex has also been reported to be carcinogenic in mice. Hens fed mirex at 300 ppm or 600 ppm in the diet lost weight; 600 ppm did not reduce egg production but hatchability and survival of the chicks was reduced.

Mirex accumulates in the tissues of fish as expected, but it is virtually nontoxic to species such as bluegill, sunfish, and goldfish, although the residues are very persistent and remain constant for a long period after exposure to the pesticide has ceased. Unfortunately, crustaceans such as crabs, shrimp, and crayfish are very susceptible to poisoning by mirex and crayfish can accumulate particularly high levels of it from the surrounding water. Possible adverse effects on crayfish populations are of particular concern in those areas where the rearing of this crustacean is an industry, although a recent investigation in Louisiana concludes that the use of mirex in the southcentral part of the state is not a threat to the industry there.[709a]

Two recent investigations describe the fate of [14]C-mirex in rats and one of them provides some information about its uptake by plants.[709b,c] In the experiments of Dorough et al.,[709b] rats given a single oral dose of [14]C-mirex (0.2 mg/kg) eliminated 18% of the administered radiocarbon within 7 days. Of this total, 85% was excreted in the feces within the first 48 hr and only traces of radioactivity were found in the urine. The excreted material was essentially unchanged mirex and the lack of urinary excretion, together with the small increment in fecal excretion after the first 48 hr suggests that the initial excretion represents material not absorbed by the gut and that mirex is poorly metabolized in the rat. As expected, the pesticide was stored in fatty tissues and no evidence was found for the presence of metabolites in the tissues.

In chlorinated molecules, the replacement of chlorine by hydrogen is likely to make the molecule more biodegradable, and a monohydro-derivative ($C_{10}Cl_{11}H$) of mirex was investigated by Dorough et al. Its behavior in rats resembles that of mirex, although there was evidence for an increased metabolic rate compared with the $Cl_{12}$ precursor in the form of somewhat higher proportions of unextractable and water soluble material in the feces.

In the study of Fishbein et al.,[709c] a single oral dose of [14]C-mirex (1.5 or 6 mg/kg) was given to rats and the results obtained are generally similar to those described above. At the higher dose, 55%

of the administered radiocarbon was excreted within 48 hr, with a further small increase to 55.8% after 7 days. During this time, only 0.69% of the administered radiocarbon appeared in the urine and the material excreted by either route consisted only of mirex. After 7 days, 34% of the administered dose was retained by tissues and organs, the largest proportions being in fat (27.8%) and muscle (3.2%). No metabolites were found in the tissues. These results indicate a first half-life for a small dose of mirex of about 38 hr (probably material not absorbed by the gut) and a second half-life of more than 100 days. Periods of incubation of up to 36 hr with fortified liver preparations of rat, mouse or rabbit that are known to attack chloroalkanes failed to produce metabolites from mirex.

In the plant study,[709c] young (2-week-old) pea and bean plants were grown for 48 hr in water containing 1 to 10 ppm of mirex, after which various parts of the plants were analyzed for radiocarbon. Mirex was absorbed by the roots in proportion to the concentration in the water and rather similar amounts were present in each species, except for a relative increase in the case of beans at the highest exposure. In either case 7 to 8% of the mirex in the roots was translocated to the shoots. This seems surprising in view of the low water solubility of this compound; the authors make no comment on the nature of the material in either roots or shoots. Incubations for up to 24 hr with root preparations from these species gave no evidence of metabolite formation from mirex.

In summary, these observations confirm the suspicion that this undoubtedly useful and selective control agent for the fire ant has considerable biological stability. Its accumulation in animal tissues is to be expected and the half-times for elimination when exposure ceases will probably be longer than for most of the other organochlorine compounds already discussed. The roots of plants grown in mirex treated areas may acquire residues of this compound. It remains to be seen whether there is the possibility of, for example, reductive removal of chlorine atoms by soil or other bacteria under anaerobic conditions. Such processes would be important, since they might expose the remaining molecule to more extensive biological degradation, as appears to be possible in the case of DDT.

# Chapter 4

# RESIDUES AND PROSPECTS

Previous sections have dealt with the details of organochlorine insecticide behavior and it is appropriate to conclude this book with some general considerations pertinent to the use of these (and other) pest control chemicals. Although the present pace of scientific advance makes it possible that some new and revolutionary nonchemical means of pest control may be just around the corner, it is wise to assume that until such a method appears and is proven to be a practical substitute, the use of chemicals must and will continue. The requirement for increasingly efficient food production to support the rapidly expanding world population will ensure that this is so. Any suggestion that we might revert to the agricultural practices of former years is out of the question; a recent paper points out, for example, that to maintain the present level of agricultural production in the United States using the practices of 1930 would require a 50 to 60% increase in the land available for crop growing.[719] Even assuming this were possible, one wonders what the effect would be on the wildlife of that country.

Given that the use of pesticides will continue in the foreseeable future, the aim must be to minimize the undesirable environmental consequences of such use, which are summarized in Table 22. Achievement of the necessary balance between beneficial and adverse effects requires the careful management of (i) the interaction between pesticide and pest, (ii) the interaction between pesticide and nontarget organisms (the biosphere in general), and (iii) the relationship between man's health, food, and fiber, and the chemicals used to protect them.[720] For the first item, the most important requirement is to avoid, as far as possible, using the toxicant in such a way that the pest is exposed to intense selection for resistance to it. This requires a good deal of knowledge of the pest, particularly with regard to its detoxicative capabilities toward various poisons and the incidence of resistant individuals in the population. Since resistance is frequently due to enhanced metabolic detoxication by pathways already present to a low degree in normal populations, information on these two questions may give some warning of the type of resistance to

be expected and the speed with which it may develop. Accordingly, there is considerable justification for continuous surveys to detect the presence of resistant individuals and these should be associated with appropriate biochemical studies. Such work can be started at the laboratory level as soon as the chemical appears likely to be a promising control agent for a particular pest.

Correct management of the third relationship, from which pest control requirements originate, is clearly crucial to the environmental considerations of item (ii). At the present time, this management, at international level, is the result of the activities of the Joint Meeting of the Food and Agriculture Organization (F.A.O.) Working Party of Experts on Pesticide Residues with the World Health Organization (W.H.O.) Expert Committee on Pesticide Residues, which has reported on the safety of pesticide residues in food since 1963. In 1965, the Codex Alimentarius Commission of the F.A.O. and W.H.O. set up the Codex Committee on Pesticide Residues, which was instructed to use the recommendations of the Joint F.A.O./W.H.O. Meeting above as a basis for the development of international standards on residues. The Codex Committee at present consists largely of members from those developed countries that have had some years of experience in operating systems of pesticide registration and is a committee of government delegations containing only a few working scientists, although the information for decision making must eventually come from scientific sources. Discussing this situation, Hurtig[721] points out that there are many difficulties in devising mutually acceptable international standards, but if regulatory systems are to be evolved, scientists must participate from the beginning to ensure that the decisions are based on the best available scientific information rather than on emotional or political considerations.

At the scientific level, there have been arguments about the need to pursue residue analysis to its limits, since this involves the tedious identification and toxicological evaluation of trace metabolites and degradation products of pesticides. Arguments against such needs often seem quite valid to those scientists who from personal

TABLE 22

Undesirable Environmental Consequences of Pesticide Usage[720]

| Abiotic environment | Target organism | Plants | Animals | Man | Food |
|---|---|---|---|---|---|
| Residues present in soil, air, water | Appearance of resistance to toxicant | Presence of tissue residues | Presence of residues in tissues | Residues in tissues and organs | Presence of residues |
| | | Phytotoxicity | | | |
| | | Vegetation changes due to herbicide usage | Mortality in some wildlife species, sublethal effects in others | Effects of occupational exposure | |
| | | | Mortality in beneficial insects or in pest predators and parasites: leads to population changes such as the appearance of secondary pests | | |

experience may be in a position to assess the likely hazards of these products, even without precise knowledge of their identity. However, it is a different matter to convince the general public that all is well, and since public opinion is the mainspring of administrative action, the only way to prevent arbitrary and perhaps extreme decisions is by presentation of hard facts about pesticide residue behavior. It must also be admitted that the discovery of the extremely toxic compound tetrachlorodioxin as a trace component of the herbicide 2,4,5-trichlorophenoxyacetic acid is salutary and throws a new light on the need for information about pesticide transformation products.

The deliberations of the F.A.O./W.H.O. Joint Meeting have led to a number of definitions that are important for pesticide management. A pesticide residue is defined as a residue in or on food of any chemical used for the control of pests. The term includes derivatives of the parent chemical and is expressed in parts per million (ppm) by weight of the contaminated commodity. Fundamental to the regulation of residue levels is the concept of *"good agricultural practice,"* which is defined as that *recommended usage* of a pesticide which is necessary and essential for the control of

a pest under all practical conditions. Recommended usage involves following the procedures of formulation, dose rates, application frequencies, and minimum preharvest application intervals (designed to ensure maximum residue dissipation before harvest) arrived at by supervised trials, taking into account variations in crop husbandry, climatic conditions, incidence of pests, etc., in the various countries in which the chemical may be used. Good agricultural practice usually results in a residue of the parent chemical or its transformation products, and the *residue tolerance* is the maximum concentration (expressed as ppm) permitted in or on a food at a specified stage in the harvesting, storage, transport, marketing, or preparation of the commodity, up to the final point of consumption. Thus, the tolerance is simply a legal upper limit, which if exceeded can result in action by regulatory authorities. The situation is complicated because "good agricultural practice" may differ between geographical locations, so that even with careful usage the residues resulting at some particular stage of crop may differ for this reason.

Here one sees a source of arguments between countries concerning acceptable tolerances, with consequent repercussions for international trade in food commodities. Furthermore, the residue at some particular commodity stage may differ both *quantitatively* and qualitatively from that in the final food processed for consumption, the initial residues being usually dissipated in various ways during processing. The residue at this point is the *terminal residue* and is the contaminant(s) actually ingested by the consumer. Since the tolerance allowable at some earlier point in the process is arrived at by reconciling both the requirement of good agricultural practice and any contribution to toxic hazard by the terminal residue in the diet, information is required about the levels of terminal residues and the toxicological properties of the chemicals involved. Without such data, it must be assumed that the whole of the residue at the earlier tolerance stage reaches the consumer unaltered and the allowed tolerance will accordingly be lower than it might have been with better information.

In addition to the residues arising in this way, there are so-called *unintentional residues* arising not through deliberate pesticide application to a particular crop but from the existing levels of environmental contamination. Thus, residues may appear in crops through earlier treatment of the soil for some different purpose, and contamination of animal products (meat, eggs, milk, etc.) arises because the animals consume contaminated fodder. The occurrence of this type of residue presents problems because it is often the result of the bioconcentration phenomenon and is unrelated to any agricultural practice that can be controlled. Such residues can only be reduced by controlling the sources of contamination so as to lower the background residue levels in the environment as a whole. At present there is good evidence that such changes are occurring in some areas as a result of restrictions placed on the use of organochlorine insecticides (Vol. II, Chap. 3B.1c). To meet this situation, the term *practical residue limit* has been defined as the maximum unintentional residue allowed in a specific commodity; it recognizes the unavoidable presence of residues in some commodities and allows a degree of flexibility in regulatory approach.

The actual intake of, say, organochlorine residues, is compounded from the terminal residues in all the individual components of the diet and is clearly not simply related to the tolerances recommended for these components, even if they are set right at the point of consump-

tion. The *acceptable daily intake* (A.D.I.) is defined as the daily intake (expressed as mg/kg) which, on the basis of all the known facts (from toxicological evaluations of a chemical and its residues), appears to be without appreciable risk during a lifetime. The figure arrived at takes into account any special risks to the young and aged and such factors as the interactions discussed in section B.3c of Chapter 3 in this volume. It incorporates the usual large safety factors included when extrapolating from experimental animals to man. Therefore, the ultimate aim is to set a tolerance for a particular pesticide at the required stage of crop processing such that its contribution of terminal residue, together with the terminal residues of the same chemical from all other dietary components, will not in total exceed the A.D.I. Summaries of recommendations concerning tolerances and practical residue limits in raw agricultural products moving in commerce, as well as A.D.I's, are published in reports of the F.A.O./ W.H.O. Joint Meeting (as issues of the *W.H.O. Technical Report Series*). Tolerances automatically include large safety factors because they assume that all food of a particular class will be treated with the chemical concerned and will carry residues at the maximum permissible level, which is by no means the case in practice.

In the case of organochlorine compounds, the treatment of plant products after harvest both in the home and in commercial processes favors reduction of the residues.[722] Two properties of residues which are particularly useful in this respect are their tendency to co-distil with water and to localize in the outer layers and peels of fruits and vegetables, so that steaming, washing, or peeling of these items can effect a substantial initial removal of residues. This affinity for the peel or rind is evident in root crops such as potatoes, beets, and carrots (with some differences between varieties in the last case) for aldrin, heptachlor, and chlordane; these compounds may be from 20 to 300 times more concentrated in potato peel than in the pulp, and 70% or more of the residues in colored varieties of carrots normally appear in the peel. Cooking usually effects some further reduction. Thus, peeling followed by boiling removes nearly all the DDT residues in potatoes and the chlordane residues from potatoes, beets, and turnips. The effect of washing depends on the nature of the surface, and removal of the residues is usually facilitated if a little detergent is present.

Many of the commercial processes used in food preparation remove substantial amounts of organochlorine residues.[722] The canning process for spinach or green beans, for example, removes 80 to 90% of the DDT residues present. Since the organochlorine insecticides are highly lipophilic, it would be logical to suppose that vegetable oils for human consumption make a substantial contribution to the human intake of residues. In fact, there is considerable evidence that the refining process applied to these oils removes residues of the main organochlorines through volatilization in steam during deodorization and probably through dechlorination or adsorption on catalysts during the hydrogenation steps. The vegetable tissues after oil extraction are frequently used in animal feeds, and since they contain lower residues than the parent crop, the processing is beneficial in this respect also. Surveys in the United States show that the dietary components from these sources contribute only 2.5 to 3.7% of the total daily organochlorine intake from all foods.[722]

The situation is rather different for dairy and other animal products since the organochlorine residues remain with the fat during processing. Thus, cheese, butter, or cream tend to have the same residue levels in fat as the original milk. However, steam deodorization reduces the residues in some milk products. Much of the reduction in residue levels effected by cooking meat is associated with the removal of fat, although the cooking itself has some effect. Numerous investigations show that the processing loss of residues from foods of animal origin occurs less readily than from those of plant origin and the former undoubtedly provide the major source of residues in the human diet. The presence of these residues in foods of animal origin is thought to be largely due to the use of the waste products of agricultural and food processes (just those portions, such as rinds and peels, which contain most residues) as components of animal feeds. Thus, some of the residues discarded in the preparation of human food may yet reach the consumer by this route.[722]

How does the actual daily human intake of residues compare with the A.D.I? The actual intake has been determined in two ways, and the agreement between them for organochlorine insecticides appears to be satisfactory. In the first method, the residues in the diet are determined from those in food items comprising the total diet which are purchased from retail sources (market basket samples), processed as for consumption and then individually analyzed.[720] Additionally, cooked meals from restaurants have been analyzed for their residue content. The second method involves the determination of residues in samples of blood or adipose tissue and the use of the mathematical relationships between tissue content and daily intake discussed in Chapter 3, section B.1c of this volume. The results of these investigations for the U.K. and the U.S. are well documented and show that the actual daily intakes of most organochlorine insecticides are only small fractions of the respective A.D.I's. Highest intakes are of DDT group residues (DDT plus DDE plus DDD), taking up one-tenth to one-twentieth of the A.D.I. (set at 0.005 mg/kg/day), and aldrin-dieldrin residues, lying on or just below the A.D.I (set at 0.0001 mg/kg/day). It must be remembered that the A.D.I itself incorporates generous safety factors and that there has been a tendency for organochlorine residues, including those of aldrin and dieldrin, to decline in recent years following restrictions placed on the use of these compounds. In the U.K. for example, the daily intake of dieldrin fell from 19.9 $\mu$g/man/day in 1965 to 6.6 $\mu$g/man/day (about 0.9 of the A.D.I for a 70 kg man) in 1968 − 69;[54] even the higher figure is some 160 times lower than the dietary no-toxic-effect intake rate for dieldrin in the rat, dog, and monkey.[303]

The adipose tissue concentrations arising from these exposures in the U.S. and the U.K. are of a similar order and resemble those found from surveys in continental Europe.[481] Dieldrin levels are at about 0.2 ppm in the U.K. and in northern Europe and at 0.2 to 0.3 ppm in the U.S. DDT group residues in the U.K. (about 3 ppm) are lower than those in southern and eastern Europe, which resemble those in the U.S. (6 to 12 ppm). These levels have tended to remain constant or even decline in recent years, predicted for constant or declining intakes by the pharmacokinetic principles discussed in Chapter 3, section B.1c. While it would clearly be ideal if there were no residues in the diet, the conclusion to be reached from the available information is that man is rather well protected by food preparation processes and by tolerance levels, from the consequences of residue concentration along food chains (Figure 19). With

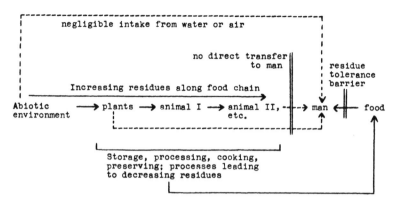

FIGURE 19.   Relationship between man, food chains, and residues in food.

many years of human exposure to these small tissue levels of organochlorines already behind us, there is no evidence that they have any biological effect, but this is no excuse for complacency. Constant vigilance must and certainly will be maintained. Since the tolerance system is based on good agricultural practice, it should act as a brake to prevent the excessive and indiscriminate use of pesticides in crop protection. However, it cannot prevent the loss of a proportion of the applied chemical to the atmosphere (if volatile, as in the case of organochlorines), nor its spreading beyond the point of application or effects on nontarget organisms.

The W.H.O. has emphasized that the spraying of individual dwellings (as in malaria control programs) contributes much less to environmental contamination than do crop protection treatments, but the aerial spraying of large areas for disease vector control, which must have occurred quite frequently in the past, undoubtedly contributes to the global burden of contamination. Another important source is the large-scale spraying of forests against pest attack. Wildlife cannot be protected from exposure to chemicals distributed in this fashion and, apart from direct exposure to the pesticide at the time of its application, the possibility of bioconcentration or accumulation along food chains from consumed to consumer (Figure 19) is now well recognized and extensively discussed. A comprehensive account of persistent pesticides in the environment has been given by Edwards[59] and a useful discussion of the bioconcentration phenomenon by Kenaga.[723] In simple terms, the bioconcentration factor of pesticide residue in a particular organism is the ratio of the level in a selected tissue to the level in the

ambient environment (soil, water, or air) or in the various organisms of the diet. The complexity of this concept may be illustrated by reference to fish, for which bioconcentration factors depend on the tissue chosen and on whether the point of reference is the chemical's concentration in the surrounding water or in the food consumed by the fish.

Although the actual maximum residues of, for example, the DDT group compounds found in aquatic organisms do not usually differ very greatly from those found in terrestrial ones, by far the largest bioconcentration factors for organochlorine insecticides are found in aquatic species. Perhaps this is not surprising, since although the concentrations in water may be minute, contact between organism and water, present in relatively infinite amount, is very efficient and is assisted by rapid circulatory processes in many cases. Thus, DDT residues in water are taken up very rapidly by the gill surfaces of fish (reflecting both the aqueous/lipid phase partitioning discussed in Chapter 3, section B. and the tendency of DDT to adsorb on any nonaqueous surface available), and the blood circulation rate appears sufficient to ensure rapid equilibration with adipose tissue. Bioconcentration factors of $10^6$ (relative to the surrounding water) have been found with aquatic organisms, but are usually less than 100 for those feeding on or in soil. Although this kind of accumulation certainly occurs, it must be remembered that we owe our existence to the ability to concentrate *nutrient* substances originating in the environment. Furthermore, it is not an invariable rule that these residues are passed on to other organisms along food chains in an ever accumulating fashion because the creatures in the various successive trophic levels have elimination processes (Vol. II, Chap. 3B.) which can greatly alter distribution patterns.

Data in reference to residues collected by Edwards[59] show that the bioconcentration factors for fish are frequently similar to those of the aquatic invertebrates on which they feed, suggesting that direct uptake from water is more important than ingestion with food. Laboratory experiments have confirmed this for at least one species, *Cottus perplexus*. When groups of this fish were kept in HEOD (dieldrin) contaminated water, those fed on HEOD contaminated worms for 3 weeks contained similar residues to others given clean worms, and the total HEOD given in the food constituted no more than 16% of the total tissue residue found. In the Farne Island marine ecosystem discussed by Robinson,[724] there is a general increase in tissue residues of both HEOD and DDE as the trophic levels are ascended, but HEOD residues in cod (0.009 ppm) are only about half those in the sand eel (its principal food) and are greater in planktonic crustacea (0.016 ppm) than in any of the fish examined. Yet another example is the pond ecosystem described by Hurtig, in which dieldrin residues in plankton and several higher levels of the food chain declined rapidly soon after its application to the water.[725]

With so many species and differing environmental factors to consider, generalizations about organochlorine insecticide behavior and effects are impossible. However, it may be the case that the levels achieved in certain aquatic environments (such as estuarine waters receiving the outflow of contaminated rivers) are dangerously high for some species.

One of the most controversial aspects of the environmental toxicology of organochlorine insecticides is the effect of their residues on birds (see section B.1c in Chapter 3 of this volume). In general, predatory birds contain higher residues of these compounds than herbivorous ones, with insectivorous species occupying an intermediate position, although the concentrations vary greatly. In Great Britain, particularly high organochlorine residues have been found in the tissues and eggs of great crested grebes and herons;[726] residues in the latter predator are similar to those found in bald eagles and brown pelicans in the United States, showing that birds with similar habits in different geographical locations tend to accumulate similar residues.[59] Despite their burden of residues (about

14 ppm of mainly DDE plus dieldrin), heron eggs produce viable young and there is no evidence of a decline in the heron population, although there have been declines in the populations of some bird-eating predators which accumulate rather similar residues in their eggs. In fact, the amount of dieldrin in British heron eggs (about 6 ppm) is about half of that which allowed production of viable young when present in chicken, quail, and homing pigeon eggs in laboratory trials. Dieldrin residues in predator eggs are generally lower than the levels in these experiments and frequently lower than those found for other species whose reproductive success seems little affected. In such comparisons, difficulties might arise through interspecific differences in sensitivity to the toxicant or through unexpected effects of the mixed residues which are usually present in tissues.

Although flesh-eating birds generally contain higher organochlorine residues than non-flesh-eating species, some dead grain-eaters found in Britain in 1960-61 had much higher total body burdens of these compounds (especially dieldrin and heptachlor epoxide) than did dead predators of that time. The probability that these particular grain-eating species were poisoned is increased by the similarity between their tissue levels (in brain or liver) of dieldrin plus heptachlor epoxide, and the levels of these compounds found in experimentally poisoned birds.[727] Using these same criteria for toxicity, it can also be concluded that few of the dead birds examined between 1961 and 1965 died from dieldrin poisoning and that the deaths of not more than 5 to 10% of the predators examined in this period can be attributed to this cause, unless they are much more sensitive to its effects than other birds. These poisoning episodes of 1960–61 seem to represent a clear cut effect of organochlorine residues on birds.

A second example is the dramatic decline in the numbers and reproductive success of Western grebes at Clear Lake in California following the extensive application of DDD to control gnats. These events have been associated with high levels of DDD in breast muscle (41.4 ppm), brain (47.9 ppm) and eggs (298.7 ppm of DDD plus additional DDE), which persisted from the first use of this compound in the early 1950's until 1967–69, when a sharp fall in egg residues seems to have occurred. There also appears to have been a return to nearly normal breeding success by 1970, although the mean egg residues (117.4 ppm of

DDD plus 47.9 ppm of DDE) reported for 1969 by Rudd[728] are still rather high. Although the precise cause of this change is uncertain, it is believed that there has been a dilution of DDD residues in the biomass owing to the proliferation of a small Atherinid fish introduced in 1967, the residues acquired by the grebes in their food being smaller in consequence. If this proves to be the case, the experience points to a way of manipulating residues in the various trophic levels of highly contaminated ecosystems when the circumstances are appropriate.

With the information at our disposal today, it must surely be possible to avoid excessive contaminations of the sort just described, and there is no certain evidence that the organochlorine residues more commonly encountered in tissues are responsible for population changes in either predatory birds or other creatures. One of the greatest difficulties in assessing the effects of chemicals on wild bird populations arises from lack of baseline data on normal levels and their fluctuations, and this problem certainly extends to terrestrial mammals.[59] From the limited data available, tissue concentrations of organochlorine insecticides are smaller in mammals than in birds and there is little evidence that predators accumulate larger residues than other species. Apart from occasional incidences of acute poisoning, there is no evidence of gross effects on wild animals, and although more information is needed to assess the long term effects of continuous sublethal exposure, extrapolation from the abundant data obtained using laboratory animals and from the lack of effect of the residues normally found in domestic ones, tend to suggest that the levels thus far encountered will not be harmful. Unfortunately, these inferences may be confounded by species differences in sensitivity, so that there is no ground for complacency about the present situation. Indeed, there is a continuing need for the systematic collection of information regarding the behavior of organochlorine insecticides and other xenobiotics in wildlife, so that there is a pool of knowledge available for the future. This is a large task, but it may prove to be no more formidable than others to be faced by mankind in the next 50 years or so.

Many of the problems discussed above are exceedingly complex, and it is clearly prudent to avoid as many of them as possible. Educational programs are needed to ensure that the public fully understands the need for the use of environmental chemicals in certain circumstances, while appreciating that they are not without hazard and must be treated with respect. In this way, it should be possible to avoid poisoning incidents resulting from accidental gross pollution of the environment. Measures to reduce the background of environmental contamination by persistent organochlorine insecticides are in hand and their gradual replacement by more degradable compounds in appropriate circumstances will clearly continue. However, it is important to remember that with a few exceptions, these compounds are less acutely toxic than are many of the more recent, less persistent types, and, due to controversies in recent years, we probably know more about their behavior in living organisms than is the case for many other toxic substances. Some of the problems attending the use of these and other residual compounds may be solved by improved application techniques such as microencapsulation, which has been investigated at the Rothamsted Experimental Station in England. By enclosing the toxicant in porous microcapsules from which it is slowly released, the amount used for control may be restricted. Alternatively, the life of a useful but highly biodegradable compound might be prolonged to make it more efficient under practical conditions.

In well-regulated environments, such as are found in glasshouses, the careful integration of biological control methods with the use of chemicals, including organochlorines such as lindane, can reduce the need for repeated chemical treatments. Chemicals which do not persist beyond their required period of effectiveness find favor in these, as well as other applications, but their cost is frequently greater and their efficiency lower than for the more persistent types.

If predictions about the growth of world population are fulfilled, which seems unavoidable unless some natural calamity decimates the human race, the recently coined term "spaceship earth" may become very real to our descendants. In these circumstances, careful control of the environment will become increasingly crucial for man's survival. As suggested by Rudd,[729] the future role of the entomologist is likely to be that of an "ecological counsellor," prescribing and regulating the use of insecticides much as the physician prescribes medicines.

# REFERENCES

1. Department of Health, Education, and Welfare, Report of the Secretary's Commission on Pesticides and Their Relationship to Environmental Health, Parts I and II, U.S. Govt. Print. Off., Washington, D.C., 1969, 44.
2. O'Brien, R. D., *Insecticides, Action and Metabolism*, Academic Press, New York, 1967, 165.
3. Perkow, W., *Die Insektizide*, Huther Verlag, Heidelberg, 1956, 360.
4. von Rumker, R., Guest, H. R., and Upholt, W. M., The search for safer, more selective and less persistent pesticides, *Bioscience*, 20, 1004, 1970.
5. *Agrochemical Industry in Japan*, published by Japan Agricultural Chemicals Overseas Development Commission (JACODEC), Tokyo, 1971.
6. Pesticides and Health, Royal Dutch Shell Group Briefing Service, February, 1967.
7. Lauger, P., Martin, H., and Muller, P., Uber Konstitution und toxische wirkung von naturlichen und neuen synthetischen insektentotenden stoffen, *Helv. Chim. Acta*, 27, 892, 1944.
8. Zeidler, O., Verbindungen von chloral mit brom-und chlorbenzol, *Chem. Ber.*, 7, 1180, 1874.
9. Lauger, P., Uber neue, sulfogruppenhaltige mottenschutzmittel, *Helv. Chim. Acta*, 27, 71, 1944.
10. Chattaway, F. D. and Muir, R. J. K., The formation of carbinols in the condensation of aldehydes with hydrocarbons, *J. Chem. Soc. (Lond.)*, 701, 1934.
11. Muller, P., Uber zusammenhange zwischen konstitution und Insektizider wirkung I, *Helv. Chim. Acta*, 29, 1560, 1946.
12. West, T. F. and Campbell, G. A., *DDT and Newer Persistent Insecticides*, revised 2nd ed., Chapman and Hall, London, 1950.
13. Busvine, J. R., New synthetic contact insecticides, *Nature (Lond.)*, 158, 22, 1946.
14. Barlsch, E., Eberle, D., Ramsteiner, K., Tomann, A., and Spindler, M., The carbinole acaricides: Chlorobenzilate and chloropropylate, *Residue Rev.*, 39, 1, 1971.
15. Martin, H., Ed., *Pesticide Manual*, 2nd ed. British Crop Protection Council, 1971, 495.
16. Busvine, J. R., Mechanism of resistance to insecticides in house flies, *Nature (Lond.)*, 168, 193, 1951.
17. Miles, J. W., Goette, M. B., and Pearce, G. W., A new high temperature test for predicting the storage stability of DDT water-dispersible powders, *Bull. W.H.O.*, 27, 309, 1962.
18. Hadaway, A. B. and Barlow, F., Some physical factors affecting the efficiency of insecticides, *Trans. R. Soc. Trop. Med. Hyg.*, 46, 236, 1952.
19. van Tiel, N., On 'Supona', a new type of DDT suspension, *Bull. Entomol. Res.*, 43, 187, 1952.
20. Philips, N. V., Gloeilampen fabricken (Stabiliser for aqueous pesticide emulsions), Netherland Patent 6809975, June 1970.
21. Wiesmann, R., cited in *DDT, the Insecticide Dichlorodiphenyltrichloroethane and its significance*, Muller, P., Ed., Birkhauser Verlag, Basel, 1955, 14.
22. van Tiel, N., Influence of Resins on Crystal Size and Toxicity of Insecticidal Residues, Ph.D. Thesis, University of Leyden, 1955.
23. Hayhurst, H., The action on certain insects of fabrics impregnated with DDT, *J. Soc. Chem. Ind.*, 64, 296, 1945.
24. Brown, A. W. A., *Insect Control by Chemicals*, John Wiley, New York, 1951.
25. Carruth, L. A. and Howe, M. L., Chlorinated hydrocarbons and cucurbit foliage, *J. Econ. Entomol.*, 41, 352, 1948.
26. Cullinan, F. P., Effect of DDT on plants, *Agric. Chem.*, 2, 18, 1947.
27. Wilson, J. K. and Choudhri, R. S., DDT and soil microflora, *J. Econ. Entomol.*, 39, 537, 1946.
28. Edwards, C. A., Soil pollutants and soil animals, *Sci. Am.*, 220, 88, 1969.
29. Linsley, E. G. and MacSwain, J. W., DDT dusts in alfalfa, and bees, *J. Econ. Entomol.*, 40, 358, 1947.
30. Stahl, C. F., Relative susceptibility of *Protoparce sexta* and *P. quinquemaculata*, *J. Econ. Entomol.*, 39, 610, 1946.
31. Dowden, P. B., DDT to control *Lymantria dispar*, cited in Brown, A. W. A., *Insect Control by Chemicals*, John Wiley, New York, 1951, 603.
32. Hofmaster, R. N. and Greenwood, D. E., Organic insecticides for fall armyworm, *J. Econ. Entomol.*, 42, 502, 1949.
33. Blanchard, R. A. and Chamberlin, T. R., Insecticides for corn earworm and armyworm, *J. Econ. Entomol.*, 41, 928, 1948.
34. Bishopp, F. C., New insecticides to date, *J. Econ. Entomol.*, 39, 449, 1946.
35. Weinman, C. J., Chlorinated hydrocarbons for codling moth control, *J. Econ. Entomol.*, 40, 567, 1947.
36. Sun, Y-P., Rawlins, W. A., and Norton, L. B., Toxicity of chlordane, BHC and DDT, *J. Econ. Entomol.*, 41, 91, 1948.
37. Plumb, G. H., DDT sprays against Dutch elm disease, *J. Econ. Entomol.*, 43, 110, 1950.
38. Kapoor, I. P., Metcalf, R. L., Nystrom, R. F., and Sangha, G. K., Comparative metabolism of methoxychlor, methiochlor and DDT in mouse, insects, and in a model ecosystem, *J. Agric. Food Chem.*, 18, 1145, 1970.
39. E.I. du Pont de Nemours & Co., Industrial and Biochemicals Department, Technical Data Sheets, 1969, etc.
40. Martin, H., Ed., *Pesticide Manual*, 2nd ed., British Crop Protection Council, 1971, 428.
41. Metcalf, R. L., *Organic Insecticides*, John Wiley (Interscience) New York, 1955.
42. Brown, A. W. A., Insect resistance in arthropods, *W.H.O. Monogr. Ser.*, No. 38, 1958.

43. Brown, A. W. A. and Pal, R., Insect Resistance in Arthropods, 2nd ed., W.H.O., Geneva, 1971.
44. Insecticide resistance and vector control, *W.H.O. Monogr. Ser.*, No. 443, 1970.
45. Niswander, R. E. and Davidson, R. H., Chlordane and lindane versus DDT for roaches, *J. Econ. Entomol.*, 41, 652, 1948.
46. Busvine, J. R., DDT and analogues for Pediculus and Cimex, *Nature (Lond.)*, 156, 169, 1945.
47. Busvine, J. R., Review of lousicides, *Br. Med. Bull.*, 3, 215, 1945.
48. Busvine, J. R. and Pal, R., The impact of insecticide-resistance on control of vectors and vector-borne diseases, *Bull. W.H.O.*, 40, 731, 1969.
49. Busvine, J. R., DDT and BHC for the bedbug and louse, *Ann. Appl. Biol.*, 33, 271, 1946.
50. Pampana, E., *A Textbook of Malaria Eradication*, 2nd ed., Oxford University Press, 1969.
51. Peffly, R. L. and Gahan, J. B., Residual toxicity of DDT analogues, *J. Econ. Entomol.*, 42, 113, 1949.
52. Metcalf, R. L., Kapoor, I. P., and Hirwe, A. S., Biodegradable analogues of DDT, *Bull. W.H.O.*, 44, 363, 1971.
53. Bruce, W. N. and Decker, G. C., Chlordane, DDD and DDT for fly control, *J. Econ. Entomol.*, 40, 530, 1947.
54. Brooks, G. T., Pesticides in Britain, in *Environmental Toxicology of Pesticides*, Matsumura, F., Boush, G. M., and Misato, T., Eds., Academic Press, New York and London, 1972, 61.
55. The place of DDT in operations against malaria and other vector-borne diseases, *Off. Rec. W.H.O.*, No. 190, 1971, 176.
56. Food and Agriculture Organization, *F.A.O. Production Year Book*, 1969, 23, 496.
56a. Carson, R., *Silent Spring*, Houghton-Mifflin Co., Boston, 1962, 368.
57. Department of Health, Education, and Welfare, *Report of the Secretary's Commission on Pesticides and Their Relationship to Environmental Health*, Pts. I and II, U.S. Govt. Print. Off., Washington, D.C., 1969.
58. Matsumura, F., Current pesticide situation in the United States, in *Environmental Toxicology of Pesticides*, Matsumura, F., Boush, G. M., and Misato, T., Eds., Academic Press, New York and London, 1972, 33.
59. Edwards, C. A., CRC Uniscience Series, *Persistent Pesticides in the Environment*, Butterworths, London, 1970.
60. Department of Education and Science, Further review of certain persistent organochlorine pesticides used in Great Britain, *Report by the Advisory Committee on Pesticides and Other Toxic Chemicals*, HMSO London, 1969.
61. Muller, P., Physike und chemie des DDT-insektizides, in *DDT, the Insecticide Dichlorodiphenyltrichloroethane and its Significance*, Muller, P., Ed., Birkhauser Verlag, Basel, 1955.
62. Haller, H. L., The chemical composition of technical DDT, *J. Am. Chem. Soc.*, 67, 1591, 1945.
63. Forrest, J., Stephenson, O., and Waters, W. A., Chemical investigations of the insecticide 'DDT' and its analogues, *J. Chem. Soc. (Lond.)*, 333, 1946.
64. Cochrane, W. P. and Chau, A. S. Y., Chemical derivatization techniques for confirmation of organochlorine residue identity, in Pesticides Identification at the Residue Level, *Adv. Chem. Ser.*, No. 104, 11, 1971.
65. Rogers, E. F., Brown, H. D., Rasmussen, I. M., and Heal, R. E., The structure and toxicity of DDT insecticides, *J. Am. Chem. Soc.*, 75, 2991, 1953.
66. Skerrett, E. J. and Woodcock, D., Insecticidal activity and chemical composition. II. Synthesis of some analogues of DDT, *J. Chem. Soc. (Lond.)*, 2718, 1950.
67. Metcalf, R. L. and Fukuto, T. R., The comparative toxicity of DDT and analogues to susceptible and resistant houseflies and mosquitos, *Bull. W.H.O.*, 38, 633, 1968.
68. Holan, G., 1,1-Bis(p-ethoxyphenyl) dimethylalkane Insecticides, German patent 1,936,494, 1970; *C.A.*, 72, 90045, 1970.
69. Kaluszyner, A., Reuter, S., and Bergmann, E. D., Synthesis and biological properties of diaryl (trifluoromethyl) carbinols, *J. Am. Chem. Soc.*, 77, 4165, 1955.
70. Wiles, R. A. (to Allied Chemical Corporation), U.S. patent 3,285,811, 1966; *C.A.*, 60, 18572, 1967.
71. Holan, G., Diphenyldichlorocyclopropane Insecticides, Australian patent 283,356, 1968; *C.A.*, 71, 12815, 1969.
72. Muller, P. (to Geigy AG), Insecticidal Compounds, U.S. patent 2,397,802; *C.A.*, 40, 3849, 1946.
73. Haas, H. B., Neher, M. B., and Blickenstaff, R. T., *Ind. Eng. Chem.*, 43, 2875, 1951.
74. Holan, G., 1,1-Bis(p-Ethoxyphenyl)dimethylalkane Insecticides, German patent 1,936,495, 1970; *C.A.*, 72, 100259, 1970.
75. Grummitt, O., Buck, A. C., and Becker, E. I., 1,1-Di(p-chlorophenyl)ethane, *J. Am. Chem. Soc.*, 67, 2265, 1945.
76. Bergmann, E. D. and Kaluszyner, A., Di(p-chlorophenyl)trichloromethylcarbinol and related compounds, *J. Org. Chem.*, 23, 1306, 1958.
77. Hinsberg, O., Uber die bildung von saureestern und saure amiden bei gegenwart von wasser und alkali, *Chem. Ber.*, 23, 2962, 1890.
78. Huisman, H. O., Uhlenbroek, J. H., and Meltzer, J., Preparation and acaricidal properties of substituted diphenylsulfones, diphenylsufides and diphenylsulfoxides, *Recl. Trav. Pays-Bas Belg.*, 77, 103, 1958.
79. Monroe, E. and Hand, C. R., Some bis-(aryloxy)methanes, *J. Am. Chem. Soc.*, 72, 5345, 1950.
80. Hilton, B. D. and O'Brien, R. D., A simple technique for tritiation of aromatic insecticides, *J. Agric. Food Chem.*, 12, 236, 1964.
80a. Kapoor, I. P., Metcalf, R. L., Hirwe, A. S., Coats, J. R., and Khalsa, M. S., Structure activity correlations of biodegradability of DDT analogs, *J. Agric. Food Chem.*, 21, 310, 1973.

81. Fields, M., Gibbs, J., and Walz, D. E., The synthesis of 1,1,1-trichloro-2,2-bis(4'-chlorophenyl-4'-$C^{14}$) ethane, *Science*, 112, 591, 1950.
82. Dachauer, A. C., Cocheo, B., Solomon, M. G., and Hennessy, D. J., The synthesis of tertiary carbon deuterated DDT and DDT analogs, *J. Agric. Food Chem.*, 11, 47, 1963.
83. Sumerford, W. T., A synthesis of DDT, *J. Am. Pharm. Assoc.*, 34, 259, 1945.
84. Rueggeberg, W. H. C. and Torrans, D. J., Production of DDT, *Ind. Eng. Chem.*, 38, 211, 1946.
85. Mosher, S. H., Cannon, M. R., Conroy, E. A., Van Strien, R. E., and Spalding, D. P., Preparation of technical DDT, *Ind. Eng. Chem.*, 38, 916, 1946.
86. *Specifications for Pesticides used in Public Health*, 3rd ed., W.H.O., Geneva, 1967
87. Gunther, F. A., cited in West, T. F. and Campbell, G. A., *DDT and Newer Persistent Insecticides*, 2nd ed., Chapman and Hall, London, 1950, 43.
88. Gooden, E. L., Optical crystallographic properties of DDT, *J. Am. Chem. Soc.*, 67, 1616, 1945.
89. Wild, H. and Brandenberger, E., Zur krystallstruktur des p,p'-dichlorodiphenyl trichlorathan, *Helv. Chim. Acta*, 28, 1692, 1945.
90. Balson, E. W., Vapour pressures of DDT, BHC, and DNOC, *Trans. Faraday Soc.*, 43, 54, 1947.
90a. Spencer, W. F. and Claith, M. M., Volatility of DDT and related compounds, *J. Agric. Food Chem.*, 20, 645, 1972.
90b. Claith, M. M. and Spencer, W. F., Dissipation of pesticides from soil by volatalization of degradation products. I. Lindane and DDT, *Environ. Sci. Technol.*, 6, 910, 1972.
91. Kenaga, E. E., Factors related to bioconcentration of pesticides, in *Environmental Toxicology of Pesticides*, Matsumura, F., Boush, G. M., and Misato, T., Eds., Academic Press, New York and London, 1972, 193.
92. Riemschneider, R., Chemical structure and activity of DDT analogues with special consideration of their spatial structures, *Adv. Pest. Control Res.*, 2, 307, 1958.
93. Bradlow, H. and Van der Werf, C., Composition of Gix, *J. Am. Chem. Soc.*, 69, 662, 1947.
94. Schneller, G. H. and Smith, G. B. L., 1-Trichloro -2,2-bis(p-methoxyphenyl)ethane (methoxychlor), *Ind. Eng. Chem.*, 41, 1027, 1949.
95. Harris, L. W., Condensation of aldehydes, U.S. patent 2,572,131; *C.A.*, 46, 5088, 1952.
96. Adams, C. E., British patent 709,564; *C.A.*, 49, 10376, 1955.
97. Grummitt, O., Di (p-chlorophenyl) methylcarbinol, a new miticide, *Science*, 111, 361, 1950.
98. Barker, J. S. and Maughan, F. R., Acaricidal properties of Rohm and Haas FW-293, *J. Econ. Entomol.*, 49, 458, 1956.
99. Gatzi, K. and Stammbach, W., Über die natur der nebenprodukte in technischen p,p'- dichlorodiphenyl-trichlorathan, *Helv. Chim. Acta*, 29, 563, 1946.
100. Schechter, M. S. and Haller, H. L., Colorimetric tests for DDT, and related compounds, *J. Am. Chem. Soc.*, 66, 2129, 1944.
101. Balaban, I. E. and Sutcliffe, F. K., Thermal stability of DDT, *Nature (Lond.)*, 155, 755, 1945.
102. Scholefield, P. G., Bowden, S. T., and Jones, W. J., The thermal decomposition of 1,1,1-trichloro-2,2-bis-(p-chlorophenyl)ethane (DDT) and some of its phase equilibria, *J. Soc. Chem. Ind.*, 65, 354, 1946.
103. Fleck, E. E. and Haller, H. L., Catalytic removal of hydrogen chloride from some substituted α- trichloroethanes, *J. Am. Chem. Soc.*, 66, 2095, 1944.
104. Gunther, F. A., Blinn, R. C., and Kohn, G. K., Labile organo-halogen compounds and their gas chromatographic detection and determination in biological media, *Nature (Lond.)*, 193, 537, 1962.
105. Fleck, E. E., Preston, R. K., and Haller, H. L., Sym-tetraphenylethane from DDT and related compounds, *J. Am. Chem. Soc.*, 67, 1419, 1945.
106. Cristol, S. J., Kinetic study of the dehydrochlorination of substituted 2,2-diphenylchloroethanes related to DDT, *J. Am. Chem. Soc.*, 67, 1494, 1945.
107. England, B. D. and McLennan, D. J., Kinetics and mechanism of the reaction of DDT with sodium benzenethiolate and other nucleophiles, *J. Chem. Soc. (Lond.)*, B7, 696, 1966.
108. Cristol, S. J. and Haller, H. L., Dehydrochlorination of 1-trichloro-2-o-chlorophenyl-2-p-chlorophenylethane (o,p'-DDT isomer), *J. Am. Chem. Soc.*, 67, 2222, 1945.
109. McKinney, J. D., Boozer, E. L., Hopkins, H. P., and Suggs, J. E., Synthesis and reactions of a proposed DDT metabolite, 2,2-bis(p-chlorophenyl) acetaldehyde, *Experientia*, 25, 897, 1969.
110. Plimmer, J. R., Klingebiel, U. I., and Hummer, B. E., Photooxidation of DDT and DDE, *Science*, 167, 67, 1970.
111. Institut für ökologische Chemie, *Jahresbericht 1970*, Klein, W., Weisgerber, I., and Bieniek, D., Eds., Gesellschaft für Strahlen und Umweltforschung mbH, München, 1971, 48.
112. McKinney, J. D. and Fishbein, L., DDE formation, dehydrochlorination or dehypochlorination, *Chemosphere*, [2], 67, 1972.
113. Zweig, G., Ed., *Analytical Methods for Pesticides, Plant Growth Regulators and Food Additives*, Vols. I–V, Academic Press, New York and London, 1963, et seq.
114. Gunther, F. A., Ed., *Residue Reviews*, Springer-Verlag, New York.
115. Raw, G. R., Ed., *CIPAC Handbook, Vol. I, Analysis of Technical and Formulated Pesticides*, Collaborative International Pesticides Analytical Council Ltd., Harpenden, U. K., 1970.
116. Shell Chemical Corporation, *Handbook of Aldrin, Dieldrin and Endrin Formulations*, 2nd ed., Agricultural Chemicals Division, New York, 1959, et seq.

117. Castillo, J. C. and Stiff, H. A., Application of the xanthydrol-KOH-pyridine method to the determination of 2,2-bis(p-chlorophenyl)-1,1,1-trichloroethane (DDT) in water, *Mil. Surg.*, 97, 500, 1945; *C.A.*, 2561, 1946.

118. Claborn, H. V., Determination of DDT in the presence of DDD (1,1-dichloro-2,2-bis(p-chlorophenyl)ethane (also called TDE), *J. Assoc. Offic. Agric. Chem.*, 29, 330, 1946.

119. Bradbury, F. R., Higgons, D. J., and Stoneman, J. P., Colorimetric method for the estimation of 2,2-bis(p-chlorophenyl)-1,1,1-trichloroethane (DDT), *J. Soc. Chem. Ind.*, 66, 65, 1947.

120. Miskus, R., Analytical methods for DDT, in *Analytical Methods for Pesticides, Plant Growth Regulators and Food Additives*, Vol. II, Zweig, G., Ed., Academic Press, New York and London, 1964, 97.

121. Claborn, H. V. and Beckman, H. F., Determination of 1,1,1-trichloro-2,2-bis(p-methoxyphenyl) ethane in milk and fatty materials, *Anal. Chem.*, 24, 220, 1952.

122. Gillett, J. W., Van der Geest, L. P. S., and Miskus, R. P., Column partition chromatography of DDT and related compounds, *Anal. Biochem.*, 8, 200, 1964.

123. Jones, L. R. and Riddick, J. A., Colorimetric determination of 2-nitro-1,1-bis(p-chlorophenyl)alkanes, *Anal. Chem.*, 23, 349, 1951.

124. Blinn, R. C. and Gunther, F. A., The utilisation of infrared and ultra violet spectrophotometric procedures for assay of pesticide residues, *Residue Rev.*, 2, 99, 1963.

125. Gilbert, G. G. and Sader, M. H., Fluorimetric determination of pesticides, *Anal. Chem.*, 41, 366, 1969.

126. Allen, P. T., Polarographic methods for pesticides and additives, in *Analytical Methods for Pesticides, Plant Growth Regulators and Food Additives*, Vol. V, Zweig, G., Ed., Academic Press, New York, 1966, 67.

127. Gajan, R. J., Analysis of pesticide residues by polarography, *J. Assoc. Offic. Agric. Chem.*, 48, 1027, 1965.

128. Gajan, R. J. and Link, J., An investigation of the oscillography of DDT and certain analogues, *J. Assoc. Offic. Agric. Chem.*, 47, 1118, 1964.

129. Cisak, A., Study of the electroreduction mechanism of halogen derivatives of cyclohexane, *Rocz. Chem.*, 37, 1025, 1963.

130. Cisak, A., Polarographic properties of aldrin, isodrin, a-chlordane, β-chlordane, dieldrin and endrin, *Rocz. Chem.*, 40, 1717, 1966.

131. Gajan, R. J., Applications of polarography for the detection and determination of pesticides and their residues, *Residue Rev.*, 5, 80, 1964.

132. Gajan, R. J., Recent developments in the detection and determination of pesticides and their residues by oscillographic polarography, *Residue Rev.*, 6, 75, 1964.

133. Casida, J. E., Radiotracer studies on metabolism, degradation, and mode of action of insecticide chemicals, *Residue Rev.*, 25, 149, 1969.

134. Gorsuch, T. T., Radioactive Isotope Dilution Analysis, Review No. 2, The Radiochemical Centre, Amersham, U.K.

135. Goodwin, E. S., Goulden, R., Richardson, A., and Reynolds, J. G., The analysis of crop extracts for traces of chlorinated pesticides by gas-liquid partition chromatography, *Chem. Ind. (Lond.)*, 1220, 1960.

136. Goodwin, E. S., Goulden, R., and Reynolds, J. G., Rapid identification and determination of residues of chlorinated pesticides in crops by gas-liquid chromatography, *Analyst*, 86, 697, 1961.

137. Widmark, G., Modern trends of analytical chemistry as regards environmental chemicals, in *Environmental Quality and Safety; Chemistry, Toxicology and Technology*, Vol. I, Coulston, F. and Korte, F., Eds., Georg Thieme Verlag, Stuttgart, Academic Press, New York, 1972, 78.

138. Mitchell, L. C., Separation and identification of chlorinated organic pesticides by paper chromatography, *J. Assoc. Offic. Agric. Chem.*, 41, 781, 1958.

139. McKinley, W. P., Paper chromatography, in *Analytical Methods for Pesticides, Plant Growth Regulators and Food Additives*, Vol. I, Zweig, G., Ed., Academic Press, New York, 1963, 227.

140. Winteringham, F. P. W., Harrison, A., and Bridges, R. G., Radioactive tracer-paper chromatography techniques, *Analyst*, 77, 19, 1952.

141. Abbott, D. C. and Thomson, J., The application of thin-layer chromatographic techniques to the analysis of pesticide residues, *Residue Rev.*, 11, 1, 1965.

142. Mangold, H. K., Isotope technique, in *Thin Layer Chromatography, a Laboratory Handbook*, 2nd ed., Stahl, E., Ed., Allen and Unwin, London, 1969.

143. Burchfield, H. P. and Storrs, E. E., *Biochemical Applications of Gas Chromatography*, Academic Press, New York and London, 1962.

144. Coulson, D. M. Cavanagh, L. A., de Vries, J. E., and Walther, B., Microcoulometric gas chromatography of pesticides, *J. Agric. Food Chem.*, 8, 399, 1960.

145. Cassil, C. C., Pesticide residue analysis by microcoulometric gas chromatography, *Residue Rev.*, 1, 37, 1962.

146. Lovelock, J. E. and Lipsky, S. R., Electron affinity spectroscopy — a new method for the identification of functional groups in chemical compounds separated by gas chromatography, *J. Am. Chem. Soc.*, 82, 431, 1960.

147. de Faubert Maunder, M. J., Egan, H., and Roburn, J., Some practical aspects of the determination of chlorinated pesticides by electron-capture gas chromatography, *Analyst*, 89, 157, 1964.

148. Clark, S. J., Quantitative determination of pesticide residues by electron absorption chromatography: characteristics of the detector, *Residue Rev.*, 5, 33, 1964.

149. Burke, J. A. and Holswade, W., A gas chromatographic column for pesticide residue analysis: retention times and response data, *J. Assoc. Offic. Agric. Chem.*, 49, 374, 1966.
150. Ott, D. E. and Gunther, F. A., DDD as a decomposition product of DDT, *Residue Rev.*, 10, 70, 1965.
151. Robinson, J., Organochlorine compounds in man and his environment, *Chem. Br.*, 7, 472, 1971.
152. Bowman, M. C. and Beroza, M., Extraction p values of pesticides and related compounds in six binary solvent systems, *J. Assoc. Offic. Agric. Chem.*, 48, 943, 1965.
153. Bache, C. A. and Lisk, D. J., Selective emission spectrometric determination of nanogram quantities of organic bromine, chlorine, iodine, phosphorus and sulfur compounds in a helium plasma, *Anal. Chem.*, 39, 787, 1967.
154. Keith, L. H., Alford, A. L., and Garrison, A. W., The high resolution NMR spectra of pesticides II; DDT-type compounds, *J. Assoc. Offic. Agric. Chem.*, 52, 1074, 1969.
155. Baldwin, M. K., Robinson, J., and Carrington, R. A. G., Metabolism of HEOD (dieldrin) in the rat: examination of the major faecal metabolite, *Chem. Ind. (Lond.)*, 595, 1970.
156. Cochrane, W. P., Forbes, M., and Chau, A. S. Y., Cyclodiene chemistry. IV. Assignment of configuration of two nonachlors via synthesis and derivatisation, *J. Assoc. Offic. Agric. Chem.*, 53, 769, 1970.
157. Mackenzie, K. and Lay, W. P., Fragmentation of tetrahydromethanonaphthalenes and their diaza- and benzo-analogues, *Tetrahedron Lett.*, No. 37, 3241, 1970.
158. McKinney, J. D., Keith, L. H., Alford, A., and Fletcher, C. E., The proton magnetic resonance spectra of some chlorinated polycyclodiene pesticide metabolites, rapid assessment of stereochemistry, *Can. J. Chem.*, 49, 1993, 1971.
159. Keith, L. H. and Alford, A. L., Long-range couplings in the chlorinated polycyclodiene pesticides, *Tetrahedron Lett.*, 2489, 1970.
160. Nagl, H. G., Klein, W., and Korte, F., Uber das reaktions verhalten von dieldrin in losung und in der gasphase, *Tetrahedron*, 26, 5319, 1970.
161. Roll, D. B. and Biros, F. J., Nuclear quadrupole resonance spectrometry of some chlorinated pesticides, *Anal. Chem.*, 41, 407, 1969.
162. Damico, J. N., Barron, R. P., and Ruth, J. M., The mass spectra of some chlorinated pesticidal compounds, *Org. Mass Spectr.*, 1, 331, 1968.
163. Scopes, N. E. A. and Lichtenstein, E. P., The use of *Folsomia fimetaria* and *Drosophila melanogaster* as test insects for the detection of insecticide residues, *J. Econ. Entomol.*, 60, 1539, 1967.
164. Sun, Y-P., Bioassay-insects, in *Analytical Methods for Pesticides, Plant Growth Regulators and Food Additives*, Vol. I, Zweig, G., Ed., Academic Press, New York and London, 1963, 399.
165. Alder, K. and Stein, G., The course of the diene synthesis, *Angew Chem.*, 50, 510, 1937.
166. Bergmann, F. and Eschinazi, H. E., Sterical course and the mechanism of the diene reactions, *J. Am. Chem. Soc.*, 65, 1405, 1943.
166a. Gill, G. B., The application of the Woodward-Hoffman orbital symmetry rules to concerted organic reactions, *Q. Rev. Chem. Soc. (Lond.)*, 22, 338, 1968.
167. Buchel, K. H., Ginsberg, A. E., Fischer, R., and Korte, F., β-Dihydroheptachlor, ein insektizid mit sehr niedriger warmblüter-toxizitat, *Tetrahedron Lett.*, No. 33, 2267, 1964.
168. Soloway, S. B., Stereochemistry of Bridged Polycyclic Compounds, Ph.D. Thesis, University of Colorado, 1955.
169. Benson, W. R., Note on nomenclature of dieldrin and related compounds, *J. Assoc. Offic. Agric., Chem.*, 52, 1109, 1969.
169a. Bedford, C. T., Von Baeyer — IUPAC Names and Recommended Trivial Names of Dieldrin, Endrin, and 24 Related Compounds, Shell Research Ltd., Tunstall Laboratory Research Report, November 1972.
170. Bird, C. W., Cookson, R. C., Crundwell, E., Cyclisations and rearrangements in the isodrin-aldrin series, *J. Chem. Soc. (Lond.)*, 4809, 1961.
171. Parsons, A. M. and Moore, D. J., Some reactions of dieldrin and the proton magnetic resonance spectra of the products, *J. Chem. Soc. (Lond.)*, C., 2026, 1966.
172. Eckroth, D. R., A method for manual generation of correct von Baeyer names of polycyclic hydrocarbons, *J. Org. Chem.*, 32, 3362, 1967.
172a. International Union of Pure and Applied Chemistry, *IUPAC Nomenclature of Organic Chemistry*, 3rd ed., Butterworth, London, 1971.
173. Straus, F., Kollek, L., and Heyn, W., Uber den ersatz positiven wasserstoffs durch halogen, *Chem. Ber.*, 63B, 1868, 1930.
174. Kleiman, M., Chlorination of cyclopentadiene, U.S. patent 2,658,085, 1953; *C.A.*, 48, 12798, 1954.
175. Lidov, R. E., Hyman, J., and Segel, E., Diels-Alder adducts of hexahalocyclopentadiene with quinones, U.S. patent 2,584,139, 1952; *C.A.*, 46, 9591, 1952.
176. Frensch, H., Entwicklung und chemie der dien-gruppe, einer neuen klasse biozider wirkstoffe, *Med. Chem.*, 6, 556, 1957.
177. Ungnade, H. E. and McBee, E. T., The chemistry of perchlorocyclopentenes and cyclopentadienes, *Chem. Rev.*, 58, 249, 1958.
178. Gilbert, E. E. and Giolito, S. L., Pesticides, U.S. patent 2,616,928; *C.A.*, 47, 2424, 1953.
179. McBee, E. T., Roberts, C. W., Idol, J. D., and Earle, R. H., An investigation of the chlorocarbon, $C_{10}Cl_{12}$, mp485° and the ketone, $C_{10}Cl_{10}O$, mp349°, *J. Am. Chem. Soc.*, 78, 1511, 1956.

180. Gilbert, E. E., Lombardo, P., Rumanowski, E. J., and Walker, G. L., Preparation and insecticidal evaluation of alcoholic analogs of kepone, *J. Agric. Food Chem.*, 14, 111, 1966.

181. Allen, W. W., The effectiveness of various pesticides against resistant two-spotted spider mites on green house roses, *J. Econ. Entomol.*, 57, 187, 1964.

182. McBee, E. T. and Smith, D. K., The reduction of hexachlorocyclopentadiene, 1,2,3,4,5-pentachlorocyclopentadiene, *J. Am. Chem. Soc.*, 77, 389, 1955.

183. McBee, E. T., Meyers, R. K., and Baranauckas, C. F., 1,2,3,4-Tetracyclopentadiene. 1. The preparation of the diene and its reaction with aromatics and dienophiles, *J. Am. Chem. Soc.*, 77, 86, 1955.

184. Soloway, S. B., Correlation between biological activity and molecular structure of the cyclodiene insecticides, *Adv. Pest Control Res.*, 6, 85, 1965.

185. Volodkovich, S. D., Melnikov, N. N., Plate, A. F., and Prianishnikova, M. A., Reaction of 1,1-difluorotetrachloro-cyclopentadiene with some unsaturated compounds, *J. Gen. Chem. (U.S.S.R.)*, 28, 3153, 1958.

186. Riemschneider, R. and Kuhnl, A., Zur chemie von polyhalocyclopentadienen und verwandten verbindungen, *Monatsh. Chem.*, 86, 879, 1953.

187. Banks, R. E., Harrison, A. C., Haszeldine, R. N., and Orrell, K. G., Diels-Alder reactions involving perfluorocyclopentadiene, *Chem. Commun.*, 3, 41, 1965.

188. Riemschneider, R., The chemistry of the insecticides of the diene group, *World Rev. Pest Control*, 2, 29, 1963.

189. Riemschneider, R. and Nehring, R., Ein Thiodan-Analoges aus trimethyltrichlorocyclopentadienen, *Z. Naturforsch. Teil. B.*, 17b, 524, 1962.

190. Bluestone, H., 1,2,3,4,10,10-Hexachloro-6,7-Epoxy-1,4,4a,5,6,7,8,8a-Octahydro-1,4,5,8-Dimethanonaphthalene and Insecticidal Compositions Thereof, U.S. patent 2,676,132, 1954; *C.A.*, 48, 8874, 1954.

191. Mackenzie, K., The reaction of hexachlorobicyclo [2,2,1] heptadiene with potassium ethoxide, *J. Chem. Soc. (Lond.)*, 86, 457, 1962.

192. Brooks, G. T. and Harrison, A., The effect of pyrethrin synergists, especially sesamex, on the insecticidal potency of hexachlorocyclopentadiene derivatives ('cyclodiene' insecticides) in the adult housefly, *Musca domestica*, L., *Biochem. Pharmacol.*, 13, 827, 1964.

193. Riemschneider, R., Gallert, H., and Andres, P., Uber der herstellung von 1,4,5,6,7,7-hexachlorobicyclo [2.2.1] hepten-5-bishydroxymethylen-2,3, *Monatsh. Chem.*, 92, 1075, 1961.

194. Riemschneider, R. and Kotzsch, H. J., Hexachlorocyclopentadiene and allylglykolather, *Monatsh. Chem.*, 91, 41, 1960.

195. Riemschneider, R., Herzel, F., and Koetsch, H. J., 2-Methylen-1,4,5,6,7,7-hexachlorbicyclo [2.2.1] hepten-(5), *Monatsh. Chem.*, 92, 1070, 1961.

196. Riemschneider, R., Penta- und hexachlorocyclopentadien als philodiene komponenten, *Botyu-Kagaku*, 28, 83, 1965.

197. Forman, S. E., Durbetaki, A. J., Cohen, M. V., and Olofson, R. A., Conformational equilibria in cyclic sulfites and sulfates. The configurations and conformations of the two isomeric thiodans, *J. Org. Chem.*, 30, 169, 1965.

198. Feichtinger, H. and Linden, H. W., Telodrin, its synthesis and derivatives, *Chem. Ind. (Lond.)*, 1938, 1965.

199. Gross, H., Darstellung und reaktionen des 2,5-dichlor-tetrahydrofurans, *Chem. Ber.*, 95, 83, 1962.

200. Hyman, J., Improvements In or Relating to a Method of Forming Halogenated Organic Compounds and the Products Resulting Therefrom, British patent 618.432, 1949.

201. Herzfeld, S. H. and Ordas, E. P., 1-Hydroxy-4, 7-methano-3a,4,7,7a-tetrahydro-4, 5,6,7,8,8-hexachloroindene and method of preparing same, U.S. patent 2,528,656, 1950.

202. Goldman, A., Kleiman, M., and Fechter, H. G., Production of Halogenated Polycyclic Alcohols, U.S. patent 2,750,397, 1956.

203. Arvey Corporation, Improvements in or Relating to Halogenated Dicyclopentadiene Epoxides, British patent 714,869, 1954.

204. Buchel, K. H., Ginsberg, A. E., and Fischer, R., Isomierung und chlorierung von dihydroheptachlor, *Chem. Ber.*, 99, 416, 1966.

205. Buchel, K. H., Ginsberg, A. E., and Fischer, R., Synthese und struktur von heptachlor-methano-tetrahydroindanen, *Chem. Ber.*, 99, 405, 1966.

206. Buchel, K. H., Ginsberg, A. E., and Fischer, R., Synthese und struktur von isomeren des chlordans, *Chem. Ber.*, 99, 421, 1966.

207. March, R. B., The resolution and chemical and biological characterisation of some constituents of technical chlordane, *J. Econ. Entomol.*, 45, 452, 1952.

208. Cochrane, W. P., Forbes, M., and Chau, A. S. Y., Assignment of configuration of two nonachlors via synthesis and derivatisation, *J. Assoc. Offic. Agric. Chem.*, 53, 769, 1970.

209. Hyman, J., Freireich, E., and Lidov, R. E., Bicyclo-heptadienes and process of preparing the same, British patent 701,211, 1953.

210. Plate, A. F. and Pryanishnikova, M. A., Preparation of bicyclo [2.2.1] hepta-2,5-diene by the condensation of cyclopentadiene with acetylene, *Izvest. Akad. Nauk SSSR., Otdel Khim. Nauk*, 741, 1956.

211. Lidov, R. E. and Soloway, S. B., Improvements In or Relating to Methods of Preparing Insecticidal Compounds and the Insecticidal Compounds Resulting From Said Methods, British patent 692,546, 1953.

212. Lidov, R. E. and Soloway, S. B., Process of Preparing Polycyclic Compounds and Insecticidal Compositions Containing the Same, British patent 692,547, 1953.

213. Winteringham, F. P. W. and Harrison, A., Mechanisms of resistance of adult houseflies to the insecticide dieldrin, *Nature, (Lond.)*, 184, 608, 1959.

214. Kuderna, J. G., Sims, J. W., Wilkstrom, J. R., and Soloway, S. B., The preparation of some insecticidal chlorinated bridged phthalazines, *J. Am. Chem. Soc.*, 81, 382, 1959.

215. Kleiman, M., Pesticidal Compounds, U.S. patent 2,655,513, 1953.

216. Bellin, R. H., Endrin Stabilization Using Inorganic and Organic Nitrite Salts, U.S. patent 2,768,178, 1956.

217. Nagl, H. G., Klein, W., and Korte, F., Uber das reaktionsverhalten von dieldrin in losung und in der gasphase, *Tetrahedron*, 26, 5319, 1970.

218. Busvine, J. R., The insecticidal potency of γ-BHC and the chlorinated cyclodiene compounds and the significance of resistance to them, *Bull. Entomol. Res.*, 55, 271, 1964.

219. Adams, C. H. M. and Mackenzie, K., Dehalogenation of isodrin and aldrin with alkoxide base, *J. Chem. Soc. (Lond.)*, C, 480, 1969.

220. Bienieck, D., Moza, P. N., Klein, W. and Korte, F., Reduktive dehalogenierung von chlorierten cyclischen kohlen wasserstoffen, *Tetrahedron Lett.*, 4055, 1970.

221. Volodkovich, S. D., Vol'Fson, L. G., Kuznetsova, K. V., and Mel'nikov, N. N., Synthesis of α-oxides by the oxidation of polycyclic haloderivatives with hydrogen peroxide, *J. Gen. Chem. (U.S.S.R.)*, 29, 2797, 1959.

222. Brooks, G. T., The synthesis of [14]C-labelled 1:2:4:10:10-hexachloro-6:7-epoxy-1:4:4a:5:6:7:8:8a-octahydro-*exo*-1:4-exo-5:8-dimethanonaphthalene, *J. Chem. Soc. (Lond.)*, 3693, 1958.

223. Korte, F., Metabolism of Chlorinated Insecticides, Paper Presented to the Joint FAO/IAEA Division of Atomic Energy in Agriculture, Panel on the uses of Radioisotopes in the detection of Pesticide Residues, Vienna, April 1965.

224. Korte, F. and Rechmeier, G., Mikrosynthese von Aldrin-[14]C und Dieldrin-[14]C, *Justus Liebigs Ann. Chem.*, 656, 131, 1962.

225. McKinney, R. M. and Pearce, G. W., Synthesis of carbon-14-labeled aldrin and dieldrin, *J. Agric. Food Chem.*, 8, 457, 1960.

226. Thomas, D. J. and Kilner, A. E., The synthesis of the insecticides aldrin and dieldrin labelled with carbon-14 at high specific activity, in *Radioisotopes in the Physical Sciences and Industry*, International Atomic Energy Agency, Vienna 1962.

227. Korte, F. and Stiasni, M., Mikrosynthese von [14]C-markiertem Telodrin, *Justus Liebigs Ann. Chem.*, 656, 140, 1962.

228. Black, A. and Morgan, A., Herstellung einiger Cl-38-markierten chlorierten Kohlen wasserstoffe durch neutronenbestrahlung und gaschromatographie, *Int. J. Appl. Radiat. Isot.*, 21, 5, 1970.

229. Dailey, R. E., Walton, M. S., Beck, V., Leavens, C. L., and Klein, A. K., Excretion, distribution, and tissue storage of a [14]C-labeled photoconversion product of [14]C-dieldrin, *J. Agric. Food Chem.*, 18, 443, 1970.

230. Riemschneider, R., Zur chemie von polyhalocyclopentadienen und verwandten verbindungen, IV. Thermische spaltung und oxydation des adduktes $C_{10}H_6Cl_6$, *Chem. Ber.*, 89, 2697, 1956.

231. Lidov, R. E. and Bluestone, H., Cyclobutano-compounds, U.S. patent 2,714,617; *C.A.*, 50, 5756, 1956.

232. Cochrane, W. P. and Forbes, M. A., Practical Applications of Chromous Chloride to the Confirmation of Organochlorine Pesticide Residue Identity, Paper presented at the 2nd International Congress of Pesticide Chemistry, Tel Aviv, Israel, 1971.

233. Davidow, B. and Radomski, J. L., Isolation of an epoxide metabolite from fat tissues of dogs fed heptachlor, *J. Pharmacol. Exp. Ther.*, 107, 259, 1953.

234. Brooks, G. T. and Harrison, A., The toxicity of α-DHC and related compounds to the housefly (*M. domestica*) and their metabolism by housefly and pig liver microsomes, *Life Sci.*, (Oxford), 6, 1439, 1967.

235. Brooks, G. T., Insect opoxide hydrase inhibition by juvenile hormone analogues and metabolic inhibitors, *Nature (Lond.)*, 245, 382, 1973.

236. Chau, A. S. Y. and Cochrane, W. P., Cyclodiene chemistry. I. Derivative formation for the identification of heptachlor, heptachlor epoxide, cis-chlordane, trans-chlordane and aldrin pesticide residues by gas chromatography, *J. Assoc. Offic. Agric. Chem.*, 52, 1092, 1969.

237. Cochrane, W. P., Cyclodiene chemistry II. Identification of the derivatives employed in the confirmation of the heptachlor, heptachlor epoxide, cis-chlordane, and trans-chlordane residues, *J. Assoc. Offic. Agric. Chem.*, 52, 1100, 1969.

238. Chau, A. S. Y. and Cochrane, W. P., Cyclodiene chemistry III. Derivative formation for the identification of heptachlor, heptachlor epoxide, cis-chlordane, trans-chlordane, dieldrin, and endrin pesticide residues by gas chromatography, *J. Assoc. Offic. Agric. Chem.*, 52, 1220, 1969.

239. Cochrane, W. P. and Forbes, M. A., Isomerisation of 1-exo-4,5,6,7,8,8-heptachloro-2,3-endo-epoxy-3a,4,7,7a-tetrahydro-4,7-methanoindane with base, *Can. J. Chem.*, 49, 3569, 1971.

240. Polen, P. B., personal communication, 1970.

241.  Rosen, J. D., Conversion of pesticides under environmental conditions, in *Environmental Quality and Safety*, Vol. 1, Coulston, F. and Korte, F., Eds., Georg Thieme Verlag, Stuttgart; Academic Press, New York, 1972.

242.  McGuire, R. R., Zabik, M. J., Schuetz, R. D., and Flotard, R. D., Photolysis of 1,4,5,6,7,8,8-heptachloro-3a,4,7,7a-tetrahydro-4,7-methanoindene (cage formation vs. photodechlorination), *J. Agric. Food Chem.*, 18, 319, 1970.

243.  Brooks, G. T. and Harrison, A., Structure-activity relationships among insecticidal compounds derived from chlordene, *Nature (Lond.)*, 205, 1031, 1965.

244.  Institut fur Okologische Chemie, Jahresbericht 1969, Klein, W. and Drefahl, B., Eds., Gesellschaft für Strahlen und Umweltforschung mbH, München, 1970.

244a. Ivie, G. W., Knox, J. R., Khalifa, S., Yamamoto, I., and Casida, J. E., Novel photoproducts of heptachlor epoxide, *trans*-chlordane, and *trans*-nonachlor, *Bull. Environ. Contam. Toxicol.*, 7, 376, 1972.

245.  Vollner, L., Klein, W., and Korte, F., Photoumlagerung der komponenten des technischen chlordans, *Tetrahedron Lett.*, 2967, 1969.

246.  Schwemmer, B., Cochrane, W. P., and Polen, P. B., Oxychlordane, animal metabolite of chlordane: isolation and synthesis, *Science*, 169, 1087, 1970.

247.  Polen, P. B., Hester, M., and Benziger, J., Characterisation of oxychlordane, animal metabolite of chlordane, *Bull. Environ. Contam. Toxicol.*, 5, 521, 1971.

248.  Maier-Bode, H., Properties, effect, residues and analytics of the insecticide endosulfan, *Residue Rev.*, 22, 1, 1968.

249.  Chau, A. S. Y. and Cochrane, W. P., Cis-opening of the dieldrin oxirane ring, *Chem. Ind. (Lond.)*, 1568, 1970.

250.  Adams, C. H. M. and Mackenzie, K., Dehalogenation of isodrin and aldrin with alkoxide base, *J. Chem. Soc. (Lond.)*, C, 480, 1969.

251.  Robinson, J., Richardson, A., Bush, B., and Elgar, K., A photoisomerisation product of dieldrin, *Bull. Environ. Contam. Toxicol.*, 1, 127, 1966.

252.  Benson, W. R., Photolysis of solid and dissolved dieldrin, *J. Agric. Food Chem.*, 19, 66, 1971.

253.  Cookson, R. C. and Crundwell, E., Transannular reactions in the isodrin series, *Chem. Ind. (Lond.)*, 703, 1959.

254.  Soloway, S. B., Damiana, A. M., Sims, J. W., Bluestone, H., and Lidov, R. E., Skeletal rearrangements in reactions of isodrin and endrin, *J. Am. Chem. Soc.*, 82, 5377, 1960.

255.  Weil, E. D., Colson, J. G., Hoch, P. E., and Gruber, R. H., Toxic chlorinated methanoisobenzofuran derivatives, *J. Heterocycl. Chem.*, 6, 643, 1969.

256.  Bruck, P., Thompson, B., and Winstein, S., Dechlorination of isodrin and related compounds, *Chem. Ind. (Lond.)*, 405, 1960.

257.  Phillips, D. D., Pollard, G. E., and Soloway, S. B., Thermal isomerisation of endrin and its behaviour in gas chromatography, *J. Agric. Food Chem.*, 10, 217, 1962.

258.  Rosen, J. D., Sutherland, D. J., and Lipton, G. R., The photochemical isomerisation of dieldrin and endrin and effects on toxicity, *Bull. Environ. Contam. Toxicol.*, 1, 133, 1966.

259.  Zabik, M. J., Schuetz, R. D., Burton, W. L., and Pape, B. E., Studies of a major photolytic product of endrin, *J. Agric. Food Chem.*, 19, 308, 1971.

260.  Brooks, G. T., Degradation of Organochlorine Insecticides, Problems and Possibilities, in Proceedings of the 2nd International IUPAC Congress of Pesticide Chemistry 1971, Vol. 6; *Fate of Pesticides in Environment*, Tahori, A. S., Ed., Gordon and Breach, London, 1972, 223.

261.  Martin, H. and Wain, R. L., Insecticidal action of DDT, *Nature (Lond.)*, 154, 512, 1944.

262.  Prill, E. A., Diels-Alder syntheses with hexachlorocyclopentadiene, *J. Am. Chem. Soc.*, 69, 62, 1947.

263.  Riemschneider, R., Neue kontakt-insetizide der halogenkohlen wasserstoffe-klasse (M410, M344), *Mitt. Physiol. Chem. Inst. Univ. Berlin*, R8, March 1947; *C.A.*, 48, 2973, 1954.

264.  Riemschneider, R. and Kuhnl, A., *Mitt. Physiol. Chem. Inst. Univ. Berlin*, R.11, October 1947 (see also reference 186).

265.  Kearns, C. W., Ingle, L., and Metcalf, R. L., New chlorinated hydrocarbon insecticide, *J. Econ. Entomol.*, 38, 661, 1945.

266.  Hyman, J., personal communication, 1972.

267.  Velsicol Corporation, Improvements in the Production of New Diels-Alder Adducts, British patent 614,931, 1948; *C.A.*, 43, 4693, 1949.

268.  Hyman, J., Improvements in or relating to method of forming halogenated organic compounds and the products resulting therefrom, British patent 618,432, 1949; *C.A.*, 43, 5796, 1949.

269.  Ingle, L., *A Monograph On Chlordane. Toxicological and Pharmacological Properties*, University of Illinois, 1965.

270.  Velsicol Chemical Corporation, *Chlordane Formulation Guide*, Bulletin No. 502−35R, 1970.

271.  Velsicol Chemical Corporation, Standard for Technical Chlordane, August 1971.

272.  Polen, P. B., Chlordane: Composition, Analytical Considerations and Terminal Residues, paper presented to a meeting of the IUPAC Commission on Terminal Residues, Geneva, Switzerland, 1966.

273.  Riemschneider, R. and Graviz, B. B., Uber den raumlichen Bau einiger Addukte aus hexachlorocyclopentadien und ungesattigen verbindungen (Dien-gruppe), *Botyu-kagaku*, 25, 123, 1960.

274.  Vogelbach, C., Isolation of new crystalline substances from technical grade chlordan, *Angew Chem.*, 63, 378, 1951.

275.  Harris, C. R., Factors influencing the biological activity of technical chlordane and some related components in soil, *J. Econ. Entomol.*, 65, 343, 1972.

276. **Saha, J. G. and Lee, Y. W.**, Isolation and identification of the components of a commercial chlordane formulation, *Bull. Environ. Contam. Toxicol.*, 4, 285, 1969.
277. *Belt-plus Insecticide*, Velsicol Chemical Corporation Development Bulletin no. 01-054-601, April 1971.
278. **Freeman, S. M. D.**, cited in Ingle, L., *A Monograph on Chlordane. Toxicological and Pharmacological Properties*, University of Illinois, 1965, 12.
279. **Herzfeld, S. H. and Ordas, E.**, Improvements In or Relating To Processes For the Production of Derivatives of Diels-Alder Adducts of Hexachlorocyclopentadiene and Cyclopentadiene, British patent 652,300, 1951.
280. **Whaley, R. E.**, Heptachlor Insecticide, U.S. patent 3,541,163, 1970; *C.A.*, 74, 53177, 1971.
281. Velsicol Chemical Corporation, Heptachlor Formulation Guide, Bulletin no. 504-16, 1964.
282. **Busvine, J. R.**, The newer insecticides in relation to pests of medical importance, in symposium on insecticides, *Trans. R. Soc. Trop. Med. Hyg.*, 46, 245, 1952.
283. Velsicol Chemical Corporation, Chlordane Federal Label Acceptances, Bulletin no. 502-32, March 1968.
284. **Miles, J. R. W., Tu, C. M., and Harris, C. R.**, Metabolism of heptachlor and its degradation products by soil microorganisms, *J. Econ. Entomol.*, 62, 1334, 1969.
285. **Carter, F. L., Stringer, C. A., and Heinzelman, D.**, 1-Hydroxy-2,3-epoxychlordene in Oregon soil previously treated with technical heptachlor, *Bull. Environ. Contam. Toxicol.*, 6, 249, 1971.
286. **Bonderman, D. P. and Slach, E.**, Appearance of 1-hydroxychlordene in soil, crops and fish, *J. Agric. Food Chem.*, 20, 328, 1972.
287. **Miles, J. R. W., Tu, C. M., and Harris, C. R.**, Degradation of heptachlor epoxide and heptachlor by a mixed culture of soil microorganisms, *J. Econ. Entomol.*, 64, 839, 1971.
288. **Harris, C. R.**, personal communication, 1972.
289. **Davidow, B.**, A spectrophotometric method for the quantitative estimation of technical chlordan, *J. Assoc. Offic. Agric. Chem.*, 33, 886, 1950.
290. **Ordas, E. P., Smith, V. C., and Meyer, C. F.**, Spectrophotometric determination of heptachlor and technical chlordane on food and storage crops, *J. Agric. Food Chem.*, 4, 444, 1956.
291. Velsicol Chemical Corporation, Evaluation of Technical Heptachlor Formulations, Bulletin No. 504-17, undated.
292. **Dorn, W.**, Statement of the State Secretary on a request of MdB(MP) Prinz Zu Sayn Wittgenstein et al. regarding the fish kill in the Rhine, given at 20.2. 1970 in the Deutsche Bundestag (Parliament), 1970. Cited in *Pesticides in the Modern World*, Cooperative Programme of Agro-Allied Industries, with FAO and other U.N. Organisations at the Newgate Press Ltd., London, 1972, 5.
293. Farbwerke Hoechst, AG, personal communications, 1972.
294. **Feichtinger, H. and Tummes, H.**, Verfahren Zur Hestellung Von 4,5,6,7,10,10-Hexachlor-4,7-Endomethylen-4,7,8,9-Tetrahydrophthalan, German patent 960,284, 1957.
295. **Whetstone, R. R.**, Chlorinated derivatives of cyclopentadiene, in Kirk-Othmer: *Encyclopedia of Chemical Technology*, Vol. 5, 2nd ed., McKetta, J. J. and Othmer, D. F., Eds., John Wiley (Interscience), 1964, 240.
296. Farbwerke Hoechst, AG, *Thiodan and the Environment*, technical bulletin translated by Hoechst U.K., 1971; *Thiodan*, [an undated bulletin on formulations and applications of Thiodan], Farbwerke Hoechst AG, Frankfurt a. Main.
297. **Frensch, H., Goebel, H., and Czech, M.**, Agents for killing undesired fish, U.S. patent 2,799,685; *C.A.*, 3852, 1962.
298. **Gorbach, S. and Knauf, W.**, Endosulfane and the environment in *Environmental Quality and Safety*, Vol. I, Coulston, F. and Korte, F., Eds., Georg Thieme Stuttgart, Academic Press, New York, 1972, 250.
299. **Schoettger, R. A.**, Toxicology of Thiodan in several fish and aquatic invertebrates, *Investigations in Fish Control 35*, United States Department of the Interior, Bureau of Sport Fisheries and Wildlife, U.S. Govt. Print. Off., Washington D.C., 1970
300. **Zweig, G. and Archer, T. E.**, Quantitative determination of Thiodan by gas chromatography, *J. Agric. Food Chem.*, 8, 190, 1960.
301. Shell Research Ltd., cited in Phillips, F. T., The rates of loss of dieldrin and aldrin by volatilisation from glass surfaces, *Pestic. Sci.*, 2, 255, 1971.
302. **Kearns, C. W., Weinman, C., and Decker, G. C.**, Insecticidal properties of some new chlorinated organic compounds, *J. Econ. Entomol.*, 42, 127, 1949.
303. **Jager, K. W.**, *Aldrin, Dieldrin, Endrin and Telodrin, an Epidemiological and Toxicological Study of Long-term Occupational Exposure*, Elsevier, Amsterdam, 1970.
304. **Park, K. S. and Bruce, W. N.**, The determination of water solubility of aldrin, dieldrin, heptachlor and heptachlor epoxide, *J. Econ. Entomol.*, 61, 770, 1968.
304a. **Delacy, T. P. and Kennard, C. H. L.**, Crystal structures of endrin and aldrin, *J. Chem. Soc. (Lond.)*, Perkin II, 2153, 1972.
305. Aldrin and Dieldrin, An Appraisal of Data Submitted to International Regulatory Agencies, Shell International Chemical Company Ltd., Regulatory Affairs Division, January 1972.
306. Summary of Technical Information, Registered Label Uses and Label Uses Withdrawn Since 1964, in; Aldrin, Dieldrin, Endrin, a Status Report, Shell Chemical Company, Agricultural Chemicals Division, September 1967.
307. *Endrin, the Foliage Insecticide*, Shell International Chemical Company Ltd., Bulletin, 1964.
308. *The Safe Handling and Toxicology of Endrin*, Shell International Chemical Company Ltd., undated bulletin.

309. Skerrett, E. J. and Baker, E. A., A new colour reaction for dieldrin and endrin, *Chem. Ind. (Lond.)*, 539, 1959.
310. Third World Food Survey, Basic Study No. 11, Food and Agriculture Organisation, 1963.
311. Pesticides and Health, Shell briefing service, Royal Dutch Shell Group, February 1967.
312. The Use of Pesticides, President's Scientific Advisory Committee Report, Weisner, J. B., Chairman, Govt. Print. Off., Washington, D.C., May 15, 1963.
313. Glasser, R. F., The actions taken by Shell in response to recommendations of advisory committees and regulatory agencies, in Symposium on the Science and Technology of Residual Insecticides in Food Production with Special Reference to Aldrin and Dieldrin, Shell Oil Company, 1968, 225.
314. Hardie, D. W. F., Benzene Hexachloride, in *Kirk-Othmer Encyclopedia of Chemical Technology*, Vol. 5, 2nd ed., McKetta, J. J. and Othmer, D. F., Eds., John Wiley (Interscience), 1964, 267.
315. Lambermont, F., L'Hexachlorcyclohexane et son isomère gamma, paper presented at the 8th International Congress on Agriculture, Brussels, July 12, 1950.
316. Van der Linden, T., Uber die benzol-hexachloride und ihren zerfall in trichlor-benzole, *Chem. Ber.*, 45, 231, 1912.
317. Bender, H., Chlorinating benzene, toluene, etc., U.S. patent 2,010,841; 1935. *C.A.*, 29, 6607, 1935.
318. Stephenson, H. P., Curtis, A. L., Grant, A. E., and Hardie, T., cited in Haller, H. L., and Bowen, C. V., Basic facts about benzene hexachloride, *Agric. Chem.*, 2, 15, 1947.
319. Dupire, A. and Raucourt, M., Un insecticide nouveau: l'hexachlorure de benzene, *C. R. Acad. Agric. Fr.*, 29, 470, 1943.
320. Bourne, L. B., Hexachlorocyclohexane as an insecticide, *Nature (Lond.)*, 156, 85, 1945.
321. Slade, R. E., The γ-isomer of hexachlorocyclohexane (gammexane), Hurter Memorial Lecture, *Chem. Ind. (Lond.)*, 64, 314, 1945.
322. Holmes, E., Recent British developments in taint-free use of BHC, *Agric. Chem.*, 6,[12], 31, 1951.
323. Ishikura, H., Impact of pesticide usage on the Japanese environment, in *Environmental Toxicology of Pesticides*, Matsumura, F., Boush, G. M., and Misato, T., Eds., Academic Press, New York and London, 1972, 1.
324. Kenaga, E. E. and Allison, W. E., Commercial and experimental organic insecticides (1971 revision), *Bull. Entomol. Soc. Am.*, 16, 68, 1970.
325. Ishii, K., Chlorinated pesticides, in *Japan Pesticide Information*, No. 3, Japan Plant Protection Association, 1970, 11.
326. Shindo, N., The present situation of agricultural chemicals in the chemical industry of Japan, in *Japan Pesticide Information*, No. 7, Japan Plant Protection Association, 1971, 10.
327. Kurihara, N., Sanemitsu, Y., Kimura, T., Kobayashi, M., and Nakajima, M., Stepwise synthesis of 3,4,5,6-tetrachlorocyclohexene-1 (BTC) isomers, *Agric. Biol. Chem.* (Tokvo), 34, 784, 1970.
328. Cited in Hardie, D. W. F., Benzene hexachloride, in *Kirk-Othmer Encyclopedia of Chemical Technology*, Vol. 5, 2nd ed., Mcketta, J. J. and Othmer, D. F., Eds., John Wiley (Interscience), 1964, 274.
329. Gunther, F. A., Chlorination of benzene, *Chem. Ind. (Lond.)*, 399, 1946.
330. Melnikov, N. N., *Chemistry of Pesticides*, Gunther, F. A. and Gunther, J. D., Eds., *Residue Rev.*, Vol. 36, Springer-Verlag, New York, 1971.
331. Cooke, W. H. and Smart, J. C., Insecticide, British patent 586,439, 1944; *C.A.*, 41, 7640, 1947.
332. Hay, J. K. and Webster, K. C., Insecticide, British patent 586,442, 1944; *C.A.*, 7640, 1947.
333. Balson, E. W., Studies in vapour pressure measurement, *Trans. Faraday Soc.*, 43, 54, 1947.
334. Kanazawa, J., Yashima, T., and Kiritani, K., Contamination of ecosystem by pesticides II, *Kagaku (Science)*, 41, 383, 1971.
335. Furst, H. and Praeger, K., Isolation of pure γ-isomer from crude hexachlorocyclohexane mixtures, German patent 1,093,791, 1959; *C.A.*, 55, 19124, 1961.
336. Bishopp, F. C., The insecticide situation, *J. Econ. Entomol.*, 39, 449, 1946.
337. Rohwer, S. A., Effect of individual BHC isomers on plants, *Agric. Chem.*, 4, 75, 1949.
338. Burnett, G. F., Trials of residual insecticides against anophelines in African-type huts, *Bull. Entomol. Res.*, 48, 631, 1957.
339. Hocking, K. S., Armstrong, J. A., and Downing, F. S., cited in Pampana E., *A Textbook of Malaria Eradication*, 2nd ed., Oxford University Press, 1969, 133.
340. Cross, H. F. and Snyder, F. M., Impregnants against chiggers, *J. Econ. Entomol.*, 41, 936, 1948.
340a. Ulmann, E., Ed., *Lindane, Monograph of an Insecticide*, Verlag K. Schillinger, Freiburg im Breisgau, 1972.
341. Hassell, O., Stereochemistry of cyclohexane, *Q. Rev. Chem. Soc. (Lond.)*, 7, 221, 1953.
342. Orloff, H. D., The stereoisomerism of cyclohexane derivatives, *Chem. Rev.*, 54, 347, 1954.
343. Riemschneider, R., Spat, M., Rausch, W., Bottger, E., Zur chemie von polyhalocyclohexanen XXIV. 1,2,3,4,5,6-hexachlorocyclohexan von schmp. 88-89°, *Monatsch. Chem.*, 84, 1068. 1953.
344. Kauer, K., Du Vall, R., and Alquist, F., Epsilon isomer of 1,2,3,4,5,6-hexachlorocyclohexane, *Ind. Eng. Chem., Ind. Ed.*, 39, 1335, 1947.
345. Visweswariah, K. and Majumder, S. K., A new insecticidal isomer of hexachlorocyclohexane (BHC), *Chem. Ind. (Lond.)*, 379, 1969.
346. Cristol, S. J., The structure of α-benzene hexachloride, *J. Am. Chem. Soc.*, 71, 1894, 1949.
347. Bastiansen, O., Ellefson, O., and Hassel, O., Structure of α, β, γ, δ, and ε benzene hexachloride, *Research*, 2, 248, 1948.

348. Bastiansen, O., Ellefson, O., and Hassell, O., Electron diffraction investigation of α, β, γ, δ and ε benzene hexachloride, *Acta Chem. Scand.,* 3, 918, 1949.

349. Norman, N., The crystal structure of the epsilon isomer of 1,2,3,4,5,6-hexachlorocyclohexane, *Acta Chem. Scand.,* 4, 251, 1950.

350. Hetland, E., Dipole moments of α, β, γ, and δ hexachlorocyclohexane and tetrachlorocyclohexane (mp 174°), *Acta Chem. Scand.,* 2, 678, 1948.

351. Kolka, A., Orloff, H., and Griffing, M., Two new isomers of benzene hexachloride, *J. Am. Chem. Soc.,* 76, 3940, 1954.

352. Whetstone, R. R., Davis, F., and Ballard, S., Interconversion of hexachlorocyclohexane isomers, *J. Am. Chem. Soc.,* 75, 1768, 1953.

353. Guilhon, M. J.. The insecticidal and toxic properties of some sulfur derivatives of hexachlorocyclohexane, *C. R. Acad. Agric. Fr.,* 23, 101, 1947; *C.A.,* 41, 6363, 1947.

354. Cristol, S., The kinetics of the alkaline dehydrochlorination of the benzene hexachloride isomers. The mechanism of second order elimination reactions, *J. Am. Chem. Soc.,* 69, 338, 1947.

355. Gunther, F. and Blinn, R. C., Alkaline degradation of benzene hexachloride, *J. Am. Chem. Soc.,* 69, 1215, 1947.

356. La Clair, J. B., Determination of BHC by hydrolysable chlorine method, *Anal. Chem.,* 20, 241, 1948.

357. Riemschneider, R., Konstitution und wirkung von insektiziden. Mitt XVI, *Z. Angew. Ent.,* 48, 423, 1961.

358. Kurihara, N., BHC – its toxicity and its penetration, translocation and metabolism in insects and mammals, *Bochu Kagaku,* 35(II), 56, 1970.

359. Riemschneider, R., Polyhalocyclohexanen und verwandten verbindungen. Mitt XXX: Relativ konfigurationsbestimmung. Konfigurations-bezeichnungen von polyhalocyclohexanen und cyclohexenen, *Osterreichische Chemiker-Zeitung,* 55, 102, 1954.

360. Kurihara, N., Sanemitsu, Y., Tamura, Y., and Nakajima, M., Studies on BHC isomers and their related compounds. Part II. Isomerisation of 3,4,5,6-tetrachlorocyclohexene-1 (BTC) isomers, *Agric. Biol. Chem. (Tokyo),* 34, 790, 1970.

361. Kurihara, N., Sanemitsu, Y., Nakajima, M., McCasland, G. E., and Johnson, L. F., Studies on BHC isomers and their related compounds. III. Proton magnetic resonance studies at 220 MHz of two new iodopentachlorocyclohexanes derived from tetrachlorocyclohexene (BTC), *Agric. Biol. Chem. (Tokyo),* 35, 71, 1971.

362. Bridges, R. G., Retention of γ-benzene hexachloride by wheat and cheese, *J. Sci. Food Agric.,* 431, 1958.

363. Schechter, M. S. and Hornstein, I., Colorimetric determination of benzene hexachloride, *Anal. Chem.,* 24, 544, 1952.

364. Physical and Chemical Properties of Hercules Toxaphene; Hercules Incorporated Agricultural Chemicals Technical Data Bulletin AP−103A; various Hercules technical bulletins.

365. Klein, A. K. and Link, D. J., Field weathering of toxaphene and chlordan, *J. Assoc. Offic. Agric. Chem.,* 50, 586, 1967.

366. Desalbres, L. and Rache, J., Les terpènes polychlorés et leurs propriétés insecticides, *Chim. Ind. (Paris),* 59, 236, 1948.

367. Khanenia, F. S. and Zhuravlev, S. V., cited in Busvine, J. R., The insecticidal potency of γ-BHC and the chlorinated cyclodiene compounds and the significance of resistance to them, *Bull. Entomol. Res.,* 55, 271, 1964.

368. Weinman, C. and Decker, G. C., Chlorinated hydrocarbon insecticides used alone and in combinations for grasshopper control, *J. Econ. Entomol.,* 42, 135, 1949.

369. Schread, J. C., Persistence of DDT, parathion, etc., for *Popillia japonica* control, *J. Econ. Entomol.,* 42, 383, 1949.

370. The Use of Toxaphene Low Volume Formulations for Cotton Insect Control, Hercules Incorporated Bulletin, Wilmington, Delaware, 1969, 24 pp.

371. Beckman, H. F., Ibert, E. R., Adams, B. B., Skoolin, D. O., Determination of total chlorine in pesticides by reduction with a liquid anhydrous ammonia-sodium mixture, *J. Agric. Food Chem.,* 6, 104, 1958.

372. Graupner, A. J. and Dunn, C. L., Determination of toxaphene by a spectrophotometric diphenylamine procedure, *J. Agric. Food Chem.,* 8, 286, 1960.

373. Brown, A. W. A., Insect resistance, *Farm Chemicals,* September 1969 et seq.

374. Brown, A. W. A. and Pal, R., *Insect Resistance in Arthropods,* 2nd ed., W.H.O., Geneva, 1971, 446.

375. Busvine, J. R., *A Critical Review of the Techniques Used for Testing Insecticides,* 2nd ed., Commonwealth Agricultural Bureaux, 1971.

376. Bliss, C. I., Calculation of dose/mortality curve, *Ann. Appl. Biol.,* 22, 134, 1935.

377. Dyte, C. E., Insecticide resistance in stored-product insects with special reference to *Tribolium castaneum, Trop. Stored Prod. Inf.,* (20), 13, 1970.

378. Georghiou, G. P., Genetics of resistance to insecticides in houseflies and mosquitoes, *Exp. Parasitol.,* 26, 224, 1969.

379. Davidson, G., Insecticide resistance in *Anopheles gambiae* Giles, a simple case of Mendelian inheritance, *Nature (Lond.),* 178, 861, 1956.

380. Davidson, G., Studies on insecticide resistance in anopheline mosquitoes, *Bull. W.H.O.,* 18, 579, 1958.

381. Guneidy, A. M. and Busvine, J. R., Genetical studies on dieldrin-resistance in *Musca domestica* L. and *Lucilia cuprina* (Wied.), *Bull. Entomol. Res.,* 55, 499, 1964.

382. Georghiou, G. P., March, R. B., and Printy, G. E., A study of the genetics of dieldrin-resistance in the housefly (*Musca domestica* L.), *Bull. W.H.O.,* 29, 155, 1963.

383. Tadano, T. and Brown, A. W. A., Genetical linkage relationships of DDT resistance and dieldrin-resistance in *Culex pipiens fatigans* Wiedemann, *Bull. W.H.O.*, 36, 101, 1967.

384. Brown, A. W. A., Insecticide resistance-genetic implications and applications, *World. Rev. Pest Control*, 6, 104, 1967.

385. Busvine, J. R., Developments in pest control, *Pest Articles and News Summaries P.A.N.S.*, [A], 14, 310, 1968.

386. Tsukamoto, M., Biochemical genetics of insecticide resistance in the housefly, *Residue Rev.*, 25, 289, 1969.

387. Tsukamoto, M. and Suzuki, R., Genetic analysis of DDT-resistance in strains of the housefly *Musca domestica* L., *Botyu-Kagaku*, 29, 76, 1964.

388. Oppenoorth, F. J., Some Cases of Resistance Caused by the Alteration of Enzymes, Proc. 12th Int. Congr. Entomol., London, July 1964, 240.

389. Grigolo, A. and Oppenoorth, F. J., The importance of DDT-dehydrochlorinase for the effect of the resistance gene kdr in the housefly, *Genetica*, 37, 159, 1966.

390. Khan, M. A. Q. and Terriere, L. C., DDT-dehydrochlorinase activity in housefly strains resistant to various groups of insecticides, *J. Econ. Entomol.*, 61, 732, 1968.

391. Sawicki, R. M. and Farnham, A. W., Examination of the isolated autosomes of the SKA strain of house-flies (*Musca domestica* L.) for resistance to several insecticides with and without pretreatment with sesamex and TBTP, *Bull. Entomol. Res.*, 59, 409, 1968.

392. Hoyer, R. F. and Plapp, F. W., Jr., A gross genetic analysis of two DDT-resistant housefly strains, *J. Econ. Entomol.*, 59, 495, 1966.

393. Plapp, F. W., Jr. and Hoyer, R. F., Possible pleiotropism of a gene conferring resistance to DDT, DDT analogs, and pyrethrins in the House Fly and *Culex tarsalis*, *J. Econ. Entomol.*, 61, 761, 1968.

394. Plapp, F. W., Jr. and Hoyer, R. F., Insecticide resistance in the housefly: decreased rate of absorption as the mechanism of action of a gene that acts as an intensifier of resistance, *J. Econ. Entomol.*, 61, 1298, 1968.

395. Sawicki, R. M. and Farnham, A. W., Genetics of resistance to insecticides of the SKA strain of Musca domestica, III. Location and isolation of the factors of resistance to dieldrin, *Entomol. Exp. Appl.*, 11, 133, 1968.

396. Perry, A. S., Hennessy, D. J., and Miles, J. W., Comparative toxicity and metabolism of p, p'-DDT and various substituted DDT-derivatives by susceptible and resistant houseflies, *J. Econ. Entomol.*, 60, 568, 1967.

397. Oppenoorth, F. J., DDT-resistance in the housefly dependent on different mechanisms and the action of synergists, *Meded. Landbouwhogesch. Opzoekingsstn. Staat. Gent.*, 30, 1390, 1965.

398. El Basheir, S., Causes of resistance to DDT in a diazinon-selected and a DDT-selected strain of house flies, *Entomol. Exp. Appl.*, 10, 111, 1967.

399. Milani, R., Genetical aspects of insecticide resistance, *Bull. W.H.O.*, Supplement, 29, 77, 1963.

400. Oppenoorth, F. J. and Nasrat, G. E., Genetics of dieldrin and gamma-BHC (lindane) resistance in the housefly (*Musca domestica*), *Entomol. Exp. Appl.*, 9, 223, 1966.

401. Davidson, G., Resistance to Chlorinated Insecticides in Anopheline Mosquitoes, Proc. 12th Int. Congr. Entomol., London, July 1964, 236.

402. Oonnithan, E. S. and Miskus, R., Metabolism of $C^{14}$-dieldrin-resistant *Culex pipiens quinquefasciatus* Mosquitoes, *J. Econ. Entomol.*, 5, 425, 1964.

403. Tadano, T. and Brown, A. W. A., Development of resistance to various insecticides in *Culex pipiens fatigans* Wiedemann, *Bull. W.H.O.*, 35, 189, 1966.

404. Plapp, F. W., Jr., Chapman, G. A., and Morgan, J. W., *J. Econ. Entomol.*, 58, 1064, 1965.

405. Klassen, W. and Brown, A. W. A., Genetics of insecticide-resistance and several visible mutants in *Aedes aegypti*, *Can. J. Genet. Cytol.*, 6, 61, 1964.

406. Lockhart, W. L., Klassen, W., and Brown, A. W. A., Crossover values between dieldrin-resistance and DDT; resistance and linkage-group-2 genes in *Aedes aegypti*, *Can. J. Genet. Cytol.*, 12, 407, 1970.

407. Kimura, T. and Brown, A. W. A., DDT-dehydrochlorinase in *Aedes aegypti*, *J. Econ. Entomol.*, 57, 710, 1964.

408. Inwang, E. E., Khan, M. A. Q., and Brown, A. W. A., DDT-resistance in West African and Asian strains of *Aedes aegypti* (L), *Bull. W.H.O.*, 36, 409, 1967.

409. McDonald, I. C., Ross, M. H., and Cochran, D. G., Genetics and linkage of aldrin resistance in the German cockroach, *Blattella germanica* (L), *Bull. W.H.O.*, 40, 745, 1969.

410. Matsumura, F., Telford, J. N., and Hayashi, M., Effect of sesamex upon dieldrin resistance in the German cockroach, *J. Econ. Entomol.*, 60, 942, 1967.

411. Read, D. C. and Brown, A. W. A., Inheritance of dieldrin-resistance and adult longevity in the cabbage maggot, *Hylemya brassicae* (Bouche), *Can. J. Genet. Cytol.*, 8, 71, 1966.

412. Hooper, G. H. S. and Brown, A. W. A., Dieldrin-resistant and DDT-resistant strains of the spotted root maggot apparently restricted to heterozygotes for resistance, *J. Econ. Entomol.*, 58, 824, 1965.

413. Lineva, V. A. and Derbeneva-Uhova, V. P., cited in Zaghloul, T. M. A. and Brown, A. W. A., Effects of sublethal doses of DDT on the reproduction and susceptibility of *Culex pipiens* L., *Bull. W.H.O.*, 38, 459, 1968.

414. Zaghloul, T. M. A. and Brown, A. W. A., Effects of sublethal doses of DDT on the reproduction and susceptibility of *Culex pipiens* L., *Bull. W.H.O.*, 38, 459, 1968.

415. Moriarty, F., The sublethal effects of synthetic insecticides on insects, *Biol. Rev.*, 44, 321, 1969.

416. Hadaway, A. B., Cumulative effect of sublethal doses of insecticides on houseflies, *Nature (Lond.)*, 178, 149, 1956.

417. Ahmad, N. and Brindley, W. A., Modification of parathion toxicity to wax moth larvae by chlorcyclizine, aminopyrine or phenobarbital, *Toxicol. Appl. Pharmacol.*, 15, 433, 1969.
418. Agosin, M., Aravena, L., and Neghme, A., Enhanced protein synthesis in *Triatoma infestans* treated with DDT, *Exp. Parasitol.*, 16, 318, 1965.
419. Plapp, F. W., Jr. and Casida, J. E., Induction by DDT and dieldrin of insecticide metabolism by house fly enzymes, *J. Econ. Entomol.*, 63, 1091, 1970.
420. Walker, C. R. and Terriere, L. C., Induction of microsomal oxidases by dieldrin in *Musca domestica*, *Entomol. Exp. Appl.*, 13, 260, 1970.
421. Yu, S. J. and Terriere, L. C., Induction of microsomal oxidases in the housefly and the action of inhibitors and stress factors, *Pestic. Biochem. Physiol.*, 1, 173, 1971.
422. Melander, A. L., Strain of *Aspidiotus* resistant to lime-sulfur, *J. Econ. Entomol.*, 7, 167, 1914.
423. Brown, A. W. A., Insecticide resistance comes of age, *Bull. Entomol. Soc. Am.*, 14, 3, 1968.
424. Keiding, J., Persistence of resistant populations after the relaxation of the selection pressure, *World Rev. Pest Control*, 6, 115, 1967.
425. Busvine, J. R., Mechanism of resistance to insecticides in house-flies, *Nature (Lond.)*, 168, 193, 1951.
426. Missiroli, A., Resistenza alli insecticidi di alcuni razze di *Musca domestica*, *Riv. Parassitol.*, 12, 5, 1951.
427. Gilbert, I. H., Couch, M. D., and McDuffie, W. C., Development of resistance to insecticides in natural populations of house-flies, *J. Econ. Entomol.*, 46, 48, 1953.
428. March, R. B. and Metcalf, R. L., Insecticide-resistant flies, *Soap Sanit. Chem.*, 26, 121, 1950.
429. March, R. B., Summary of Research on Insects Resistant to Insecticides, in National Research Council, Division of Medical Sciences, *Conference on Resistance and Insect Physiology*, Washington, D.C., Dec. 8—9, 1951, 45 (National Research Council Publication 219).
430. Metcalf, R. L., Physiological basis for insect resistance to insecticides, *Physiol. Rev.*, 35, 197, 1955.
431. Georgopoulos, G. D., Extension to chlordane of the resistance to DDT observed in *Anopheles sacharovi*, *Bull. W.H.O.*, 11, 855, 1954.
432. Davidson, G., Insecticide resistance in *Anopheles gambiae* Giles, *Nature (Lond.)*, 178, 705, 1956.
433. Wright, J. W., Present status of mosquito control, *Pestic. Sci.*, 3, 471, 1972.
434. Busvine, J. R., Insecticide-resistance in mosquitoes, *Pestic. Sci.*, 3, 483, 1972.
435. Hawkins, W. B., Tests on the Resistance of Anopheles Larvae in the Region of the Tennessee Valley Authority, in Seminar on the Susceptibility of Insects to Insecticides, Panama, 23—28 June 1958, Pan American Sanitary Bureau, W.H.O., Washington D.C., 1958.
436. Cova-Garcia, P., Manifestations of behaviouristic resistance in Venezuela, in Seminar on the Susceptibility of Insects to Insecticides, Panama, 23 June 1958, Pan American Sanitary Bureau, W.H.O., Washington D.C., 1958.
437. Gjullin, C. M. and Peters, R. F., Recent studies of mosquito resistance to insecticides in California, *Mosquito News*, 12, 1, 1952.
438. Rachou, R. G., Some manifestations of behaviouristic resistance in Brazil. in Seminar on the Susceptibility of Insects to Insecticides, Panama, 23 June 1958, Pan American Sanitary Bureau, W.H.O., Washington D.C., 1958.
439. Wharton, R. H., The behaviour and mortality of *Anopheles maculatus* and *Culex fatigans* in experimental huts treated with DDT and BHC, *Bull. Entomol. Res.*, 42, 1, 1950.
440. Mosna, E., *Culex pipiens* autogenicus DDT-resistenti e loro controllo con Octaklor e esachlorocicloesano, *Riv. Parassitol.*, 9, 19, 1948.
441. Kerr, J. A., de Camargo, S., and Abedi, Z. H., cited in Busvine, J. R. and Pal, R., The impact of insecticide-resistance on control of vectors and vector-borne diseases, *Bull. W.H.O.*, 40, 731, 1969.
442. Fay, R. W., Insecticide resistance in *Aedes aegypti*, *Am. J. Trop. Med. Hyg.*, 5, 378, 1956.
443. Wright, J. W. and Brown, A. W. A., Survey of possible insecticide resistance in body lice, *Bull. W.H.O.*, 16, 9, 1957.
444. Wright, J. W. and Pal, R., Second survey of insecticide resistance in body-lice, 1958—63, *Bull. W.H.O.*, 33, 485, 1965.
445. Grayson, J. M., Effects on the German cockroach of twelve generations of selection for survival to treatments with DDT and benzene hexachloride, *J. Econ. Entomol.*, 46, 124, 1953.
446. Grayson, J. M., Differences between a resistant and a non-resistant strain of the German cockroach, *J. Econ. Entomol.*, 47, 253, 1954.
447. Whitnall, A. B. M. and Bradford, B., An arsenic resistant tick and its control with gammexane dips, *Bull. Entomol. Res.*, 40, 207, 1949.
448. Whitnall, A. B. M., Thorburn, J. A., Mellardy, W. M., Whitehead, G. B., and Meerholz, F., A BHC-resistant tick, *Bull. Entomol. Res.*, 43, 51, 1952.
449. Stone, B. F. and Meyers, R. A. J., Dieldrin-resistant cattleticks, *Boophilus microplus* (Canestrini) in Queenland, *Aust. J. Agric. Res.*, 8, 312, 1957.
450. Wharton, R. H. and Roulston, W. J., Resistance of ticks to chemicals, *Ann. Rev. Entomol.*, 15, 381, 1970.
451. Adkisson, P. L., Development of resistance by the tobacco budworm to endrin and carbaryl, *J. Econ. Entomol.*, 61, 37, 1968.
452. Plapp, F. W., Jr., Insecticide resistance in *Heliothis*: tolerance in larvae of *H. viriscens* as compared with *H. zea* to organophosphate insecticides, *J. Econ. Entomol.*, 64, 999, 1971.

453. Ishikura, H., High yield rice cultivation and the use of pesticides, *Japan Pesticide Information*, No. 5, Japan Plant Protection Association, October 1970, 5.

454. Fukaya, F., Insecticide resistance, detection methods and counter-measures, *Japan Pesticide Information*, No. 6, Japan Plant Protection Association, January 1971, 25.

455. Vernon, A. J., Control of cocoa capsids in West Africa, *Chem. Ind. (Lond.)*, 1219, 1961.

456. Rushton, M. P., Cocoa capsid control in Sierra Leone, *Span*, 6, 23, 1963.

457. Dunn, J. A., Insecticide resistance in the cocoa capsid, *Distantiella theobroma, Nature (Lond.)*, 199, 1207, 1963.

458. Harris, C. R., Mazurek, J. H., and Svec, H. J., Cross-resistance shown by aldrin-resistant seed maggot flies, *Hylemya* spp., to other cyclodiene insecticides and related materials, *J. Econ. Entomol.*, 57, 702, 1964.

459. McClanahan, R. J., Harris, C. R., and Miller, L. A., Resistance to aldrin, dieldrin, and heptachlor in the onion maggot, *Hylemya antiqua* (Meig.) in Ontario, *Rep. Entomol. Soc. Ont.*, 89, 55, 1958.

460. Harris, C. R., Manson, G. F., and Mazurek, J. H., Development of insecticidal resistance by soil insects in Canada, *J. Econ. Entomol.*, 55, 777, 1962.

461. Coaker, T. H., Mowat, D. J., and Wheatley, G. A., Insecticide resistance in the cabbage root fly in Britain, *Nature (Lond.)*, 200, 664, 1963.

462. Gostick, K. G. and Coaker, T. H., Monitoring for Insecticide Resistance in Root Flies, *Proc. 5th Br. Insectic. Fungic. Conf.*, 1969, 89.

463. Kilpatrick, J. W. and Schoof, H. F., Interrelationship of water and *Hermetia illucens* breeding to *Musca domestica* production in human excrement, *Am. J. Trop. Med. Hyg.*, 8, 103, 1959.

464. Knutson, H., Changes in reproductive potential in houseflies in response to dieldrin, *Misc. Publ. Entomol. Soc. Am.*, 1, 27, 1959.

465. Parkin, E. A., The onset of insecticide resistance among field populations of stored-product insects, *J. Stored Prod. Res.*, 1, 3, 1965.

466. Lindgren, D. L. and Vincent, L. E., The susceptibility of laboratory reared and field-collected cultures of *Tribolium confusum* and *T. castaneum*, to ethylene dibromide, hydrocyanic acid and methyl bromide, *J. Econ. Entomol.*, 58, 551, 1965.

467. Dyte, C. E. and Blackman, D. G., The spread of insecticide resistance in *Tribolium castaneum* (Herbst), *J. Stored Prod. Res.*, 6, 255, 1970.

468. Lloyd, C. J., Studies on the cross-tolerance to DDT-related compounds of a pyrethrin-resistant strain of *Sitophilus granarius* (L.), *J. Stored Prod. Res.*, 5, 337, 1969.

468a. Weisgerber, I., Klein, W., and Korte, F., Verteilung und umwandlung von heptachlor-[14]C in weisskohl und weizen, *Chemosphere*, No. 2, 89, 1972.

469. Abedi, Z. H. and Brown, A. W. A., Development and reversion of DDT-resistance in *Aedes aegypti, Can. J. Genet. Cytol.*, 2, 252, 1960.

470. Georghiou, G. P., March, R. B., and Printy, G. E., Induced regression of dieldrin-resistance in the housefly (*Musca domestica* L.), *Bull. W.H.O.*, 29, 167, 1963.

471. Georghiou, G. P. and Metcalf, R. L., Dieldrin susceptibility: partial restoration in anopheles selected with a carbamate, *Science*, 140, 301, 1963.

472. Georghiou, G. P. and Bowen, W. R., An analysis of housefly resistance to insecticides in California, *J. Econ. Entomol.*, 59, 204, 1966.

473. Stone, B. F., The inheritance of DDT-resistance in the cattle tick, Boophilus microplus, *Aust. J. Agric. Res.*, 13, 984, 1962.

474. Schnitzerling, H. J., Roulston, W. J., and Schuntner, C. A., The absorption and metabolism of [14C] DDT in DDT-resistant and susceptible strains of the cattle tick *Boophilus microplus, Aust. J. Biol. Sci.*, 23, 219, 1970.

475. Winteringham, F. P. W., Mechanisms of selective insecticidal action, *Ann. Rev. Entomol.*, 14, 409, 1969.

476. Gerolt, P., Mode of entry of contact insecticides, *J. Insect Physiol.*, 15, 563, 1969.

477. Moriarty, F. and French, M. C., The uptake of dieldrin from the cuticular surface of *Periplaneta Americana, Pestic. Biochem. Physiol.*, 1, 286, 1971.

478. Narahashi, T., Effects of insecticides on excitable tissues, in *Advances in Insect Physiology*, Vol. 8, Beament, J. W. L., Treherne, J. E., and Wigglesworth, V. B., Eds., Academic Press, London and New York, 1971.

479. Winteringham, F. P. W. and Lewis, S. E., On the mode of action of insecticides, *Ann. Rev. Entomol.*, 4, 303, 1959.

480. Soto, A. R. and Deichmann, W. B., Major metabolism and acute toxicity of aldrin, dieldrin and endrin, *Environ. Res.*, 1, 307, 1967.

481. Robinson, J., Persistent pesticides, *Ann. Rev. Pharmacol.*, 10, 353, 1970.

482. Brooks, G. T., The metabolism of diene-organochlorine (cyclodiene) insecticides, *Residue Rev.*, 27, 81, 1969.

483. Brooks, G. T., The fate of chlorinated hydrocarbons in living organisms, in *Pesticide Terminal Residues*, Invited papers from the IUPAC International Symposium on Terminal Residues, Tel Aviv, 1971, Tahori, A. S., Ed., Butterworths, London, 1971, 111.

484. Brooks, G. T., Pathways of enzymatic degradation of pesticides, in *Environmental Quality and Safety*, Vol. I, Coulston, F. and Korte, F., Eds., Georg Thieme, Stuttgart, Academic Press, New York, 1972, 106.

485. Matsumura, F., Metabolism of Pesticides in Higher Plants, in *Environmental Quality and Safety*, Vol. I, Coulston, F. and Korte, F., Eds., Georg Thieme, Stuttgart, Academic Press, New York, 1972, 96.

486.  Klein, W., Metabolism of Insecticides in Microorganisms and Insects, in *Environmental Quality and Safety*, Vol. I, Coulston, F., Korte, F., Eds., George Thieme, Stuttgart, Academic Press, 1972, 1964.

486a. Weisgerber, I., Klein, W., and Korte F., Verteilung und umwandlung von heptachlor-[14]C in weisskohl und weizen, *Chemosphere*, No. 2, 89, 1972.

487.  Schaefer, C. H. and Sun, Y. P., A study of dieldrin in the housefly central nervous system in relation to dieldrin resistance, *J. Econ. Entomol.*, 60, 1580, 1967.

488.  Sellers, L. G. and Guthrie, F. E., Localisation of dieldrin in the housefly thoracic ganglion by electron microscopic autoradiography, *J. Econ. Entomol.*, 64, 352, 1971.

489.  Telford, J. N. and Matsumura, F., Dieldrin binding in subcellular nerve components of cockroaches. An electron microscopic and autoradiographic study, *J. Econ. Entomol.*, 63, 795, 1970.

490.  Kurihara, N., Nakajima, E., and Shindo, H., Whole body autoradiographic studies on the distribution of BHC and nicotine in the American cockraoch, in *Biochemical Toxicology of Insecticides*, O'Brien, R. D. and Yamamoto, I., Eds., Academic Press, New York, 1970, 41.

491.  Sun, Y. P., Dynamics of insect toxicology – a mathematical and graphical evaluation of the relationship between insect toxicity and rates of penetration and detoxication of insecticides, *J. Econ. Entomol.*, 61, 949, 1968.

492.  Glasstone, S., *A Textbook of Physical Chemistry*, 2nd ed., MacMillan, London, 1948, 1075.

493.  Hewlett, P. S., Interpretation of dosage-mortality data for DDT-resistant houseflies, *Ann. Appl. Biol.*, 46, 37, 1958.

494.  Menzie, C. M., Fate of pesticides in the environment, *Ann. Rev. Entomol.*, 17, 199, 1972.

495.  Hamaker, J. W., Mathematical prediction of cumulative levels of pesticides in soil, in Organic Pesticides in the Environment, *Am. Chem. Soc., Adv. Chem. Ser.*, 60, 122, 1966.

496.  Decker, G. C., Bruce, N. W., and Bigger, J. H., The accumulation and dissipation of residues resulting from the use of aldrin in soils, *J. Econ. Entomol.*, 58, 266, 1965.

497.  Lichtenstein, E. P., Persistence and degradation of pesticides in the environment, in *Scientific Aspects of Pest Control*, Publication 1402, National Academy of Sciences, National Research Council, Washington, D.C., 1966.

498.  Harris, C. R., Factors influencing the effectiveness of soil insecticides, *Ann. Rev. Entomol.*, 17, 177, 1972.

499.  Harris, C. R. and Sans, W. W., Behaviour of dieldrin in soil: microplot field studies on the influence of soil type on biological activity and absorption by carrots, *J. Econ. Entomol.*, 65, 333, 1972.

500.  Harris, C. R., Factors influencing the biological activity of technical chlordane and some related compounds in soil, *J. Econ. Entomol.*, 65, 341, 1972.

501.  Strickland, A. H., Some estimates of insecticide and fungicide usage in agriculture and horticulture in England and Wales 1960–64. Pesticides in the environment and their effects on wildlife, *J. Appl. Ecol.*, 3[suppl.], 3, 1966.

502.  *Third Report of the Research Committee on Toxic Chemicals*, Agricultural Research Council, HMSO London, 1970, 69.

503.  Plapp, F. W., Jr., On the molecular biology of insecticide resistance, in *Biochemical Toxicology of Insecticides*, O'Brien, R. D. and Yamamoto, I., Eds., Academic Press, New York, 1970, 179.

504.  Lichtenstein, E. P. and Corbett, J. R., Enzymatic conversion of aldrin to dieldrin with subcellular components of pea plants, *J. Agric. Food Chem.*, 17, 589, 1969.

505.  Yu, S. J., Kiigemagi, U., and Terriere, L. C., Oxidative metabolism of aldrin and isodrin by bean root fractions, *J. Agric. Food Chem.*, 19, 5, 1971.

506.  Lyr, H. and Ritter, G., Zum wirkungsmechanismus von Hexachlorocyclohexan – Isomeren in Hefezellen, *Z. Allge. Mikrobiol.*, 9, 545, 1969.

507.  Tu, C. M., Miles, J. R. W., and Harris, C. R., Soil microbial degradation of aldrin, *Life Sci.*, [Oxford] 7, 311, 1968.

508.  Korte, F., Metabolism of [14]C-labelled insecticides in microorganisms, insects and mammals, *Botyu-kagaku*, 32, 46, 1967.

508a. Rice, C. P. and Sikka, H. C., Uptake and metabolism of DDT by six species of marine algae, *J. Agric. Food Chem.*, 21, 148, 1973.

508b. Rice, C. P. and Sikka, H., Fate of dieldrin in selected species of marine algae, *Bull. Environ. Contam. Toxicol.*, 9, 116, 1973.

509.  Robinson, J., The burden of chlorinated hydrocarbon pesticides in man, *Can. Med. Assoc. J.*, 100, 180, 1969.

510.  Quaife, M. L., Winbush, J. S., and Fitzhugh, O. G. Survey of quantitative relationships between ingestion and storage of aldrin and dieldrin in animals and man, *Food Cosmet. Toxicol.*, 5, 39, 1967.

511.  Department of Health, Education, and Welfare, Report of the Secretary's Commission on Pesticides and their Relationship to Environmental Health, Parts I and II, U.S. Govt. Print. Off., Washington, D.C., 1969, 263.

512.  Bitman, J., Cecil, H. C., Harris, S. J., and Fries, G. F., Comparison of DDT effect on pentobarbital metabolism in rats and quail, *J. Agric. Food Chem.*, 19, 333, 1971.

513.  Robinson, J., Roberts, M., Baldwin, M., and Walker, A. I. T., The pharamacokinetics of HEOD [dieldrin] in the rat, *Food Cosmet. Toxicol.*, 7, 317, 1969.

514.  Brown, V. K. H., Robinson, J., and Richardson, A., Preliminary studies on the acute and subacute toxicities of a photoisomerisation product of HEOD, *Food Cosmet. Toxicol.*, 5, 771, 1967.

515.  Robinson, J. and Richardson, A. R., personal communication.

516.  Hunter, C. G. and Robinson, J., Pharmacodynamics of dieldrin [HEOD]. I. Ingestion by human subjects for 18 months, *Arch. Environ. Health*, 15, 614, 1967.

517. Klein, W., Muller, W., and Korte, F., Ausscheidung, verteilung und stoffwechsel von endrin-$^{14}$C in ratten, *Justus Liebigs Ann. Chem.*, 713, 180, 1968.

518. Kaul, R., Klein, W., and Korte, F., Metabolismus und kinetic der verteilung von β-dihydroheptachlor-$^{14}$C in mannlichen ratten, *Tetrahedron*, 26, 99, 1970.

519. Robinson, J. and Roberts, M., Accumulation, distribution and elimination of organochlorine insecticides by vertebrates, *Soc. Chem. Ind. (Lond.) Monogr.*, No. 29, 106, 1968.

520. Abbott, D. C., Collins, G. B., and Goulding, R., Organochlorine pesticide residues in human fat in the United Kingdom 1969—71, *Br. Med. J.*, 553, 1972.

521. Coulson, J. C., Deans, I. R., Potts, G. R., Robinson, J., and Crabtree, A. N., Changes in organochlorine contamination of the marine environment of Eastern Britain monitored by Shag eggs, *Nature (Lond.)*, 236, 454, 1972.

522. Harrison, H. L., Loucks, O. L., Mitchell, J. W., Parkhurst, D. F., Tracy, C. R., Watts, D. G., and Yannacone, V. J., Jr., Systems studies of DDT transport, *Science*, 170, 503, 1970.

523. Woodwell, M., Craig, P. P., and Johnson, H. A., DDT in the biosphere: where does it go? *Science*, 174, 1101, 1971.

524. Kapoor, I. P., Metcalf, R. L., Hirwe, A. S., Po-Yung Lu, Coats, J. R., and Nystrom, R. F., Comparative metabolism of DDT, methylchlor, and ethoxychlor in mouse, insects, and in a model ecosystem, *J. Agric. Food Chem.*, 20, 1, 1972.

525. Ivie, G. W. and Casida, J. E., Enhancement of photoalteration of cyclodiene insecticide chemical residues by rotenone, *Science*, 167, 1620, 1970.

526. Bailey, S., Bunyan, P. J., Rennison, B. D., and Taylor, A., The metabolism of 1,1-di[p-chlorophenyl]-2,2-dichloro-ethylene and 1,1-di[p-chlorophenyl]-2-chloroethylene in the pigeon, *Toxicol. Appl. Pharmacol.*, 14, 23, 1969.

527. Bailey, S., Bunyan, P. J., and Taylor, A., The metabolism of p,p'-DDT in some avian species, in *Environmental Quality and Safety*, Vol. I, Coulston, F. and Korte, F., Eds., Georg Thieme, Stuttgart, Academic Press, New York, 1972, 244.

528. Datta, P. R., In vivo detoxication of p,p'-DDT via p,p'-DDE to p,p'-DDA in rats, *Ind. Med.*, 39, 49, 1970.

529. Peterson, J. E. and Robinson, W. H., Metabolic products of p,p'-DDT in the rat, *Toxicol. Appl. Pharmacol.*, 6, 321, 1964.

530. Morgan, D. P. and Roan, C. C., Absorption, storage and metabolic conversion of ingested DDT and DDT metabolites in man, *Arch. Environ. Health*, 22, 301, 1971.

531. Agosin, M., Michaeli, D., Miskus, R., Nagasawa, S., and Hoskins, W. M., A new DDT metabolising enzyme in the German cockroach, *J. Econ. Entomol.*, 54, 340, 1961.

532. Rowlands, D. G. and Lloyd, C. J., DDT metabolism in susceptible and pyrethrin resistant *Sitophilus granarius* [L.], *J. Stored Prod. Res.*, 5, 413, 1969.

533. Leeling, N. C., personal communication, 1972.

534. Oppenoorth, F. J. and Houx, N. W. H., DDT resistance in the housefly caused by microsomal degradation, *Entomol. Exp. Appl.*, 11, 81, 1968.

535. McKinney, J. D. and Fishbein, L., DDE formation; dehydrochlorination or dehypochlorination, *Chemosphere*, No. 2, 67, 1972.

536. Hassall, K. A., Reductive dechlorination of DDT; the effect of some physical and chemical agents on DDD production by pigeon liver preparations, *Pestic. Biochem. Physiol.*, 1, 259, 1971.

537. Hathway, D. E., Biotransformations, in *Foreign Compound Metabolism in Mammals*, Hathway, D. E., Brown, S. S., Chasseaud, L. F., and Hutson, D. H., Reporters, *Specialist Periodical Report of the Chemical Society*, Vol. 1, London, 1970, 295.

538. Wedemeyer, G., Dechlorination of 1,1,1-trichloro-2,2-bis[p-chlorophenyl] ethane by *Aerobacter aerogenes*, *Appl. Microbiol.*, 15, 569, 1967.

539. Alexander, M., Microbial degradation of pesticides, in *Environmental Toxicology of Pesticides*, Matsumura, F., Boush, G. M., and Misato, T., Eds., Academic Press, New York and London, 1972, 365.

539a. Pfaender, F. K. and Alexander, M., Extensive microbial degradation of DDT *in vitro* and DDT metabolism by natural communities, *J. Agric. Food Chem.*, 20, 842, 1972.

539b. Albone, E. S., Eglinton, G., Evans, N. C., and Rhead, M. M., Formation of bis(p-chlorophenyl) acetonitrile (p,p'-DDCN) from p,p'-DDT in anaerobic sewage sludge, *Nature (Lond.)*, 240, 420, 1972.

539c. Jensen, S., Gothe, R., and Kindstedt, M. O., Bis-(p-chlorophenyl)-acetonitrile (DDCN), a new DDT derivative formed in anaerobic digested sewage sludge and lake sediment, *Nature (Lond.)*, 240, 421, 1972.

540. Oppenoorth, F. J., Resistance to gamma-hexachlorocyclohexane in *Musca domestica* L., *Arch. Neerl. Zool.*, 12, 1, 1956.

541. Reed, W. T. and Forgash, A. J., Lindane: metabolism to a new isomer of pentachlorocyclohexene, *Science*, 160, 1232, 1968.

542. Reed, W. T. and Forgash, A. J., Metabolism of lindane to organic-soluble products by houseflies, *J. Agric. Food Chem.*, 18, 475, 1970.

543. Grover, P. L. and Sims, P., The metabolism of γ-2,3,4,5,6-pentachlorocyclohex-1-ene and γ-hexachlorocyclohexane in rats, *Biochem. J.*, 96, 521, 1965.

544.   Saha, J., cited in Hurtig, H., Significance of conversion products and metabolites of pesticides in the environment, in *Environmental Quality and Safety*, Coulston, F. and Korte, F., Eds., Georg Thieme, Stuttgart and Academic Press, New York, 1972, 67.

544a.  Freal, J. F. and Chadwick, R. W., Metabolism of hexachlorocyclohexane to chlorophenols and effect of isomer pretreatment on lindane metabolism in rats, *J. Agric. Food Chem.*, 21, 424, 1973.

545.   Duxbury, J. M., Tiedje, J. M., Alexander, M., and Dawson, J. E., 2,4-D metabolism; enzymatic conversion of chloromaleylacetic acid to succinic acid, *J. Agric. Food Chem.*, 18, 199, 1970.

546.   Bowman, M. C., Acree, F., Jr., Lofgren, C. S., and Beroza, M., Chlorinated insecticides; Fate in aqueous suspensions containing mosquito larvae, *Science*, 146, 1480, 1964.

547.   Brooks, G. T., Harrison, A., and Lewis, S. E., Cyclodiene epoxide ring hydration by microsomes from mammalian liver and houseflies, *Biochem. Pharmacol.*, 19, 255, 1970.

548.   Kaul, R., Klein, W., and Korte, F., Verteilung, ausscheidung und metabolismus von Telodrin und heptachlor in ratten und mannlichen kaninchen, endprodukt des warmblutermetabolismus von heptachlor, *Tetrahedron*, 26, 331, 1970.

549.   Matsumura, F. and Nelson, J. O., Identification of the major metabolic product of heptachlor epoxide in rat faeces, *Bull. Environ. Contam. Toxicol.*, 5, 489, 1971.

550.   Schwemmer, B., Cochrane, W. P., and Polen, P. B., Oxychlordane, animal metabolite of chlordane: isolation and synthesis, *Science*, 169, 1087, 1970.

551.   Lawrence, J. H., Barron, R. P., Chen, J. Y. T., Lombardo, P., and Benson, W. R., Note on identification of a chlordane metabolite found in milk and cheese, *J. Assoc. Offic. Agric. Chem.*, 53, 261, 1970.

552.   Street, J. C. and Blau, S. E., Oxychlordane: accumulation in rat adipose tissue on feeding chlordane isomers or technical chlordane, *J. Agric. Food Chem.*, 20, 395, 1972.

553.   Davidow, B., Hagan, E., and Radomski, J. L., A metabolite of chlordane in tissues of animals, *Fed. Proc. Abstr.*, 10, 291, 1951.

554.   Georgacakis, E. and Khan, M. A. Q., Toxicity of the photoisomers of cyclodiene insecticides to freshwater animals, *Nature (Lond.)*, 233, 120, 1971.

555.   Gorbach, S., Haaring, R., Knauf, W., and Werner, H. J., Residue analysis and biotests in rice fields of East Java treated with Thiodan, *Bull. Environ. Contam. Toxicol.*, 6, 193, 1971.

556.   Brooks, G. T., Perspectives of cyclodiene metabolism, in Proceedings of the Symposium on the Science and Technology of Residual Insecticides in Food Production with Special Reference to Aldrin and Dieldrin, Shell Chemical Co. New York, Library of Congress Cat. No. 68-27527, 1968, 89.

557.   Wong, D. T. and Terriere, L. C., Epoxidation of aldrin, isodrin and heptachlor by rat liver microsomes, *Biochem. Pharmacol.*, 14, 375, 1965.

558.   Nakatsugawa, T., Ishida, M., and Dahm, P. A., Microsomal oxidation of cyclodiene insecticides, *Biochem. Pharmacol.*, 14, 1853, 1965.

559.   Heath, D. F., [36]Cl-dieldrin in mice, in *Radioisotopes and Radiation in Entomology*, Proc. Bombay Symposium 1960, International Atomic Energy Agency, Vienna, 1962, 83.

560.   Heath, D. F. and Vandekar, M., Toxicity and metabolism of dieldrin in rats, *Br. J. Ind. Med.*, 21, 269, 1964.

561.   Korte, F. and Arent, H., Isolation and identification of dieldrin metabolites from urine of rabbits after oral administration of dieldrin-[14]C, *Life Sci.*, [Oxford], 4, 2017, 1965.

562.   Richardson, A., Baldwin, M. K., and Robinson, J., Metabolites of dieldrin [HEOD] in the urine and faeces of rats, *Chem. Ind. (Lond.)*, 588, 1968.

563.   Klein, A. K., Link, J. D., and Ives, N. F., Isolation and purification of metabolites found in the urine of male rats fed aldrin and dieldrin, *J. Assoc. Offic. Agric. Chem.*, 51, 895, 1968.

564.   Feil, V. J., Hedde, R. D., Zaylskie, R. G., and Zachrison, C. H., Identification of *trans*-6,7-dihydroxydihydroaldrin and 9-[*syn*-epoxy]hydroxy-1,2,3,4,10,10-hexachloro-6,7-epoxy-1,4,4a,5,6,7,8,8a-octahydro-1,4-endo-5,8-exo-dimethanonaphthalene, *J. Agric. Food Chem.*, 18, 120, 1970.

564a.  Bedford, C. T. and Harrod, R. K., Synthesis of 9-hydroxy-HEOD, a major mammalian metabolite of HEOD (dieldrin), *Chem. Commun.*, 735, 1972.

565.   Baldwin, M. K., The metabolism of the chlorinated insecticides aldrin, dieldrin, endrin and isodrin, Ph.D. Thesis, University of Surrey, 1971.

565a.  Richardson, A. and Robinson, J., Identification of a major metabolite of HEOD (dieldrin) in human faeces, *Xenobiotica*, 1, 213, 1971.

566.   Oda, J. and Muller, W., Identification of a mammalian breakdown product of dieldrin, in *Environmental Quality and Safety*, Vol. 1, Coulston, F. and Korte, F., Eds., George Thieme, Stuttgart, Academic Press, New York, 1972, 248.

567.   McKinney, J. D., personal communication, June, 1972.

568.   McKinney, J. D., Matthews, H. B., and Fishbein, L., Major faecal metabolite of dieldrin in rat, structure and chemistry, *J. Agric. Food Chem.*, 20, 597, 1972.

569.   Rosen, J. D., Conversion of pesticides under environmental conditions, in *Environmental Quality and Safety*, Vol. I, Coulston, F. and Korte, F., Eds., George Thieme, Stuttgart, Academic Press, New York, 1972, 85.

570.   Matsumura, F., Patil, K. C., and Boush, G. M., Formation of photodieldrin by microorganisms, *Science*, 170, 1206, 1970.

571. Baldwin, M. K. and Robinson, J., Metabolism in the rat of the photoisomerisation product of dieldrin, *Nature (Lond.)*, 224, 283, 1969.

572. Khan, M. A. Q., Sutherland, D. J., Rosen, J. D., and Carey, W. F., Effect of sesamex on the toxicity and metabolism of cyclodienes and their photoisomers in the housefly, *J. Econ. Entomol.*, 63, 470, 1970.

573. Matthews, H. B. and Matsumura, F., Metabolic fate of dieldrin in the rat, *J. Agric. Food Chem.*, 17, 845, 1969.

574. Klein, A. K., Dailey, R. E., Walton, M. S., Beck, V., and Link, J. D., Metabolites isolated from urine of rats fed ¹⁴C-photodieldrin, *J. Agric. Food Chem.*, 18, 705, 1970.

575. Dailey, R. E., Klein, A. K., Brouwer, E., Link, J. D., and Braunberg, R. C., Effect of testosterone on metabolism of ¹⁴C-photodieldrin in normal, castrated and oophorectomised rats, *J. Agric. Food Chem.*, 20, 371, 1972.

576. El Zorgani, G. A., Walker, C. H., and Hassall, K. A., Species differences in the *in vitro* metabolism of HEOM, a chlorinated cyclodiene epoxide, *Life Sci.*, [Oxford], 9. (Part II), 415, 1970.

576a. Robinson, J. and Brown, V. K. H., Pharmacodynamics of dieldrin in pigeons *Nature (Lond.)*, 213, 734, 1967.

577. Korte, F., Ludwig, G., and Vogel, J., Umwandlung von Aldrin-¹⁴C and Dieldrin-¹⁴C durch mikroorganismen, leberhomogenate und moskito-larven, *Justus Liebigs Ann. Chem.*, 656, 135, 1962.

578. Matsumura, F., Boush, G. M., and Tai, A., Breakdown of dieldrin by a soil microorganism, *Nature (Lond.)*, 219, 965, 1968.

578a. Khan, M. A. Q., Khamal, A., Wolin, R. J., and Runnels, J., *In vivo* and *in vitro* epoxidation of aldrin by aquatic food chain organisms, *Bull. Environ. Contam. Toxicol.*, 8, 219, 1972.

579. Mehendale, H. M., Skrentny, R. F., and Dorough, H. W., Oxidative metabolism of aldrin by subcellular root fractions of several plant species, *J. Agric. Food Chem.*, 20, 398, 1972.

580. Giannotti, O., Estudos sobre o mecanismo de acao do aldrin, dieldrin e endrin em *Periplaneta americana* [L.]. *Arq. Inst. Biol. Sao Paulo*, 25, 253, 1958.

581. Brooks, G. T., Mechanisms of resistance of the adult housefly [*Musca domestica*] to 'cyclodiene' insecticides, *Nature (Lond.)*, 186, 96, 1960.

582. Richardson, A., Robinson, J., and Baldwin, M. K., Metabolism of endrin in the rat, *Chem. Ind. (Lond.)*, 502, 1970.

583. Pimentel, D., *Ecological Effects of Pesticides on Non-target Species*, Office of Science and Technology, U.S. Govt. Print. Off., Washington, D.C., June, 1972.

584. Brooks, G. T. and Harrison, A., Relations between structure, metabolism and toxicity of the cyclodiene insecticides, *Nature (Lond.)*, 198, 1169, 1963.

585. Schupfan, I., Sajko, B., and Ballschmiter, K., The chemical and photochemical degradation of the cyclodiene insecticides aldrin, dieldrin, endosulfan and other hexachloronorbornene derivatives, *Z. Naturforsch. Teil B*, 27b, 147, 1972.

586. Schupfan, I. and Ballschmiter, K., Metabolism of polychlorinated norbornenes by *Clostridium butyricum*, *Nature (Lond.)*, 237, 100, 1972.

587. Parke, D. V., *The Biochemistry of Foreign Compounds*, Pergamon Press, Oxford, 1968.

588. Estabrook, R. W., Baron, J., and Hildebrandt, A., A new spectral species associated with cytochrome $P_{450}$ in liver microsomes, *Chem.-Biol. Interactions*, 3, 260, 1971.

589. Tsukamoto, M., Metabolic fate of DDT in *Drosophila melanogaster;* identification of a non-DDE metabolite, *Botyu-kagaku*, 24, 141, 1959.

590. Sun, Y. P. and Johnson, E. R., Synergistic and antagonistic actions of insecticide-synergist conbinations and their mode of action, *J. Agric. Food Chem.*, 8, 261, 1960.

591. Ray, J. W., Insect microsomal cytochromes, in *Pest Infestation Research 1965*, Agricultural Research Council, H.M.S.O., London, 1966, 59.

592. Lewis, S. E., Effect of carbon monoxide on metabolism of insecticides *in vivo*, *Nature (Lond.)*, 215, 1408, 1967.

593. Hodgson, E. and Plapp, F. W., Jr., Biochemical characteristics of insect microsomes, *J. Agric. Food Chem.*, 18, 1048, 1970.

594. Casida, J. E., Mixed-function oxidase involvement in the biochemistry of insecticide synergists, *J. Agric. Food Chem.*, 18, 753, 1970.

595. Krieger, R. I. and Wilkinson, C. F., Localisation and properties of an enzyme system effecting aldrin epoxidation in larvae of the southern armyworm [*Prodenia eridania*], *Biochem. Pharmacol.*, 18, 1403, 1969.

596. Krieger, R. I., Feeny, P. P., and Wilkinson, C. F., Detoxication enzymes in the guts of caterpillars: an evolutionary answer to plant defenses? *Science*, 172, 579, 1971.

597. Wilkinson, C. F. and Brattsten, L. B., Microsomal drug metabolizing enzymes in insects, *Drug Metab. Rev.*, 1, 153, 172. 1972.

598. Evans, W. C., Smith, B. S. W., Moss, P., and Fernley, H. N., Bacterial metabolism of 4-chlorophenoxyacetate, *Biochem. J.*, 122, 509, 1971, et. seq.

599. Rowlands, D. G., The metabolism of contact insecticides in stored grain, *Residue Rev.*, 17, 105, 1967.

600. Glass, B. L., Relation between the degradation of DDT and the iron redox system in soils, *J. Agric. Food Chem.*, 20, 324, 1972.

601. Brooks, G. T., Progress in metabolic studies of the cyclodiene insecticides and its relevance to structure-activity correlations, *World Rev. Pest Control*, 5, 62, 1966.

602. Oesch, F., Jerina, D. M., and Daly, J. W., Substrate specificity of hepatic epoxide hydrase in microsomes and in a purified preparation: evidence for homologous enzymes, *Arch. Biochem. Biophys.*, 144, 253, 1971.

603. Boyland, E., in *Special Publication No. 5*, Williams, R. T., Ed., The Biochemical Society, London, 1950, 40.

604. Oesch, F., Kaubisch, N., Jerina, D. M., and Daly, J. W., Hepatic epoxide hydrase. Structure-activity relationships for substrates and inhibitors, *Biochemistry*, 10, 4858, 1971.

605. Boyland, E. and Chasseaud, L. F., Glutathione S-aralkyltransferase, *Biochem. J.*, 115, 985. 1969.

606. Schonbrod, R. D., Khan, M. A. Q., Terriere, L. C., and Plapp, F. W., Jr., Microsomal oxidases in the housefly: a survey of fourteen strains, *Life Sci.* (Oxford), 7 (Part 1), 681, 1968.

607. Khan, M. A. Q., Some biochemical characteristics of the microsomal cyclodiene epoxidase system and its inheritance in the housefly, *J. Econ. Entomol.*, 62, 388, 1969.

608. Matthews, H. B. and Casida, J. E., Properties of housefly microsomal cytochromes in relation to sex, strain, substrate specificity, and apparent inhibition and induction by synergist and insecticide chemicals, *Life Sci.* (Oxford), 9 (Part 1), 989, 1970.

609. Ishida, M., Comparative studies on BHC metabolising enzymes, DDT dehydrochlorinase and glutathione S-transferases, *Agric. Biol. Chem. (Tokyo)*, 32, 947, 1968.

610. Goodchild, B. and Smith, J. N., The separation of multiple forms of housefly 1,1,1-trichloro-2,2-bis-(p-chlorophenyl)ethane (DDT) dehydrochlorinase from glutathione S-aryltransferase by electrofocusing and electrophoresis, *Biochem. J.*, 117, 1005, 1970.

611. Dinamarca, M. L., Levenbook, L., and Valdes, E., DDT-dehydrochlorinase II. Subunits, sulfhydryl groups, and chemical composition, *Arch. Biochem. Biophys.*, 147, 374, 1971.

612. Sternburg, J., Kearns, C. W., and Moorefield, H., DDT-dehydrochlorinase, an enzyme found in DDT-resistant flies, *J. Agric. Food Chem.*, 2, 1125, 1954.

613. Lipke, H. and Kearns, C. W., DDT-dehydrochlorinase, in *Advances in Pest Control Research*, Vol. 3, Metcalf, R. L., Ed., Interscience, New York, 1960, 253.

614. Mullins, L. J., The structure of nerve cell membranes, in *Molecular Structure and Functional Activity of Nerve Cells*, Grenell, R. G. and Mullins, L. J., Eds., Publication No. 1, American Institute of Biological Sciences, Washington, D.C., 1956.

615. Hennessy, D. J., Fratantoni, J., Hartigan, J., Moorefield, H. H., and Weiden, M. H. J., Toxicity of 2-(2-halogen-4-chlorophenyl)-2-(4-chlorophenyl)-1,1,1-trichloroethanes to normal and to DDT-resistant houseflies, *Nature (Lond.)*, 190, 341, 1961.

616. Kimura, T., Duffy, J. R., and Brown, A. W. A., Dehydrochlorination and DDT-resistance in Culex mosquitos, *Bull. W.H.O.*, 32, 557, 1965.

617. Pillai, M. K. K., Hennessy, D. J., and Brown, A. W. A., Deuterated analogues as remedial insecticides against DDT-resistant *Aedes aegypti, Mosquito News*, 23, 118, 1963.

618. Pillai, M. K. K. and Brown, A. W. A., Physiological and genetical studies on resistance to DDT substitutes in *Aedes aegypti, J. Econ. Entomol.*, 58, 255, 1965.

619. Busvine, J. R. and Townsend, M. G., The significance of BHC degradation in resistant houseflies, *Bull. Entomol. Res.*, 53, 763, 1963.

620. Decker, G. C. and Bruce, W. N., Housefly resistance to chemicals, *Am. J. Trop. Med. Hyg.*, 1, 395, 1952.

621. Bridges, R. G. and Cox, J. T., Resistance of houseflies to γ-benzene hexachloride and dieldrin, *Nature (Lond.)*, 184, 1740, 1959.

622. Balabaskeran, S. and Smith, J. N., The inhibition of 1,1,1-trichloro-2,2-bis(p-chlorophenyl)ethane(DDT)dehydrochlorinase and glutathione S-aryltransferase in grass-grub and housefly preparations, *Biochem. J.*, 117, 989, 1970.

623. Brooks, G. T. and Harrison, A., The oxidative metabolism of aldrin and dihydroaldrin by houseflies, housefly microsomes and pig liver microsomes and the effect of inhibitors, *Biochem. Pharmacol.*, 18, 557, 1969.

624. Gillette, J. R., Conney, A. H., Cosmides, G. J., Estabrook, R. W., Fouts, J., and Mannering, G. J., Eds., *Microsomes and Drug Oxidations*, Academic Press, New York, London, 1969.

625. Gillett, J. W. and Chan, T. M., Cyclodiene insecticides as inducers, substrates and inhibitors of microsomal epoxidation, *J. Agric. Food Chem.*, 16, 590, 1968.

626. Kinoshita, F. K., Frawley, J. P., and DuBois, K. P., Quantitative measurement of induction of hepatic microsomal enzymes by various dietary levels of DDT and toxaphene in rats, *Toxicol. Appl. Pharmacol.*, 9, 505, 1966.

627. Poland, A., Smith, D., Kuntzman, R., Jacobson, M., and Conney, A. H., Effect of intensive occupational exposure to DDT on phenylbutazone and cortisol metabolism in human subjects, *Clin. Pharmacol. Ther.*, 11, 724, 1970.

628. Cram, R. L. and Fouts, J. R., The influence of DDT and γ-chlordane on the metabolism of hexobarbital and zoxazolamine in two mouse strains, *Biochem. Pharmacol.*, 16, 1001, 1967.

629. Stephen, B. J., Gerlich, J. D., and Guthrie, F. E., Effect of DDT on induction of microsomal enzymes and deposition of calcium in the domestic hen, *Bull. Environ. Contam. Toxicol.*, 5, 569, 1971.

630. Alary, J. G. and Brodeur, J., Studies on the mechanism of phenobarbital-induced protection against parathion in adult female rats, *J. Pharmacol. Exp. Ther.*, 169, 159, 1969.

631. Triolo, A. J., Mata, E., and Coon, J. M., Effects of organochlorine insecticides on the toxicity and *in vitro* plasma detoxication of paraoxon, *Toxicol. Appl. Pharmacol.*, 17, 174, 1970.

632. Chapman, S. K. and Leibman, K. C., The effect of chlordane, DDT and 3-methylcholanthrene upon the metabolism and toxicity of diethyl-4-nitrophenyl phosphorothionate [parathion], *Toxicol. Appl. Pharmacol.*, 18, 977, 1971.

633. Wright, A. S., Potter, D., Wooder, M. F., Donninger, C., and Greenland, R. D., The effects of dieldrin on the subcellular structure and function of mammalian liver cells, *Food Cosmet. Toxicol.*, 10, 311, 1972.

634. Gillett, J. W., Chan, T. M., and Terriere, L. C., Interactions between DDT analogues and microsomal epoxidase systems, *J. Agric. Food Chem.*, 14, 540, 1966.

635. Street, J. C., Modification of animal responses to toxicants, in *Enzymatic Oxidations of Toxicants*, Hodgson, E., Ed., North Carolina State University, 1968, 197.

636. Street, J. C., Urry, F. M., Wagstaff, D. J., and Blau, S. E., Induction by different inducers: structure-activity relationships among DDT analogues, in *Proc. 2nd Int. IUPAC Congress of Pesticide Chemistry*, Tahori, A. S., Ed., Gordon Breach, London, 1972.

637. Abernathy, C. O., Hodgson, E., and Guthrie, F. E., Structure-activity relations on the induction of hepatic microsomal enzymes in the mouse by 1,1,1-trichloro-2,2-bis(p-chlorophenyl)ethane(DDT) analogues, *Biochem. Pharmacol.*, 20, 2385, 1971.

638. Chadwick, R. W., Cranmer, M. F., and Peoples, A. J., Comparative stimulation of $\gamma$-HCH metabolism by pretreatment of rats with $\gamma$-HCH, DDT, and DDT plus $\gamma$-HCH, *Toxicol. Appl. Pharmacol.*, 18, 685, 1971.

639. Mayer, F. L., Jr., Street, J. C., and Neuhold, J. M., Organochlorine insecticide interactions affecting residue storage in rainbow trout, *Bull. Environ. Contam. Toxicol.*, 5, 300, 1970.

640. Kuntzman, R., Drugs and enzyme induction, *Ann. Rev. Pharmacol.*, 9, 21, 1969.

641. Welch, R. M., Levin, W., and Conney, A. H., Insecticide inhibition and stimulation of steroid hydroxylases in rat liver, *J. Pharmacol. Exp. Ther.*, 155, 167, 1967.

642. Bitman, J., Cecil, H. C., Harris, S. J., and Fries, G. F., Estrogenic activity of o,p'-DDT in the mammalian uterus and avian oviduct, *Science*, 162, 371, 1968.

643. Bitman, J. and Cecil, H. C., Estrogenic activity of DDT analogs and polychlorinated biphenyls, *J. Agric. Food Chem.*, 18, 1108, 1970.

644. Bitman, J., Cecil, H. C., Harris, S. J., and Fries, E. F., DDT induces a decrease in eggshell calcium, *Nature (Lond.)*, 224, 44, 1969.

645. Heath, R. G., Spann, J. W., and Kreitzer, J. F., Marked DDE impairment of Mallard reproduction in controlled studies, *Nature (Lond.)*, 224, 47, 1969.

646. Porter, R. D. and Wiemeyer, S. N., Dieldrin and DDT: effects on Sparrow Hawk eggshells and reproduction, *Science*, 165, 199, 1969.

647. Robinson, J., Birds and pest control chemicals, *Bird Study*, 17, 195, 1970.

648. Sun, Y. P., Correlation between laboratory and field data on testing insecticides, *J. Econ. Entomol.*, 59, 1131, 1966.

649. Sternburg, J. and Kearns, C. W., Metabolic fate of DDT when applied to certain naturally tolerant insects, *J. Econ. Entomol.*, 45, 497, 1952.

650. Brooks, G. T., Harrison, A., and Power, S. V., in Pest Infestation Research 1964, Annual Report of the Pest Infestation Laboratory, Agricultural Research Council, HMSO London, 1965, 58.

650a. Sagar, W. C., Monroe, R. E., and Zabik, M. J., Syntheses and resolution of optically active DDT analogs and their toxicity to the housefly, *Musca domestica* L., *J. Agric. Food Chem.*, 20, 1176, 1972.

651. Holan, G., Rational design of degradable insecticides, *Nature (Lond.)*, 232, 644, 1971.

652. Holan, G., Rational design of insecticides, *Bull. W.H.O.*, 44, 355, 1971.

652a. Hirwe, A. S., Metcalf, R. L., and Kapoor, I. P., $\alpha$-Trichloromethylbenzylanilines and $\alpha$-trichloromethylbenzyl phenyl ethers with DDT-like insecticidal action, *J. Agric. Food Chem.*, 20, 818, 1972.

653. Metcalf, R. L. and Georghiou, G. P., Cross tolerances of dieldrin-resistant flies and mosquitos to various cyclodiene insecticides, *Bull. W.H.O.*, 27, 251, 1962.

654. Brooks, G. T., Harrison, A., and Cox, J. T., Significance of the epoxidation of the isomeric insecticides aldrin and isodrin by the adult housefly *in vivo*, *Nature (Lond.)*, 197, 311, 1963.

655. Brooks, G. T., The design of insecticidal chlorohydrocarbon derivatives, in *Drug Design*, Vol. IV, Ariens, E. J., Ed., Academic Press, New York, 1973, 379.

656. Wang, C. M., Narahashi, T., and Yamada, M., The neurotoxic action of dieldrin and its derivatives in the cockroach, *Pestic. Biochem. Physiol.*, 1, 84, 1971.

657. Kurihara, N., personal communication, 1973.

658. Spector, W. S., Ed., *Handbook of Toxicology*, Vol. 1, NAS-NRS Wright Air Development Center Tech. Rep. 55-16, 1955, 408.

658a. Deichman, W. B., The debate on DDT, *Arch. Toxikol.*, 29, 1, 1972.

659. Rogers, A. J., Eagles, affluence and pesticides, *Mosquito News*, 32, 151, 1972.

660. Robinson, J., Organochlorine insecticides and bird populations in Britain, in *Chemical Fallout*, Miller, M. W., and Berg, G. C., Eds., Charles C Thomas, Springfield, Illinois, 1969, 113.

661. Gaines, T. B., Acute toxicity of pesticides, *Toxicol. Appl. Pharmacol.*, 14, 515, 1969.

662. *Summary of Toxicology of Aldrin and Dieldrin*, Shell International Chemical Company, Toxicology Division, 1971.

663. Deichmann, W. B., MacDonald, W. E., Blum, E., Bevilacqua, M., Radomski, J., Keplinger, M., and Balkus, M., Tumorigenicity of aldrin, dieldrin and endrin in the albino rat, *Ind. Med.*, 39, 37, 1970.

664. Alabaster, J. S., Evaluating Risks of Pesticides to Fish, Proc. 5th Br. Insectic. Fungic. Conf. Vol. 2, 1969, 370.

665. Hercules Incorporated, (Agricultural Chemicals Department), *Hercules Toxaphene, Summary of Toxicological Investigations,* Bulletin T-105, 1962.

666. Winteringham, F. P. W. and Barnes, J. M., Comparative response of insects and mammals to certain halogenated hydrocarbons used as insecticides, *Physiol. Rev.,* 35, 701, 1955.

667. Dale, W. E., Gaines, T. B., Hayes, W. J., Jr., and Pearce, G. W., Poisoning by DDT: relation between clinical signs and concentration in rat brain, *Science,* 142, 1474, 1963.

668. Yeager, J. F. and Munson, S. C., Site of action of DDT in Periplaneta, *Science.* 102. 305, 1945.

669. Hoffman, R. A. and Lindquist, A. W., Temperature coefficients for chlorinated hydrocarbons, *J. Econ. Entomol.,* 42, 891, 1949.

670. Guthrie, F. E., Holding temperature and insecticidal mortality; Blattella, *J. Econ. Entomol.,* 43, 559, 1950.

671. Holan, G., New halocyclopropane insecticides and the mode of action of DDT, *Nature (Lond.),* 221, 1025, 1969.

672. Eaton, J. L. and Sternburg, J. G., Temperature effects on nerve activity in DDT-treated American cockroaches, *J. Econ. Entomol.,* 60, 1358, 1967.

673. Eaton, J. L. and Sternburg, J. G., Uptake of DDT by the American cockroach central nervous system, *J. Econ. Entomol.,* 60, 1699, 1967.

674. Sternburg, J. and Kearns, C. W., The presence of toxins other than DDT in the blood of DDT-poisoned roaches, *Science,* 116, 144, 1952.

675. Casida, J. E. and Maddrell, S. H. P., Diuretic hormone release on poisoning Rhodnius with insecticide chemicals, *Pestic. Biochem. Physiol.,* 1, 71, 1971.

676. Sternburg, J. G. and Hewitt, P., In vivo protection of cholinesterase against inhibition by TEPP and its methyl homologue by prior treatment with DDT, *J. Insect Physiol.,* 8, 643, 1962.

677. Winteringham, F. P. W., Hellyer, G. C., and McKay, M. A., Effects of the insecticides DDT and dieldrin on phosphorus metabolism of the adult housefly *Musca domestica* L., *Biochem. J.,* 76, 543, 1960.

678. Matsumura, F. and O'Brien, R. D., Interactions of DDT with components of American cockroach nerve, *J. Agric. Food Chem.,* 14, 39, 1966.

679. Matsumura, F. and Patil, K. C., Adenosine triphosphatase sensitive to DDT in synapses of rat brain, *Science,* 166, 121, 1969.

680. Koch, R. B., Cutkomp, L. K., and Do, F. M., Chlorinated hydrocarbon insecticide inhibition of cockroach and honeybee ATPase, *Life Sci.* (Oxford), 8, 289, 1969.

681. Weiss, D. E., A molecular mechanism for the permeability changes in nerve during the passage of an action potential, *Aust. J. Biol. Sci.,* 22, 1355, 1969.

682. Martin, H. and Wain, R. L., Properties versus toxicity of DDT analogues, *Nature (Lond.),* 154, 512, 1944.

683. Riemschneider, R., Chemical structure and activity of DDT analogs, with special consideration of their spatial structures, in *Advances in Pest Control Research,* Vol. 2, Metcalf, R. L., Ed., Interscience, New York, 1958, 307.

684. Gunther, F. A., Blinn, R. C., Carman, G. E., and Metcalf, R. L., Mechanisms of insecticidal action. The structural topography theory and DDT-type compounds, *Arch. Biochem. Biophys.,* 50, 504, 1954.

685. Mullins, L. J., Structure-toxicity in hexachlorocyclohexane isomers, *Science,* 122, 118, 1955.

685a. Fahmy, M. A. H., Fukuto, T. R., Metcalf, R. L., and Holmstead, R. L., Structure-activity correlations in DDT analogs, *J. Agric. Food Chem.,* 21, 585, 1973.

686. Wilson, W. E., Fishbein, L., and Clements, S. T., DDT; participation in ultraviolet-detectable, charge-transfer complexation, *Science,* 171, 180, 1971.

687. O'Brien, R. D. and Matsumura, F., DDT; A new hypothesis of its mode of action, *Science,* 146, 657, 1964.

688. Baranyovits, F. L. C., cited in Winteringham, F. P. W. and Barnes, J. M., Comparative response of insects and mammals to certain halogenated hydrocarbons used as insecticides, *Physiol. Rev.,* 35, 701, 1955.

689. McNamara, B. P. and Krop, S., Pharmacology of BHC isomers, *J. Pharmacol. Exp. Ther.,* 92, 140, 1948.

690. Nakajima, E., Shindo, H., and Kurihara, N., Whole body autoradiographic studies in the distribution of a-, β- and γ-BHC in mice, *Radioisotopes,* 19, 532, 1970.

691. Telford, J. N. and Matsumura, F., Electron microscopic and autoradiographic studies on distribution of dieldrin in the intact nerve tissues of German cockroaches, *J. Econ. Entomol.,* 64, 230, 1971.

692. Cherkin, A., Mechanisms of general anaesthesia by non-hydrogen-bonding molecules, *Ann. Rev. Pharmacol,* 9, 259, 1969.

693. Wang, C. M., Narahashi, T., and Yamada, M., The neurotoxic action of dieldrin and its derivatives in the cockroach, *Pestic. Biochem. Physiol.,* 1, 84, 1971.

694. Ryan, W. H. and Shankland, D. L., Synergistic action of cyclodiene insecticides with DDT on the membrane of giant axons of the American cockroach, *Periplaneta americana, Life Sci. (Oxford),* 10, 193, 1971.

694a. Shankland, D. L. and Schroeder, M. E., Pharmacological evidence for a discrete neurotoxic action of dieldrin (HEOD) in the American cockroach, *Pestic. Biochem. Physiol.,* 3, 77, 1973.

695. Matsumura, F., Studies on the Biochemical Mechanisms of Resistance in the German Cockroach Strains, paper presented at the 2nd International Congress of Pesticide Chemistry, Tel Aviv, 1971.

696. Narahashi, T., personal communication, 1971.

697. Brooks, G. T. and Harrison, A., The metabolism of some cyclodiene insecticides in relation to dieldrin resistance in the adult housefly, *Musca domestica* L., *J. Insect Physiol.*, 10, 633, 1964.

698. Sellers, L. G. and Guthrie, F. E., Distribution and metabolism of $^{14}$C-dieldrin in the resistant and susceptible housefly, *J. Econ. Entomol.*, 65, 378, 1972.

699. Colhoun, E. H., Approaches to mechanisms of insecticidal action, *J. Agric. Food Chem.*, 8, 252, 1960.

700. Yamasaki, T. and Narahashi, T., Resistance of houseflies to insecticides and the susceptibility of nerve to insecticides. Studies on the mechanism of action of insecticides XVII, *Botyu-kagaku*, 23, 146, 1958.

701. Sun, Yun-Pei, Schaefer, C. H., and Johnson, E. R., Effects of application methods on the toxicity and distribution of dieldrin in houseflies, *J. Econ. Entomol.*, 60, 1033, 1967.

702. Ray, J. W., Insecticide absorbed by the central nervous system of susceptible and resistant cockroaches exposed to dieldrin, *Nature (Lond.)*, 197, 1226, 1963.

703. Matsumura, F. and Hayashi, M., Dieldrin resistance; biochemical mechanisms in the German cockroach, *J. Agric. Food Chem.*, 17, 231, 1969.

704. Hayashi, M. and Matsumura, F., Insecticide mode of action: effect of dieldrin on ion movement in the nervous system of *Periplaneta americana* and *Blattella germanica* cockroaches, *J. Agric. Food Chem.*, 16, 622, 1967.

705. Welsh, J. H. and Gordon, H. T., The mode of action of certain insecticides on the arthropod nerve axon, *J. Cell. Comp. Physiol.*, 30, 147, 1947.

706. Hodge, H. C., Boyce, A. M., Deichmann, W. B., and Kraybill, H. F., Toxicology and no-effect levels of aldrin and dieldrin, *Toxicol. Appl. Pharmacol.*, 10, 613, 1967.

707. Hathway, D. E. and Mallinson, A., Effect of Telodrin on the liberation and utilisation of ammonia in rat brain, *Biochem. J.*, 90, 51, 1964.

708. Hathway, D. E., Mallinson, A., and Akintonwa, D. A. A., Effects of dieldrin, picrotoxin and Telodrin on the metabolism of ammonia in brain, *Biochem. J.*, 94, 676, 1965.

709. Hathway, D. E., The biochemistry of Telodrin and dieldrin, *Arch. Environ. Health*, 11, 380, 1965.

709a. Markin, G. P., Ford, J. H., and Hawthorne, J. C., Mirex residues in wild populations of the edible red crawfish (*Procambarus Clarki*), *Bull. Environ. Contam. Toxicol.*, 8, 369, 1972.

709b. Gibson, J. R., Ivie, G. W., and Dorough, H. W., Fate of mirex and its major photodecomposition product in rats, *J. Agric. Food Chem.*, 20, 1246, 1972.

709c. Mehendale, H. M., Fishbein, L., Fields, M., and Matthews, H. B., Fate of mirex-$^{14}$C in the rat and in plants, *Bull. Environ. Contam. Toxicol.*, 8, 200, 1972.

710. Pocker, Y., Beug, W. M., and Ainardi, V. R., Carbonic anhydrase interaction with DDT, DDE and Dieldrin, *Science*, 174, 1336, 1971.

711. Gair, R., Agricultural Development and Advisory Service, Cambridge, U.K., cited in Hessayon, D. G., Tennant Memorial Lecture – *Homo Sapiens*, the species the conservationist forgot, *Chem. Ind. (Lond.)*, 407, 1972.

712. Sherma, J. and Zweig, G., *Paper Chromatography: Paper Chromatography and Electrophoresis*, Vol. 2, Zweig, G. and Whitaker, J. R., Eds., Academic Press, New York, 1971.

713. DeLacy, T. P. and Kennard, C. H. L., Crystal structure of 1,1-bis(p-chlorophenyl)-2,2-dichloropropane and 1,1-bis(p-ethoxyphenyl)-2,2-dimethylpropane, *J. Chem. Soc. (Lond.)*, Perkin II, 2141, 1972.

713a. DeLacy, T. P. and Kennard, C. H. L., Crystal structures of 1,1-bis(p-chlorophenyl-2,2,2-trichloroethane (p,p'-DDT) and 1-(o-chlorophenyl)-1-(p-chlorophenyl)-2,2,2-trichloroethane (o,p'-DDT), *J. Chem. Soc. (Lond.)*, Perkin II, 2148, 1972.

714. Hoskins, W. M. and Gordon, H. T., Arthropod resistance to chemicals, *Ann. Rev. Entomol.*, 1, 89, 1956.

715. Hill, R. L. and Teipel, J. W., Fumarase and Crotonase, in *The Enzymes*, 3rd ed., Boyer, P. D., Ed., Academic Press, New York, London, 1971, 539.

716. Moriarty, F., The sublethal effects of synthetic insecticides on insects, *Biol. Rev.*, 44, 321, 1969.

717. Kagan, Yu. S., Fudel-Ossipova, S. I., Khaikina, B. J., Kuzminskaya, U. A., and Kouton, S. D., On the problem of the harmful effect of DDT and its mechanism of action, *Residue Rev.*, 27, 43, 1969.

718. Davies, J. E., Edmundson, W. F., Eds., Community Studies on Pesticides, Dade County, Florida: *Epidemiology of DDT*, Futura Publishing, Mount Kisco, New York, 1972, Chap. 10, 11, 13.

719. Lykken, L., Relation of U.S. Food Production to Pesticides, paper presented at the Symposium on Nutrition and Public Policy in the U.S., 163rd American Chemical Society National Meeting, Boston, April 1972.

720. Van Tiel, N., Pesticides in environment and food, in *Environmental Quality and Safety*, Vol. I, Coulston, F. and Korte, F., Eds., Georg Thieme, Stuttgart, Academic Press, New York, 1972, 180.

721. Hurtig, H., A commentary on pesticides and perspective, *Chem. Ind. (Lond.)*, 888, 1969.

722. Polen, P. B., Fate of insecticidal chlorinated hydrocarbons in storage and processing of foods, in *Pesticide Terminal Residues*, Invited papers from the IUPAC International Symposium on Terminal Residues, Tel Aviv, 1971, Tahori, A. S., Ed., Butterworth's London, 1971, 137.

723. Kenaga, E. E., Factors related to the bioconcentration of pesticides, in *Environmental Toxicology of Pesticides*, Matsumura, F., Boush, G. M., and Misato, T., Eds., Academic Press, New York and London, 1972, 193.

724. Robinson, J., Richardson, A., Crabtree, A. N., Coulson, J. C., and Potts, G. R., Organochlorine residues in marine organisms, *Nature (Lond.)*, 214, 1307, 1967.

725. **Hurtig, H.,** Significance of conversion products and metabolites of pesticides in the environment, in *Environmental Quality and Safety,* Vol. I, Coulston, F. and Korte, F., Eds., Georg Thieme, Stuttgart, Academic Press, New York, *1972, 58.*

726. **Walker, C. H.,** *Environmental Pollution by Chemicals,* Hutchinson Educational, London, 1971, Chap. V.

727. **Robinson, J.,** Residues of organochlorine insecticides in dead birds in the United Kingdom, *Chem. Ind. (Lond.),* 1974, 1967.

728. **Rudd, R. L. and Herman, S. G.,** Ecosystemic transferal of pesticides residues in an aquatic environment, in *Environmental Toxicology of Pesticides,* Matsumura, F., Boush, G. M., and Misato, T., Eds., Academic Press, New York and London, 1972, 471.

729. **Rudd, R. L.,** *Pesticides and the Living Landscape,* Faber and Faber, London, 1965, 41.

# SYSTEMATIC NAME INDEX

## A

*Acris creptans*, 49
*Acris gryllus*, 49
*Aedes aegypti*, 8, 20, 21, 22, 23, 25, 41, 42, 43, 59, 88, 89, 103, 104, 123
*Aedes melanimon*, 40
*Aedes nigromaculis*, 40
*Aedes sollicitans*, 40
*Aedes taeniorynchus*, 22, 40
*Aeneolamia varia*, 51, 52
*Aerobacter aerogenes*, 80
*Agrotis ypsilon*, 51, 56
*Alabama argillacea*, 51
*Amblyomma americanum*, 46
*Anopheles albimanus*, 19, 40, 60, 123
*Anopheles albitarsis*, 39
*Anopheles aquasalis*, 39
*Anopheles aztecus*, 39
*Anopheles cruzii*, 40
*Anopheles culicifacies*, 38, 61
*Anopheles darlingi*, 40
*Anopheles funestus*, 19
*Anopheles gambiae*, 8, 9, 19, 20, 30, 38, 39, 58, 59, 60
*Anopheles maculatus*, 41
*Anopheles maculipennis*, 37
*Anopheles nuneztovari*, 40
*Anopheles pharoensis*, 19, 20, 37
*Anopheles pseudopunctipennis*, 19, 37, 39
*Anopheles punctimacula*, 40
*Anopheles quadrimaculatus*, 19, 29, 39, 123
*Anopheles sacharovi*, 19, 29, 37, 38
*Anopheles sp.*, 8, 9, 61
*Anopheles stephensi*, 8, 19, 20
*Anopheles, sundaicus*, 19, 61
*Anthonomus grandis*, 30, 51, 52, 61
*Aphis gossypii*, 51
*Apis mellifera*, 49
*Argyrotaenia velutinana*, 30, 50
*Asillus sp.*, 89
*Aspergillus flavus*, 95

## B

*Bacillus sp.*, 92
*Blattella germanica*, 8, 23, 45
*Blissus pulchellus*, 51
*Boophilus decoloratus*, 46, 47
*Boophilus microplus*, 8, 47, 49, 61
*Boophilus sp.*, 8
*Bracon mellitor*, 49
*Bacculatrix thurberiella*, 51

## C

*Calliphora erythrocephala*, 36, 100
*Callitroga macellaria*, 36
*Carassius auratus*, 89

*Carpocapsa pomonella*, 30, 50, 52
*Caryedon serratus*, 57
*Chaoborus astictopus*, 36
*Chilo suppressalis*, 51, 52, 53, 54
*Chironomus zealandicus*, 36
*Chlorella sp.*, 89
*Chlorops oryzae*, 53
*Christoneura fumiferana*, 51, 52
*Chrysomyia putoria*, 36
*Cimex hemipterus*, 44, 45
*Cimex lectularius*, 44, 45, 123
*Cimex sp.*, 8
*Clostridium butyricum*, 96
*Columbia livia*, 126
*Conoderus fallii*, 30, 52
*Conoderus vespertinus*, 52
*Costelytra zealandica*, 50, 81, 105
*Cottus perplexus*, 150
*Ctenocephalides canis*, 44
*Ctenocephalides felis*, 44
*Culex fatigans*, 8, 10, 20, 22, 23, 40, 41, 60, 104, 123
*Culex pipiens fatigans*, 20
*Culex pipiens*, 25, 41
*Culex pipiens molestus* (Culex molestus), 29, 41
*Culex pipiens pallens*, 121
*Culex quinquefasciatus*, 41
*Culex tarsalis*, 21, 40, 103, 104
*Culicoides furens*, 36
*Culicoides sp.*, 36
*Cyclops sp.*, 89
*Cyclotella nana*, 73
*Cydia pomonella*, 61
*Cyprinus carpio*, 89

## D

*Daphnea sp.*, 89
*Delia antiqua*, 56
*Delia platura*, 56
*Delphacodes striatella*, 51
*Dermestes maculatus*, 58
*Diabrotica balteata*, 52
*Diabrotica longicornis*, 52
*Diabrotica undecimpunctata*, 52
*Diabrotica vergifera*, 52
*Diatraea saccharalis*, 52
*Distantiella theobroma*, 52
*Drosophila melanogaster*, 8, 23, 26, 36, 58, 97
*Drosophila sp.*, 23
*Drosophila virilis*, 36
*Dysdercus peruvianus*, 51

## E

*Earias insulana*, 51
*Echinocystis macrocarpa*, 99
*Ephestia cautella*, 58
*Epilachna varivestis*, 101

*Epiphyas postvittana*, 30, 50
*Epitrix cucumeris*, 50
*Epitrix tuberis*, 50
*Erioischia brassicae*, 56
*Eriosoma lanigerum*, 51
*Erythroneura lawsoniana*, 50
*Erythroneura variabilis*, 50
*Esox Lucius*, 89
*Estigmene acraea*, 51
*Euxesta notata*, 8, 50, 52, 56
*Euxoa detersa*, 51
*Euxoa messoria*, 51

## F

*Fannia canicularis*, 36

## G

*Galerucella birmanica*, 51
*Galleria mellonella*, 26
*Gambusia affinis*, 49, 77
*Glossina morsitans*, 36
*Glossina* sp., 36
*Glyptotendipes paripes*, 36
*Graphognathus leucoloma*, 52
*Grapholitha molesta*, 50

## H

*Haematobia irritans*, 36
*Heliothis virescens*, 50, 51, 53
*Heliothis zea*, 50, 52, 53
*Heptagenia hebe*, 49
*Hermetia illucens*, 57
*Hippelates collusor*, 36
*Hydrella sasakii*, 53
*Hydrogenomonas* sp., 80
*Hylemia antiqua*, 24, 51, 55
*Hylemia arambourgi*, 52
*Hylemia brassicae*, 24, 51, 56
*Hylemia floralis*, 52
*Hylemia liturata*, 52, 55
*Hylemia platura*, 52
*Hylemia* sp., 8
*Hypera postica*, 52

## I

*Ixodes persulcatus*, 46

## L

*Laodelphax striatellus*, 53
*Laphygma exigua*, 50
*Lebistes reticulatus*, 89
*Lema oryzae*, 51

*Leptinotarsa decemlineata*, 50
*Leptotrombidium* sp., 46
*Lepomis machrochirus*, 49
*Leptocira hirtula*, 36
*Leptocorisa varicornis*, 51
*Leptohylemya coarctata*, 56
*Limonius californicus*, 52
*Liriomyza archboldi*, 51
*Lissorhoptrus oryzophilus*, 51
*Lucilia cuprina*, 36, 61, 104
*Lucilia sericata*, 36
*Lucilia* sp., 8
*Lymnaea palustris*, 92

## M

*Macrocentrus ancylivorus*, 49
*Malanoplus femur-rubrum*, 111
*Merodon equestris*, 52, 56
*Mus musculus*, 49
*Musca domestica*, 7, 8, 17, 23, 28, 34, 123
*Musca domestica nebulo*, 34
*Musca domestica vicina*, 9, 10, 18, 34
*Musca sorbens*, 33
*Myzus persicae*, 50

## N

*Nephotettix cincticeps*, 51, 53, 54
*Nilaparvata lugens*, 53
*Notemigonus crysoleucas*, 49

## O

*Ornithodoros coniceps*, 46
*Ornithodoros moubata*, 46, 123
*Ornithodoros* sp., 46
*Ornithodoros tholozani*, 46
*Oryzaephilus mercator*, 58
*Oryzaephilus surinamensis*, 58

## P

*Pachytilus migratorius migratorioides*, 89
*Panonychus ulmi*, 50
*Pectinophora gossypiella*, 50, 52
*Pediculus humanus capitis*, 44
*Pediculus humanus corporis*, 30, 43
*Pediculus* sp., 8
*Penicillium notatum*, 71
*Peridroma saucia*, 56
*Periplaneta americana*, 139
*Phaenicia* sp., 8
*Phlebotomus papatasii*, 35
*Phlebotomus* sp., 36
*Phormia regina*, 36
*Phthorimaea operculella*, 50, 51
*Physa* sp., 77

# INDEX

## A

Abbott's formula, 5–6
Acceptable daily intake (A.D.I.), see pesticides
Acetylcholine, 132–133, 136–137, 139–140, 143
  accumulation during organochlorine poisoning,
    132–133, 136–137, 139–140, 142
Action potential, 64–65, 131, 133, 136
  effect of DDT on, 131, 133
  effect of lindane on, 136
Adamantane, 123–124
Adipose tissue
  chlorinated insecticides in, 73–76, 77–78, 88–89, 106,
    108–109, 144, 148
Adenosine triphosphatases
  effect of chlordane on, 140
  effect of DDT on, 133–134, 142
  effect of dieldrin on, 140–142
  effect of HCH isomers on, 137
  effect of heptachlor epoxide on, 140
  effect of lindane on, 137
  inhibitors of, 133–134, 140–142
Agricultural pests
  resistance in, 48, 61
    economic consequences of, 48–49, 54–55, 58–59
    First World Survey of, 48–49
    number of species involved, 49–50
Alanine in rat brain
  increase during dieldrin poisoning, 143
  increase during isobenzan poisoning, 143
Aldrin
  biological epoxidation of, 68–71, 89–90, 92–93,
    95–99, 101–102, 105–107, 117–119, 127–128
  as a measure of enzyme induction, 106–107
  genetic control of, 101–102
  effect on cockroach respiration, 139–140
  metabolism of
    in aquatic organisms, 92
    in insects, 97–98, 101–102, 117–119
    in microorganisms, 71, 92
    in plants, 68–70, 90–93, 95–96
    in vertebrates, 89–90, 127–128
  molecular rearrangement of, 92–93
  persistence of in soil, 68, 71, 90–93
  photodechlorination of, 92–93
  residues of in plants, 70–71, 92–93
  signs of poisoning by
    in insects, 138
    in mammals, 142–143
  toxicity
    of dechlorinated analogues, 119
    to insects, 115–119, 127, 139
    to vertebrates, 127–129
Algae
  aldrin epoxidation by, 92
  metabolism
    of DDT in, 80
    of endosulfan in, 89–90
  photosynthesis of
    effect of dieldrin on, 71–73

  uptake of DDT by, 80
  uptake of dieldrin by, 71–73
Alodan, 87–88, 115
  metabolism of, 87–88
  toxicity
    to insects, 115
    to mammals, 96, 115
American cockroach, 23, 63, 66, 93–95, 132, 137–142
  action of cyclodiene insecticides on, 138–142
  action of DDT on, 132–135
  action of lindane on, 137
  effect of DDT on central nervous system of, 139,
    141–142
  effect of dieldrin on central nervous system of,
    139–142
  isodrin metabolism in, 93–95
  pharmacokinetics of HCH isomers in, 66, 137
  uptake of DDT by, 63
American partridge, 128–129
American sparrowhawk, 109–110
Amino acids, changes during DDT poisoning in insects,
  133
Aminopyrine, enzyme induction by, 26–27
Ammonia, liberation of in rat brain during cyclodiene
    poisoning, 143
Aniline, hydroxylation of, 107–108
Antipyrine, 106–107
Aromatization
  of cyclodiene insecticides, 96
  of lindane derivatives, 80–83
Arsenicals, resistance to, 46–48
Atropine, as antidote in lindane poisoning, 136–137
Azinphos methyl, 62

## B

Baboons, 124–125
Bald eagle, 150
Barbiturates, in alleviation of chlorinated insecticide
    poisoning, 136–137, 142–143
Beans, 84, 92–93, 144, 147–148
Bedbugs, 43–46
  resistance in
    geographic distribution of, 44–45
    to chlordane, 44–45
    to DDT, 44–45
    to dieldrin, 44–46
    to lindane, 45
    to methoxychlor, 45–46
Bees, 49
Beetles
  flour, 6–7, 57–58, 120–121
  Mexican bean, 67, 101–102, 111–113
Beets, 147–148
Bengalese finch, 109
Benzoic acids, from DDT, 78
Biochemical genetics, use of visibly marked mutants in,
  12

Croton betaine, in dieldrin poisoned rat brain, 142–143
Crustaceans
    planktonic, HEOD residues in, 150
    toxicity of mirex to, 144
Cushing's syndrome, 108–109
Cyclodiene insecticides
    and narcotic action, 138
    biochemical effects of in insects, 139–140, 142–143
        on acetylcholine levels in nervous system, 142
        on ATP-ases, 14
        on carbohydrate reserves, 139–140
    biodynamics of, 66–73
    chemical aromatization of, 96
    dietary exposure to, 106–107, 128
        and enzyme induction, 106–107, 128
    effect on cockroach central nervous system, 139, 141–142
    enzyme induction by, 27–30, 105, 127
    intrinsic toxicity of, 115
    isotope-labeled, 66, 68–74
    metabolism of, 89–90, 139–142
        in insects, 84–89
        in microorganisms, 70–73, 87, 89
        in plants, 68–71, 88–89
        in vertebrates, 71–73, 84–85
    natural tolerance to, 101
    occupational exposure to, 106
    persistence in soil, 67–68, 70–71, 87
    pharmacokinetics of, 73–76, 91–92, 129
        mathematical models in, 74–76
    photochemistry of, 77, 84–85, 88–89, 119–120, 123–124, 142
    physiological effects of in insects, 139–140
    reductive dechlorination of, 96, 119
    residues in plants, 68–71, 84–85, 148
        removal of during processing, 147–148
    residues in vertebrate tissues, 73–77, 84–85, 147–148
    resistance to, 7–12, 17–18, 20, 23, 29–40, 42–50, 53–56, 59–60, 104, 139–142
        number of species involved, 30
    signs of poisoning by
        alleviation of by barbiturates, 142
        in insects, 137–140, 143
        in mammals, 142–143
        reversibility of, 138, 142
    structural resemblance to lindane, 121–124
    structure-activity relations of, 115–121, 142
        aldrin-isodrin group, 117–120
        chlordane-heptachlor group, 115–119
    synergism of, 105
    toxicity
        to insects, 127, 139
        to mammals, 127
Cycloheximide, inhibition of protein synthesis by, 28
Cytochrome oxidase, effect of DDT on, 132–133
Cytochrome $P_{450}$, 97–99, 101–102, 105–108
    in fungi, 99
    in insects, 98–99, 101–102
    in microorganisms, 99
    in plants, 98–99
    in vertebrates, 97–98, 105–108

**D**

Dairy products, organochlorine residues in, 147–148
DANP (dianisyl neopentane)
    analogues of, 113–114, 133–134
        insect toxicity of, 113–114
    toxicity to insects, 111–113, 133–134, 136
DBH
    co-metabolism of by sewage microorganisms, 80
    from DDT in plant tissues, 70–71
    from dicofol, 78–79
DBP
    from DDE, 77
    from DDOH, 79
    from DDT, 78–80
    from dicofol, 78–79
DDA
    as ATP-ase inhibitor, 133–134
    from DDD in vertebrates, 78, 100–101
    from DDT
        in insects, 78
        in microorganisms, 79–80
        in plant tissues, 70–71, 79
        in vertebrates, 73–74, 77–79, 100–101
DDCN, from DDT in sewage sludge, 80
DDD
    accumulation in fish, 76–77
    accumulation in snails, 76–77
    and adrenocortical atrophy in dogs, 108–109, 125–126
    as ATP-ase inhibitor, 133–134
    biological formation from DDT, 77–80, 99–100
    $^{14}$C-labeled, 21
    dehydrochlorination of in insects, 12–13, 21, 23
    effect of on adrenal steroid metabolism, 108–109, 125–126
    effect of on cyclodiene insecticide storage, 107–108
    hydroxylation of, 78
    in bird tissues and eggs, 150–151
    in technical DDT, 114–115
    in vertebrate tissues, 73–74, 76–77, 148, 150–151
    metabolism of, 21, 26, 77–78, 111
        in humans, 77–78
        in insects, 21, 23, 111
    resistance to, 20, 29–30, 40
    toxicity of
        to insects, 34–35, 113–115
        to vertebrates, 124–126, 150–151
2,4'-DDD, effect on steroid hydroxylation, 108–109
DDD-dehydrochlorinase, 12–13, 23
DDE
    accumulation in fish, 76–77
    accumulation in snails, 76–77
    as ATP-ase inhibitor, 133–134
    effects on cyclodiene insecticide storage, 107–108
    effects on drug metabolism, 107
    environmental persistence of, 77–78, 150–151
    from DDT, 77–80, 102, 126
        in insects, 12–15, 19, 22–23, 25, 42, 61, 67
        in microorganisms, 79–80
        in plant tissues, 70–71, 79
    from deutero-DDT, 15
    from dicofol, 78–79

Epoxide ring hydration, 20, 22–23, 79, 85–87, 91–92, 95, 99–101, 104–105, 120
  and dieldrin action, 105, 120, 139
  inhibitors of, 100–101, 139–140
  mechanism of, 99–101
    and fumarase action, 100–101
Epoxides
  as epoxide hydrase inhibitors, 100–101
  as intermediates in lindane metabolism, 81–83
  deoxygenation of by microorganisms, 87, 96–97
  in aromatic hydrocarbon metabolism, 100
  role of in cyclodiene insecticide toxicity, 115–121, 139
Estrogenic activity
  in DDT analogues, 79, 109–110, 114–115
    structure-activity relationships for, 109
  in polychlorobiphenyls, 109
  in polychlorotriphenyls, 109
Ethereal sulfates, 81–83, 100–101
Ethoxychlor
  accumulation in fish and in snails, 76–77
  metabolism of, 79–80, 111–113
  toxicity to insects, 111–114

## F

Farne Island ecosystem
  DDE in trophic levels of, 150
  HEOD in trophic levels of, 150
F-DMC, as DDT synergist, 15–17
Fenitrothion, 54
Fenthion, 54
Fire ant, mirex for control of, 143–144
Fish
  bluegill sun-, 49, 127, 129, 143–144
  carp, 89, 129
  chlorinated insecticide residues in, 126–129, 143–144, 149–150
  cod, 150
  comparative toxicity of chlorinated insecticides to, 124–125, 128–130
  golden shiner, 49
  goldfish, 89, 143–144
  guppy, 89
  minnow, 127
  mosquito, 49, 76–77
  pharmacokinetics of chlorinated insecticides in, 74, 76–77, 91–92, 128–129
  pike, 89
  resistance to chlorinated insecticides in, 49, 127, 129–130
  storage interactions of chlorinated insecticides in, 108–109
  swordtail, 89
  trout, 89, 108–109, 126–127, 129
Flies,
  bean seed-, 30–31, 57
  black-, 36
  cabbage root-, 56–57
  carrot rust-, 55–56
  fruit, 23, 25–26, 57, 97–98
  horn, 36

large bulb-, 56
little house-, 36
May-, 49
number of species resistant to chlorinated insecticides, 30–31
onion-, 24, 56
pomace-, 36, 57, 97–98
sand-, 35–36
spotted root-, 55–56
stable-, 36, 120–121
tsetse-, 36, 91–92, 120–121, 129–130, 139
wheat bulb-, 56
Food and Agriculture Organization (F.A.O.), 48, 53, 145
  Working Party of Experts on Pesticide Residues, 145
  Working Party on Resistance to Pesticides, 48
Food chains, accumulation of chlorinated insecticides in, 76–77, 148–150
Forage crops, 50, 53
Frogs, resistance to chlorinated insecticides in, 49
Fruit pests, 30

## G

General anesthesia, 138
Genes for resistance, frequency of occurrence, 29–30, see also individual compounds
German cockroach, 23, 45–46, 97–98, 119–120, 139–140
  action of cyclodiene insecticides in, 139–142
  chromosomes of, 23
  effect of chlorinated insecticides on respiration of, 139–140
  effect of cyclodienes on ATP-ases of, 140
  effect of cyclodienes on central nervous system of, 141–143
  effect of DDT on central nervous system of, 139, 141–142
  effect of lindane on central nervous system of, 139
  effect of toxaphene on central nervous system of, 139
  pharmacokinetics of dieldrin in nervous system of, 66, 141–142
  resistance in
    geographic distribution of, 46
    to chlordane, 23, 46
    to DDT, 45–46
    to dieldrin, 23–24, 140–142
    spectrum of, 46
Gibberellin synthesis, microsomal oxidases in, 99
Glucosides, formation in insects, 78
Glucuronides, 81–83, 96–98, 100–101
Glucuronyl transferase of rat liver, 91, 101
Glutamine in rat brain, increase during isobenzan poisoning, 143
Glutathione (GSH)
  conjugation with, 81–83, 97, 101
  GSH-S-transferases, 81–83, 101–102, 104–105
  in insecticide metabolism, 78, 81–83, 97, 100–102, 104–105
Glycosides, formation in plants, 97, 101
Gnats, resistance in, 36

paddy stem-, 53
rice stem-, 53
root-, 30
seed corn-, 55, 57
spotted root-, 8–9
turnip, 55
Maize, 70, 84, 90–91
Malaria, 19–20, 33–34, 36–40, 42–45
  eradication of, 19–20, 33, 36–40, 42–43, 45–46, 48
    status of eradication program, 42–43
  in the United States, 38–39
  transmission of, 19–20, 36–40, 43
Malathion
  for cockroach control, 46
  for flea control, 44
  for flour beetle control, 57–58
  for head-louse control, 44
  for mosquito control, 40, 43
  resistance to, 30, 38, 46, 54, 57–58, 103–104
    mechanism of, 58
Mallard, 109–110, 124–125, 128–130
Man, 79, 90–91, 123–124, 131, 142
Mathematical models
  in analysis of environmental distribution of pesticides,
    76–77
  in chlorinated insecticide pharmacokinetics, 74–76
Mathematical relationships, between tissue concentrations
    of chlorinated insecticides, 73–74
Meat, reduction of organochlorine residues in, 148
Median lethal concentration (of insecticides), 1, 4–7
Mercapturic acids, 81–83, 97, 100–101
  from lindane and derivatives, 81–83
Mersalyl acid, as ATP-ase inhibitor, 140
Meso-inositol, 121, 137
  and lindane action, 137
  as antagonist of lindane effects on *S. cervisiae*, 137
Metabolism studies
  techniques used in
    autoradiography, 66, 80–81, 136–137, 141–142
    electron microscopy, 66
    radioisotope-labeled insecticides, 66, 68–74, 76–79,
      87–92, 95–96, 136–137, 141–142, 144
Methiochlor
  accumulation in snails, 76–77
  metabolism of, 79–80, 113
  toxicity to insects, 111–113
Methoxychlor
  accumulation of in fish and snails, 76–77
  dehydrochlorination of in insects, 12–13, 102–103
  effect of on dieldrin storage in fish, 108–109
  mammalian toxicology of, 125–126
  metabolism of, 79–80, 111
  resistance to, 12–17, 29–33, 43–45
  storage in adipose tissue, 79, 124–126
  toxicity of, 79–80, 102–103, 111–113, 124–126
    to fish, 124–126
Methylchlor
  accumulation of in fish and snails, 76–77
  metabolism of, 79–80, 111
  toxicity to insects, 111–113
    synergism of, 111–113
Methyl parathion, 53, 61

Metyrapone
  effects on adrenocortical enzymes, 108–109
  effects on steroid hydroxylation, 108–109
Mexican bean beetle, 67, 101–102, 111–113
Mice, 49–50, 88–91, 107–108, 111–113, 120–121,
    124–125, 127–130, 136–137, 143–144
  resistance to DDT and endrin in, 49–50, 128
Microencapsulation, 151
Microorganisms
  role in pesticide degradation, 71–73, 79–80, 84, 87,
      89, 91–92, 95–97, 99, 120, 144
    of aldrin, 71–73, 92
    of DDT, 80
    of dieldrin, 71–73, 91–92, 120
    of endosulfan, 89
    of endrin, 71–73, 95–96
    of heptachlor, 71, 87
    of heptachlor epoxide, 87
    of photodieldrin, 71–73
Microsomal detoxication, 14–15, 18, 27–29, 61, 76,
      78–79, 85–87, 89–92, 97–98, 107, 111
  housefly genes for, 14–17, 60
  of DDT, 14–17, 60–61, 78, 111
Microsomal mixed function oxidases, 19, 27–28, 76,
      78–79, 85–87, 89–101, 105–107, 111, 115–117,
      120–121
  cofactor requirements of, 27–28, 78, 92–93, 97–98
  electron transport system for, 97–98
  inhibition of by carbon monoxide, 97–98
  inhibitors of, 14, 19, 66, 78–79, 85–87, 91–95,
      97–98, 100–101, 107, 111–115, 139
  in insects, 19, 27–29, 78–79, 85–87, 91, 97–98, 111,
      115, 136
  in plants, 92–93
  in vertebrates, 76, 91, 97–98, 105, 128
    endogenous substrates for, 105
  nature of the enzymes, 98
  oxidative dechlorination by, 99–100
  reductive dechlorination by, 99–100
Midges, resistance in, 36
Mirex
  dechloro-derivative of, metabolism in rats, 144
  inertness of in vertebrate liver preparations, 144
  persistence of in vertebrate tissues, 143–144
  pharmacokinetics of in rats, 144
  sublethal effects of, 143–144
  toxicity of
    to aquatic organisms, 143–144
    to vertebrates, 129, 143–144
  uptake by plants, 143
Mites, trombiculid, in scrub typhus, 46
Mitochondrial enzymes, effect of dieldrin on, 142–143
Model ecosystems, 76–77
Monkeys, 73–74, 90–91, 125–126, 128, 130, 148
Mosquitos, 3,5, 12, 49–50, 76–77, 84–85, 92–93,
    117–119, 121
  anthropophily in, 3, 40
  endophily in, 3, 40
  exophily in, 3
  metabolism of DDT in, 78, see also resistance to
    DDT
  mutant marker genes in, 20–23

Milton Keynes UK
Ingram Content Group UK Ltd.
UKHW051934141024
449569UK00027B/1489